GEOARCHAEOLOGY OF LANDSCAPES IN THE GLACIATED NORTHEAST

THE UNIVERSITY OF THE STATE OF NEW YORK

Regents of the University

Robert M. Bennett, *Chancellor*, B.A., M.S.	Tonawanda
Adelaide L. Sanford, *Vice Chancellor*, B.A., M.A., P.D.	Hollis
Diane O'Neill McGivern, B.S.N., M.A., Ph.D.	Staten Island
Saul B. Cohen, B.A., M.A., Ph.D.	New Rochelle
James C. Dawson, A.A., B.A., M.S., Ph.D.	Peru
Robert M. Johnson, B.S., J.D.	Huntington
Anthony S. Bottar, B.A., J.D.	North Syracuse
Merryl H. Tisch, B.A., M.A.	New York
Geraldine D. Chapey, B.A., M.A., Ed.D.	Belle Harbor
Arnold B. Gardner, B.A., LL.B.	Buffalo
Harry Phillips, 3rd, B.A., M.S.F.S.	Hartsdale
Joseph E. Bowman, Jr., B.A., M.L.S., M.A., M.Ed., Ed.D.	Albany
Lorraine A. Cortés-Vázquez, B.A., M.P.A.	Bronx
Judith O. Rubin, A.B.	New York
James R. Tallon, Jr., B.A., M.A.	Binghamton
Milton L. Cofield, B.S., M.B.A., Ph.D.	Rochester

President of The University and Commissioner of Education
Richard P. Mills

Chief Operating Officer
Richard H. Cate

Director of the New York State Museum
Clifford A. Siegfried

The State Education Department does not discriminate on the basis of age, color, religion, creed, disability, marital status, veteran status, national origin, race, gender, genetic predisposition or carrier status, or sexual orientation in its educational programs, services and activities. Portions of this publication can be made available in a variety of formats, including braille, large print or audio tape, upon request. Inquiries concerning this policy of nondiscrimination should be directed to the Department's Office for Diversity, Ethics, and Access, Room 530, Education Building, Albany, NY 12234. **Requests for additional copies of this publication may be made by contacting the Publications Sales Desk, New York State Museum, Cultural Education Center, Room 3140, Albany, NY 12230.**

Geoarchaeology of Landscapes in the Glaciated Northeast

Proceedings of a Symposium
Held at the
New York Natural History Conference VI

Edited by
David L. Cremeens
and John P. Hart

New York State Museum
Bulletin 497
2003

The University of the State of New York
The State Education Department

Copyright © 2003 The New York State Education Department, Albany, New York

Printed in the United States of America

Copies may be ordered from:
  Publication Sales Desk
  New York State Museum
  Cultural Education Center, Room 3140
  Albany, NY 12230
  Phone: (518) 402-5344
  Fax: (518) 474-2033
  Web address: www.nysm.nysed.gov/publications.html

Library of Congress Catalog Number: 2002115979

ISSN: 0278-3355
ISBN: 1-55557-214-6

*This book is dedicated to
the memory of Robert E. Funk
an early proponent of geoarchaeology in the Northeast.*

TABLE OF CONTENTS

List of Figures ix

List of Tables xii

Preface xiii
John P. Hart

CHAPTER 1: Introduction 1
David L. Cremeens and John P. Hart

SECTION I: GEOLOGICAL BACKGROUND

CHAPTER 2: Geomorphic History of New York State 7
Donald H. Cadwell, Ernest H. Muller, and P. Jay Fleisher

CHAPTER 3: The Last Deglaciation of the Northeastern United States: A Combined Varve, Paleomagnetic, and Calibrated ^{14}C Chronology 15
John C. Ridge

SECTION II: UPLAND SETTINGS

CHAPTER 4: Geoarchaeology of Soils on Stable Geomorphic Surfaces: Mature Soil Model for the Glaciated Northeast 49
David L. Cremeens

CHAPTER 5: Bluff Top Sand Sheets in Northeastern Archaeology: A Physical Transport Model and Application to the Neville Site, Amoskeag Falls, New Hamsphire 61
Robert M. Thorson and Christian A. Tryon

CHAPTER 6: Life in a Postglacial Landscape: Settlement-Subsistence Change During the Pleistocene-Holocene Transition in Southern New England 75
Brian D. Jones and Daniel T. Forrest

CHAPTER 7: Glacial Geology and Prehistoric Sensitivity Modeling Fort Drum, New York 91
Laurie W. Rush, Randy Amici, James Rapant, Carol Cady, and Steve Ahr

CHAPTER 8: Behavioral Continuity on a Changing Landscape 103
Douglas Frink and Allen Hathaway

SECTION III: ALLUVIAL SETTINGS

CHAPTER 9: Postglacial Development of the Penobscot River Valley: Implications for Geoarchaeology 119
Alice R. Kelley and David Sanger

CHAPTER 10: Geoarchaeological and Cultural Interpretations in the Lower
   Penobscot Valley, Maine 135
David Sanger, Alice R. Kelley, and Heather Almquist

CHAPTER 11: Geochronology from Archaeology: An Example from the Connecticut River Valley 151
Kathryn Curran

CHAPTER 12: Archaeological Site Formation in Glaciated Settings, New Jersey and
   Southern New York 163
Donald M. Thieme

CHAPTER 13: Landscape Change, Human Occupation, and Archaeological Site Preservation
   at the Glacial Margin: Geoarchaeological Perspectives from the Sandts Eddy Site (36Nm12),
   Middle Delaware Valley, Pennsylvania 181
Joseph Schuldenrein

List of Contributors 211

LIST OF FIGURES

2.1.	Physiographic provinces of New York.	7
2.2.	Locations and geological features discussed in text.	9
2.3.	Central portion of glacial Lake Albany with location of deltas and the buried Colonie Channels.	10
2.4.	Representative Wisconsinan ice margins in New York.	11
3.1.	Map of western New England and adjacent New York State showing the approximate geographic positions and varve numbers of long measured varve sequences in the northeastern United States.	17
3.2.	Varve sequences of the New England varve chronology as it stands today.	18
3.3.	Time-distance plot of the time spans of all measured varve sections in the Connecticut and Passumpsic Valleys from Connecticut to northern Vermont and New Hampshire.	21–23
3.4.	Time-distance plot of time spans of all measured varve sections in the Merrimack Valley of New Hampshire.	23
3.5.	Atmospheric ^{14}C and calibrated (U-Th) time scales applied to the New England varve chronology.	26
3.6.	Correlation of late Wisconsinan paleomagnetic declination records from New England and New York State.	28
3.7.	The deglaciation chronology of the northeastern United States in atmospheric ^{14}C years B.P.	29
3.8.	The deglaciation chronology of the northeastern United States in calibrated (U-Th) years B.P.	30
3.9.	The Greenland Ice Sheet Project 2 (GISP2) ice-core record of oxygen isotopes plotted with the general inferred history of deglaciation in New York and New England.	32
4.1.	Interpretive process in evaluating archaeological context.	50
4.2.	Examples of soil profiles as arrangements of soil horizons.	54
4.3.	Traditional soil formation model.	54
4.4.	Progressive-regressive pedogenesis.	55
4.5.	Artifact distribution in immature soil profiles.	56
4.6.	Artifact distribution in moderately formed soil profiles.	57
4.7.	Artifact distribution in mature soil profiles.	57
4.8.	The mature soil model.	58
4.9.	Distribution of cultural remains as a function of the timing of prehistoric occupations relative to the development stages of soils.	58
5.1.	Sketch map of the northeastern states showing resultants for sand roses of winds exceeding 6 m sec^{-1}.	62
5.2.	Geographic terms used in this chapter.	64
5.3.	Simple hydraulic factors associated with deposition of bluff-top sand sheets.	65
5.4.	Schematic model emphasizing temporal component of bluff-edge sand sheet accretion.	67
5.5.	Schematic stratigraphic summary of the Neville site.	69
5.6.	Grain-size distribution of selected samples from the Neville site, compared with a representative sample of the Massachusetts "eolian mantle" and loess from the Tanana Valley, Alaska.	70

6.1.	Location of the Hidden Creek and Sandy Hill Sites on the Mashantucket Pequot Reservation, Ledyard, Connecticut.	75
6.2.	View of the northwest corner of the 1994 excavation block at the Hidden Creek site, 0–80 cm below ground surface.	78
6.3.	Black sand features at site 72-97 exposed by a machine cut.	80
6.4.	North-south section along the West 2-m line at the Sandy Hill site showing complex overlapping black sand feature strata.	82
6.5.	Excavation of a pit-house feature at the Sandy Hill site during the 2000 field season.	84
7.1.	Location of Fort Drum in New York State.	91
7.2.	Fort Drum physiographic zones.	92
7.3.	Fort Drum hillshade model.	94
7.4.	Clovis point.	94
7.5.	Fort Drum elevations and prehistoric archaeological site distribution.	95
7.6.	Glacial Lake Iroquois, Frontenac-level shoreline defined by 600-ft elevation line.	95
7.7.	Fort Drum glacial Lake Iroquois shoreline defined by 700-ft elevation line.	96
7.8.	Overview of Fort Drum Prehistoric Site 1093.	98
7.9.	Metate discovered at FDP 1093.	99
7.10.	Friable clay features found at FDP 1093.	99
7.11.	Rapid aeolian deposition.	100
7.12.	Overview FDP 1154.	100
8.1.	Reproduction of Loring's 1980 map of Paleoindian sites in the Champlain Valley.	104
8.2.	Plot of Paleoindian-Period site dates versus latitude.	107
8.3.	Three defined stages of Lake Vermont.	109
8.4.	Four defined stages of Champlain Sea.	109
8.5.	Lake Champlain today.	110
8.6.	Enlarged view of emergent ponds and wetlands as indicated by soils.	111
8.7.	Comparison by association with reconstructed forest communities of Paleoindian-Period sites in the Champlain Valley with all known sites in Chittendon County, Vermont.	113
9.1.	Central Maine with locations of major rivers, lakes, and archaeological sites mentioned in text.	120
9.2.	Generalized bedrock geological map of the Penobscot River drainage.	121
9.3.	Generalized surficial geological map of the Penobscot River drainage.	123
9.4.	Plano-like projectile point in fine-grained sediments immediately above gravel layer, Blackman Stream site.	126
9.5.	Idealized composite stratigraphic section at Blackman Stream site.	128
9.6.	Idealized composite stratigraphic section at Gilman Falls site.	129
10.1.	Central Maine with locations of major waterways and archaeological sites.	137
10.2.	Late Paleoindian bifaces from Eddington Bend (a) and Blackman Stream (b).	138
10.3.	Pushaw Stream and the Milford drainage basin.	140
10.4.	Bedrock map of Gilman Falls Island.	141
10.5.	Bedrock metamorphic grades in central Maine.	142
10.6.	Production sequence of rods from early to final stages of manufacture.	143
10.7.	Water levels in Mansell Pond.	146
11.1.	Location of Gill, Massachusetts, in Southern New England.	151
11.2.	Riverside section of Gill, Massachusetts.	152

11.3.	Riverside terraces.	152
11.4.	Preglacial and modern paths of the Connecticut River.	153
11.5.	Glacial Lake Hitchcock.	153
11.6.	Extent of glacial Lake Hitchcock at the 350-ft elevation line.	154
11.7.	The proto-Connecticut River flows west through White Ash Swamp.	155
11.8.	The White Ash Swamp channel is closed. The proto-Connecticut spills over the Lily Pond Barrier as two waterfalls.	156
11.9.	Barton's Cove and Barton's Field ca. 1890.	157
11.10.	The proto-Connecticut River travels around the Lily Pond Barrier through the Narrows. The plunge pools are now closed.	158
11.11.	The Connecticut River establishes a course similar to that seen today.	160
12.1.	The glaciated portions of New Jersey and southern New York showing sites discussed in text.	163
12.2.	Late Pleistocene recessional margins in New Jersey.	165
12.3.	Stratigraphic cross-section east of Penhorn Creek in the Hackensack Meadowlands, North Bergen, New Jersey.	168
12.4.	Prehistoric archaeological sites along Route 21 Corridor, Passaic County, New Jersey.	169
12.5.	Stratigraphy of the Dundee Canal site (28PA143), Passaic County, New Jersey.	171
12.6.	Composite stratigraphy for Block 1 at the Dundee Canal site (28PA143).	171
12.7.	Grain size trends in the Block 1 stratigraphic column at the Dundee Canal site (28PA143).	172
13.1.	Physiographic settings of Sandts Eddy (36Nm12), Valley and Ridge province.	182
13.2.	Aerial photo of Sandts Eddy site and surrounding terrain.	183
13.3.	Geomorphic map of Sandts Eddy site (36Nm12), Middle Delaware Valley, Pennsylvania and New Jersey.	184
13.4.	Morphology of the T-1 terrace.	185
13.5.	Plan of Sandts Eddy site excavations with contours of site.	185
13.6.	Sandts Eddy generalized stratigraphic profile.	187
13.7.	Stratigraphy and soil-sediment analysis: N31 E43, Column 1.	191
13.8.	Stratigraphy and soil-sediment analysis: N27 E65, Column 2.	192
13.9.	Stratigraphy and soil-sediment analysis: N44 E69, Column 3.	192
13.10.	Triaxial plot (sand, silt, clay) of particle-size distributions by alluvial unit, Sandts Eddy site.	194
13.11.	Photomicrograph of lamella sample, Alluvial Unit 3.	195
13.12.	Photomicrograph of lamella sample, Alluvial Unit 2.	195
13.13.	Composite Holocene stratigraphy, Sandts Eddy site: 36Nm12 (N30 axis).	197
13.14a.	Diachronic model of landform evolution, occupation and site preservation, Sandts Eddy (36Nm12), Pennsylvania.	201
13.14b.	Diachronic model of landform evolution, occupation, and site preservation, Sandts Eddy (36Nm12), Pennsylvania.	201
13.15.	Holocene sedimentation rates: Delaware Valley alluvial sites.	203
13.16.	Holocene "moist-dry cycles" versus running mean sedimentation rates: Delaware Valley alluvial sites.	204
13.17.	Schematic stratigraphy of the Middle Delaware Valley.	206

LIST OF TABLES

3.1.	^{14}C ages from Varves in the New England Varve Chronology.	25
3.2.	Abbreviations of Moraines, Ice Margins, and Glacial Readvances on Figures 3.6, 3.7, 3.8, and 3.9.	29
5.1.	Northeast Sites in Bluff-top Settings.	62
6.1.	SEM Carbonized Plant Identifications from Hidden Creek and Sandy Hill.	79
6.2.	Radiocarbon Dates from the Hidden Creek and Sandy Hill Sites.	85
8.1.	Site and Date Data Used in Figure 8.2.	106
8.2.	Native American Site Components Associated with Defined Forest Communities for Chittenden County, Vermont.	112
8.3.	Paleoindian-Period Site Components Associated with Defined Forest Communities in the Champlain Valley.	113
10.1.	Calibrated Radiocarbon Dates from Sites.	136
12.1.	Radiocarbon Dates from Geoarchaeological Investigations, New Jersey and Southern New York.	166
13.1.	Stratigraphic Correlations for Three Phases of Excavation at Sandts Eddy.	186
13.2.	Radiocarbon Dates from 36Nm12.	189
13.3.	Integrated Cultural and Sedimentary Stratigraphy, Sandts Eddy.	196
13.4.	Landscape History and Archaeological Preservation Potential by Archaeological Component, Sandts Eddy.	199
13.5.	Correlation of Prehistoric Landscape Histories of Middle Delaware Valley Sites.	205

PREFACE

This is the fourth in a current series of edited books concerned with archaeological topics in the Northeast published in the New York State Museum's Bulletin series. The first volume, published in 1999, was *Current Northeast Paleoethnobotany* (Bulletin 494), edited by John P. Hart. This book addresses the prehistoric interactions of plants and humans and the impacts of agricultural evolution on human societies in the Northeast. The second, *Nineteenth and Early Twentieth Century Domestic Site Archaeology in New York* (Bulletin 495), edited by John P. Hart and Charles L. Fisher, was published in 2000. This volume explores the importance of the archaeology of relatively recent domestic sites for understanding our history. The third, *Northeast Subsistence-Settlement Change: A.D. 700–1300* (Bulletin 496), edited by John P. Hart and Christina B. Rieth, was published in 2002. This book explores subsistence and settlement during the early Late Prehistoric Period, when many of the subsistence and settlement traits of Native American societies recorded at the time of European contact first become evident in the archaeological record.

The present volume is based on a symposium that I organized with coeditor David Cremeens for the New York Natural History Conference VI, held at the New York State Museum in April 2000. Formerly glaciated terrains of northeastern North America present a wide variety of landscapes that affected the location, formation, and preservation of prehistoric archaeological sites. Many of these landscapes, such as simple till-covered uplands, are little altered since the terminal stages of the Pleistocene. Other landscapes are more complex, for example, glaciofluvial and glaciolacustrine valley floor environments that have undergone significant modification through Holocene alluvial and colluvial processes. The symposium was organized to address current geoarchaeological work in these glaciated landscapes. The papers presented at the symposium covered a wide geographical area including New England, New York, Pennsylvania, New Jersey, and southern Ontario and addressed the development of the archaeological record on various postglacial landforms.

Following an introductory chapter by David Cremeens and me, the 12 substantive chapters in this volume summarize current knowledge of the deglaciation of the Northeast and provide geoarchaeological case studies in upland and alluvial settings. Geographically chapters cover Pennsylvania, New Jersey, New York, and New England. By themselves, the chapters show how detailed geoarchaeological investigations are critical to our understanding of the archaeological record in formerly glaciated landscapes. The volume as a whole fills an important gap in the geoarchaeological literature; until now, there have been no edited volumes devoted exclusively to the geoarchaeology of the Northeast. By filling this gap, the volume may encourage additional geoarchaeological investigations in the region.

As this book goes to press, other volumes are in preparation for the Museum's Bulletin series that treat important topics in Northeast archaeology. The chapters in *The Archaeology of Albany*, edited by Charles L. Fisher, present the results of recent historical archaeological investigations in Albany, New York, one of the oldest, continuously occupied European settlements in the United States. *Perishable Material Culture in the Northeast*, edited by Penelope B. Drooker, is a collection that reviews the archaeological record of perishable material culture in the Northeast. *Three Sixteenth-Century Mohawk Iroquois Village Sites*, by Robert E. Funk and Robert D. Kuhn, summarizes the excavations at three important Mohawk Iroquois village sites. Other volumes will follow in the coming years.

This book would not have been possible without the efforts of the chapter authors. My coeditor David Cremeens provided much of the impetus for the symposium and volume. Many thanks to the five peer reviewers who provided timely and very useful comments on the chapters. Finally, thanks, as always, go to Jack Skiba, for shepherding the book through the publication process.

John P. Hart
December 2002

CHAPTER 1

INTRODUCTION

David L. Cremeens and John P. Hart

GEOARCHAEOLOGY

Inferring past human behavior from the archaeological record can be done only by examining the contextual association of the artifacts and other traces of past human behavior and evaluating the behavioral and nonbehavioral factors that have produced that record (Rapp and Hill 1999). Geoarchaeologists address both the behavioral and nonbehavioral aspects of archaeological context from a landscape point of view. The current landscape and the artifacts and features contained within constitute the record from which past processes must be inferred. Geoarchaeology investigations use techniques and concepts from the fields of geomorphology, sedimentology, stratigraphy, and pedology to evaluate site formation processes and landscape reconstruction.

Thus geoarchaeologists attempt to answer two broad questions: (1) Why did prehistoric people pick a specific location for their activities? (2) What has happened to the record of these people since they abandoned the site? The first question is largely behavioral and involves concepts such as choice of one point on a landscape versus another: what made a particular location attractive, and what influence did that particular location have on the activities done there? This latter point is a general situation where natural and behavioral context begin to overlap. The second question deals with the more site-specific aspects of natural and human caused processes affecting the record or patterning of artifacts. Because so much of the behavioral inference is derived from the patterns of artifact distributions at a site, it is crucial that we understand the specific site formation processes that influence the patterning since site abandonment. Specific site formation processes also influence the interpretation of the ages generated from site samples (Waters 1992). Without the precise stratigraphic and sedimentological context, and the context of the postdepositional matrix, the dating of archaeological remains[1] may be potentially skewed from mixing and contamination.

As geoarchaeologists we have to work backward in time from the present site condition and take into account all that has happened to the landscape since site abandonment including more recent events. For example, in the northeastern United States (hereafter Northeast), many locations have had a developing forest ecosystem superimposed over abandoned European settler agricultural fields. To understand the original pattern of artifacts, we have to consider the effects of intensive hill-slope agriculture as practiced in the pre-Industrial Revolution Northeast. Agriculture introduces both order and disorder into the landscape. Order comes in the form of straight lines of field boundaries and roads. Disorder comes from the severe erosion and sedimentation resulting from hill-slope cultivation. After considering the effects of agriculture, the effects of a developing forest on these same landscapes needs to be evaluated. In several areas, the forest has had 200 years or more to develop on the previously cultivated landscape, long enough to introduce a new set of processes and the resultant patterns.

GLACIATED-DEGLACIATED LANDSCAPES IN THE NORTHEAST

Formerly glaciated landscapes in the Northeast are complex but have a well-defined chronology (Cadwell et al., this volume; Ridge, this volume). Glacial deposits of North America do not commonly contain archaeological remains (Waters 1992). However, humans did occupy the proglacial and periglacial environments and certainly ventured north following the retreat of the ice margins into newly created proglacial and periglacial conditions. All of these related settings have specific landscapes and resultant geomorphic contexts that can be related to prehistoric human occupation (Rapp and Hill 1999).

Finding a site and interpreting site location is a landscape scale evaluation that has to take into account what

---

Geoarchaeology of Landscapes in the Glaciated Northeast edited by David L. Cremeens and John P. Hart. New York State Museum Bulletin 497. © 2003 by the University of the State of New York, The State Education Department, Albany, New York. All rights reserved.

the landscape was at the time of occupation versus what the landscape has become since. For Woodland-Period archaeology, most landscapes, with the exception of flood plains and low terraces, have not changed much since site abandonment. Archaic- and Paleoindian-Period sites, on the other hand, have the potential to have had increasingly significant landscape scale changes because of the amount of time elapsed since occupation. For example, a lakeshore ridge of gravelly sediments may have provided access to lacustrine resources to the earliest peoples, while the same landscape in Woodland times was simply a low-lying, linear ridge in the forest, possibly a dry place to camp (Rush et al., this volume).

One aspect of the deglaciating and postglacial environments that stands out in this volume is how wet the terminal Pleistocene and initial Holocene landscape was in the Northeast (Rush et al., this volume; Frink and Hathaway, this volume). The proglacial-periglacial landscapes contained extensive and numerous lakes and poorly developed drainage networks with associated wetlands. Geoarchaeologists are faced with a series of site-specific questions to evaluate: How did this wet environment influence the earliest people? Part of the answer lies in determining the type of resources they were exploiting in the wet environment. Another part of the answer has to evaluate their various activities. Were specific tasks associated with specific landforms? What happened to the record they left behind? How did the Younger Dryas Cold Episode (10,800–10,000 B.P.) affect the exploitation of the wet environments? How did the Mid-Holocene Thermal Maximum (7000–6000 B.P.) affect the same landscapes and the associated resources for later inhabitants? The nature of the landscape dictates the type and magnitude of a variety of postoccupation processes that are a significant part of site formation. Some landscapes are more susceptible to modification via pedoturbation, while other landscapes are more resilient. How did this influence later people in what they did and where they did it and in what happened to the record they left behind?

Fortunately the deglaciation record of the Northeast has a well-defined chronology due in large part to some of the finely detailed records left by the numerous lakes (Ridge, this volume). This record, combined with pollen records of the changing vegetation patterns, points to rapidly changing conditions at the beginning of the Holocene. Many of the lakes did not exist for a very long time, while others were more persistent. This rich database should provide a good context from which to locate and evaluate early sites throughout the Northeast.

VOLUME SUMMARY

This volume derives from a symposium we organized for the New York Natural History Conference VI held at the New York State Museum in Albany in April 2000. The goal of the symposium was to address current geoarchaeological work in glaciated landscapes. Participants provided regional geological and archaeological summaries and case studies that demonstrated geoarchaeological approaches to archaeological site formation and site locations in formerly glaciated landscapes in the Northeast. The latter papers were organized to highlight processes in upland and valley floor settings.

Formerly glaciated terrains of the Northeast present a wide variety of landscapes that affected the location, formation, and preservation of prehistoric archaeological sites. Many of these landscapes, such as simple till-covered uplands, have been little altered since the terminal stages of the Pleistocene with the exception of soil formation (pedogenesis). Other landscapes such as glaciofluvial and glaciolacustrine valley floor environments that have undergone significant modification through Holocene alluvial and colluvial processes are more complex. Following the structure of the symposium, this volume is organized in three sections to address current geoarchaeological concepts and work in these glaciated landscapes. The first section presents regional overviews of the deglaciation (Cadwell et al.; Ridge) of the Northeast to provide a context for the subsequent chapters. The second section presents geoarchaeological concepts and studies in upland settings (Cremeens; Thorson and Tryon; Jones and Forrest; Rush et al.; Frink and Hathaway), and the third in valley floor settings (Kelly and Sanger; Sanger et al.; Curran; Thieme; Schuldenrein). These case studies provide a sense of the variety of geoarchaeological research being done in the Northeast on topics related to the selection of landscape features for settlement and postoccupation transformation of the archaeological record and the resulting contexts.

This is the first published volume devoted to geoarchaeology in the Northeast. Our goal was to gather together papers based on studies that reflect the broad range of geoarchaeological work being done across the region. Although the volume is not an exhaustive treatment of geoarchaeology in the Northeast, it will provide the reader with an understanding of the scope of work being done on the issues raised in this introductory chapter.

END NOTE

1. Throughout this volume dates preceded by cal. have been calibrated with CALIB 4.3 (Stuiver et al. 1998) unless an earlier version is noted by the chapter author(s). Dates without the cal. designation represent uncalibrated radiocarbon ages.

REFERENCES CITED

Rapp, G., Jr., and Hill, C.L. (1998). *Geoarchaeology: The Earth Science Approach to Archaeological Interpretation*. Yale University Press, New Haven, Conn.

Stuiver, M., Reimer, P.J., Bard, E., Beck, J.W., Burr, G.S., Hughen, K.A., Kromer, B., McCormac, F.G., van der Plicht, J., and Spark, M. (1998). INTCAL98 radiocarbon age calibration 24,000–0 cal. B.P. *Radiocarbon* **40**:1041–1083.

Waters, M.R. (1992). *Principles of Geoarchaeology: A North American Perspective*. The University of Arizona Press, Tucson.

SECTION I
GEOLOGICAL BACKGROUND

CHAPTER 2

GEOMORPHIC HISTORY OF NEW YORK STATE

Donald H. Cadwell, Ernest H. Muller, and P. Jay Fleisher

Through geological time, landscapes have evolved to form existing surface topography. Interaction between geomorphic process and rock resistance is typically so slow that vestiges of former landscapes may be recognized long after environmental conditions have changed. Such is the case in New York State. Geological processes in the remote past produced the present distribution of underlying rock types. Bedrock effected by geomorphic agents account for the distinctions mapped as present physiographic provinces (Figure 2.1).

PREGLACIAL DEVELOPMENT

Differences among physiographic provinces, as identified on the New York State Geological Map, reflect rock structure and distribution that had been developed before the end of the Mesozoic Era (ca. 65

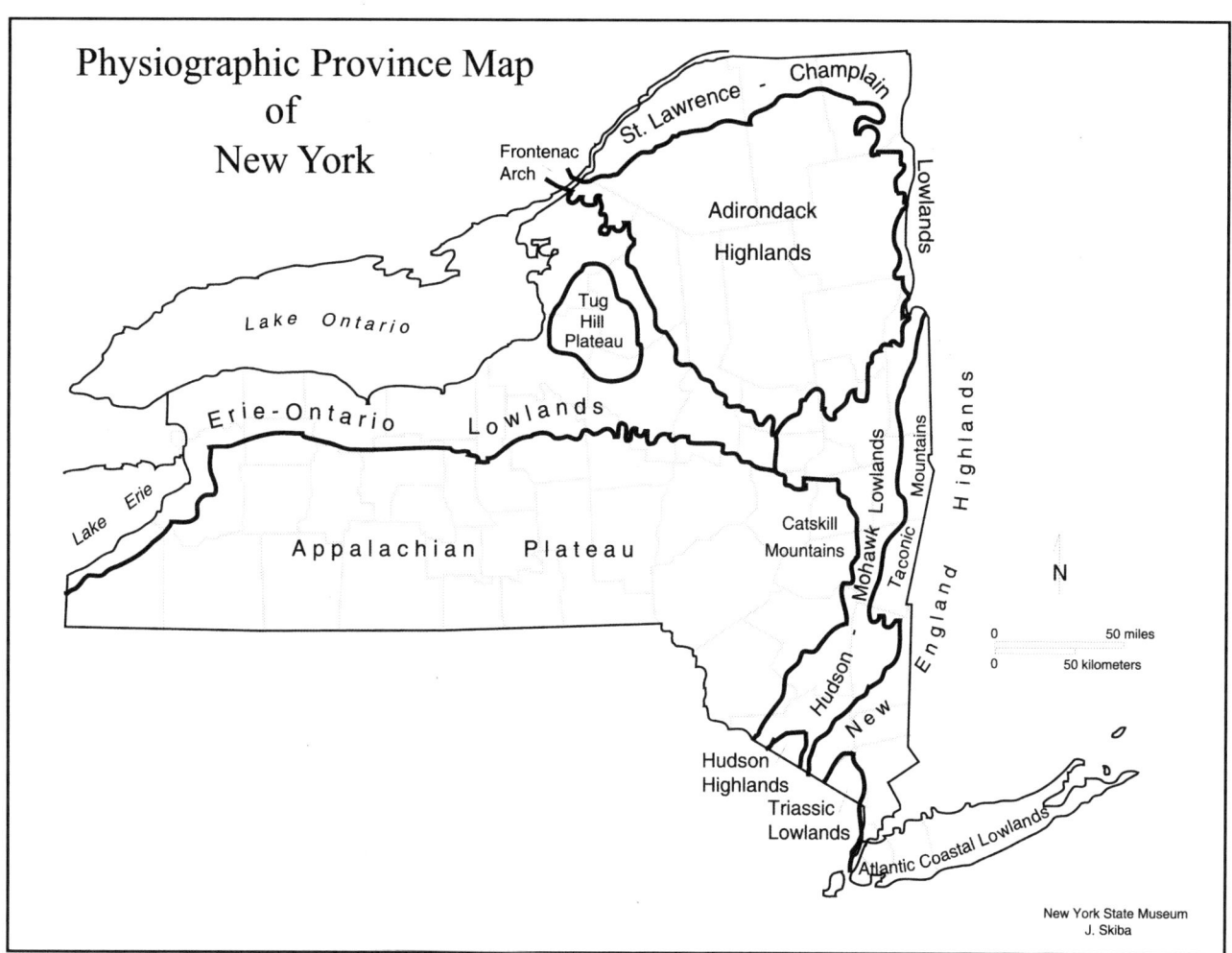

Figure 2.1. Physiographic provinces of New York.

Geoarchaeology of Landscapes in the Glaciated Northeast edited by David L. Cremeens and John P. Hart. New York State Museum Bulletin 497. © 2003 by the University of the State of New York, The State Education Department, Albany, New York. All rights reserved.

Ma [million years ago]). The New England Highlands, along the eastern border of the state, are developed on a structurally deformed, diverse association of sedimentary (shale, greywacke, and limestone) and metamorphic (phyllite, schist, quartzite, and marble) rocks (Isachsen et al. 2000). Most rugged are the Adirondack Highlands on Precambrian crystalline rocks, complex in structure and erosionally resistant. High Peaks in the eastern Adirondacks owe their prominence to the erosional resistance of Precambrian anorthosite bodies unique to the region (Isachsen et al. 2000). Northwestward, the Adirondacks are linked to the Canadian Shield by the Frontenac Arch. This area of low-lying Precambrian rocks accounts for the Thousand Islands, a threshold crossed by the St. Lawrence River with the entire outflow of the Great Lakes.

Apart from these Precambrian terranes, most of New York is underlain by minimally deformed Paleozoic (500–300 Ma) sedimentary rocks uplifted from marine and coastal depositional environments. The Erie-Ontario, St. Lawrence-Champlain and Hudson-Mohawk Lowlands are underlain by essentially undeformed and flat-lying strata of Cambrian, Ordovician, and Silurian sandstone, limestone, and shale. The Tug Hill Plateau, west of the Adirondacks, owes its elevation to a capping sandstone atop this sedimentary sequence. The most extensive physiographic province, the Appalachian Plateau, was carved on younger Devonian clastic rocks that dip gently southwestward (Isachsen et al. 2000).

Rock structure younger than Paleozoic controls topography in very limited areas of southeastern New York State. The Triassic Lowland in Rockland County is developed on a down-dropped block of shale, sandstone, and intruded volcanic basalt. The Atlantic Coastal Lowlands of Long Island consist of a mantle of glacier-derived overburden and reworked sediments draped over a gently inclined strike ridge of Cretaceous sedimentary rocks (Isachsen et al. 2000).

On foundations provided by underlying rocks, New York landscapes were shaped by differential weathering and fluvial erosion during a lengthy and complex history of denudation. Much of the state was reduced to a surface of low relief as is evidenced by uniformity of summit elevations in southwestern New York. This concept of accordant summits was originally proposed by W.M. Davis (1889) as a peneplain. It was modified by D.W. Johnson (1916) to a peneplane. Renewal of erosion etched valley and ridge topography that still characterizes Appalachian Plateau uplands in counties along New York's southern border. However, within the past 2 million years (i.e., the Pleistocene Epoch), southward expansion of the Laurentian ice sheet across New York has radically altered existing landscapes.

EVIDENCE OF MULTIPLE GLACIATIONS

The topographic expression of continental glaciation, however, is far from uniform across New York. Generally it diminishes southward. Glacial erosion and deposition are extensively expressed in the Erie-Ontario Lowland and in an adjacent east-west Appalachian Plateau that contains northern portions of the Finger Lake Basins (Figure 2.2). To the south, in a zone that includes the southern segments of the Finger Lakes, relief is moderate to high. Glacial troughs are deeply incised and through valleys are scoured across the north-south drainage divide. Along the Pennsylvania border, relief is likewise moderate to high; uplands bear only superficial evidence of glacial erosion, and valleys follow courses minimally deranged from preglacial patterns. This southward decrease in intensity of glacial modification of topography resulted because of the relatively high relief and elevation of the plateau prior to glaciation, but also because of diminishing duration and frequency of glacial expansion southward across the state.

Although an advancing ice sheet tends to efface depositional features of previous glaciations, evidence is clear that New York experienced multiple episodes of glaciation (Figure 2.2). Striking evidence of this fact involves glacially induced deflection of the Chemung River, from the broad valley it occupies upstream from Big Flats, into the canyon between Big Flats and Elmira. Glacial sediments inset within the canyon reach could have been deposited there only after initial stream derangement. Similarly, though perhaps less obviously, major scour of through valleys and deepening of the Finger Lake troughs (some of which bottom well below sea level) predate the last major glacial advance (i.e., Wisconsinan glaciation).

Preglacial streams in western New York drained northward off the Appalachian Plateaus. In the Salamanca Reentrant, where the limit of Wisconsinan glaciation has its most northerly position in New York, glacial damming of Allegheny headwaters diverted drainage southward across the present Kinzua Dam site in northwestern Pennsylvania. Isolated areas of deeply weathered outwash, kame gravel, and till are preserved south of the Allegheny River, beyond the limits of Wisconsinan glaciation. This suggests deposition during a pre-Wisconsinan glacial advance perhaps more than 30,000 years ago. The conversion of radiocarbon years ago to calendar years ago with CALIB 4.3 is valid only to 20,265 radiocarbon years; for example, 20,200 B.P. equals cal. 23,924 B.P. (Stuiver et al. 1998).

Southeast of the Adirondack Highlands is evidence of a branching interglacial river system concealed

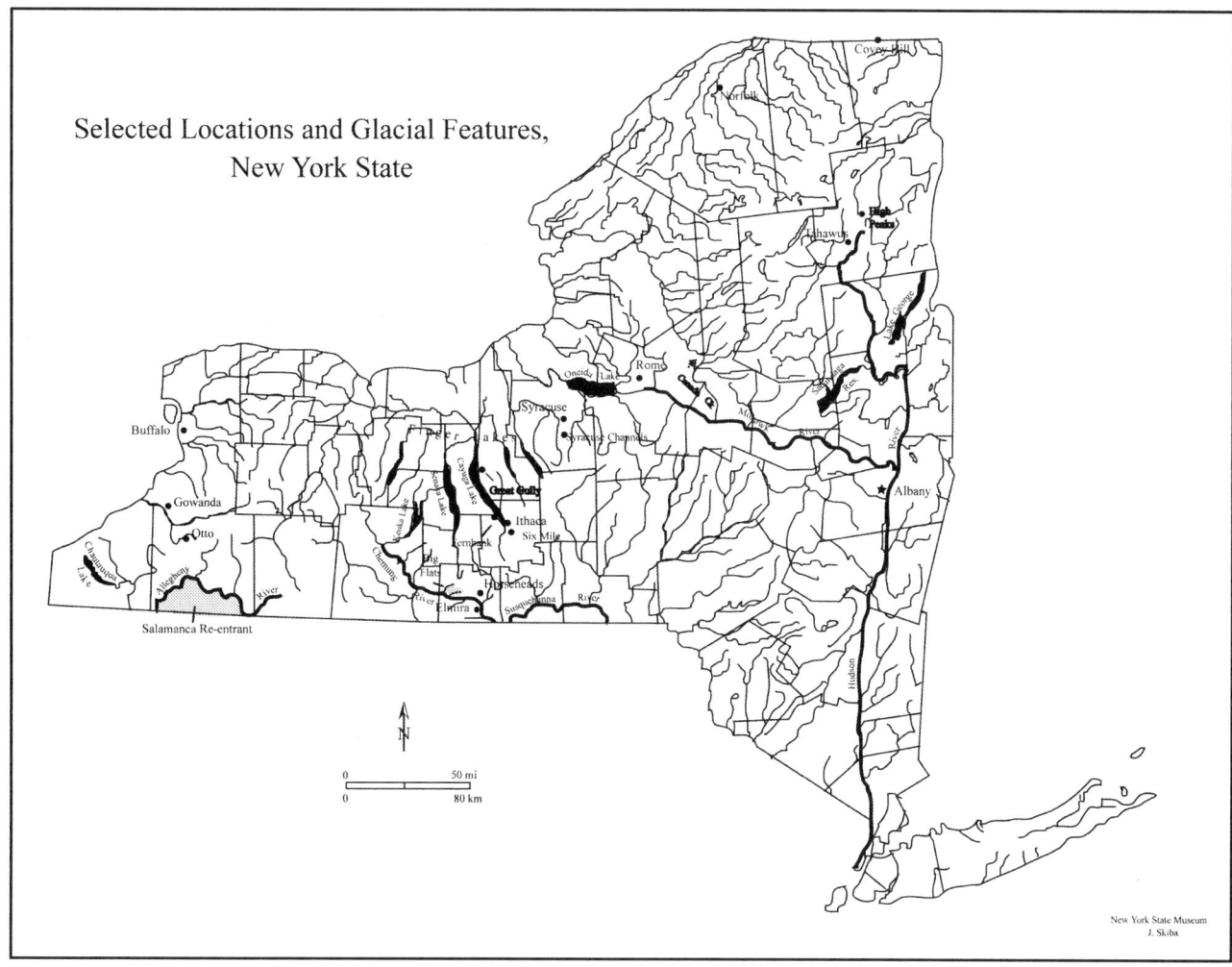

Figure 2.2. Locations and geological features discussed in text.

beneath late Wisconsinan glaciolacustrine (lake) sediments (Dineen and Hanson 1983). The western branch of this system, Mohawk Channel, was the pre-late Wisconsinan, and possibly preglacial, Mohawk River (Figure 2.3). Colonie Channel, draining the Lake George basin and the eastern Adirondacks, had a southward course that passed beneath present Saratoga and Round Lakes was confluent with the Mohawk Channel near Albany. These channels were confluent with the Battenkill-Hudson south of Albany, and were a predecessor of the Hudson River.

In widely scattered locations, remnants of older drift deposited in gullies transverse to the main glacial flow were protected from subsequent erosion and buried beneath younger glacial deposits. Such is the stratigraphic record in valleys of Six Mile Creek (Bloom 1967, 1972) and Great Gully (Shumaker 1957), east flank tributaries to Cayuga trough. Organic materials in stratified sediment between till sheets have been radiocarbon-dated at more than 30,000 years ago. At Fernbank, on the west shore of Cayuga Lake a few miles north of Ithaca, sediments contain plant debris and freshwater shells that indicate glacial impoundment in the Cayuga Valley more than 50,000 years ago (Figure 2.2). Repeated glacial episodes are also documented at Tahawus, in the central Adirondack Highlands, where wood fragments and plant debris more than 40,000 years ago have been found in nonglacial lacustrine sediment preserved between two tills. At Otto (MacClintock and Apfel 1944; Muller 1960, 1965; Suess 1954) and Gowanda (Muller 1960; Rubin and Alexander 1960) in western New York, organic remains dated from 52,000 to 64,000 years ago indicate boreal conditions are preserved between underlying and overlying glacial deposits. These sites provide clear evidence of multiple glacial episodes alternating with intervening nonglacial intervals. Nonetheless, because each glacial advance tends to

Chapter 2 *Geomorphic History of New York State* **9**

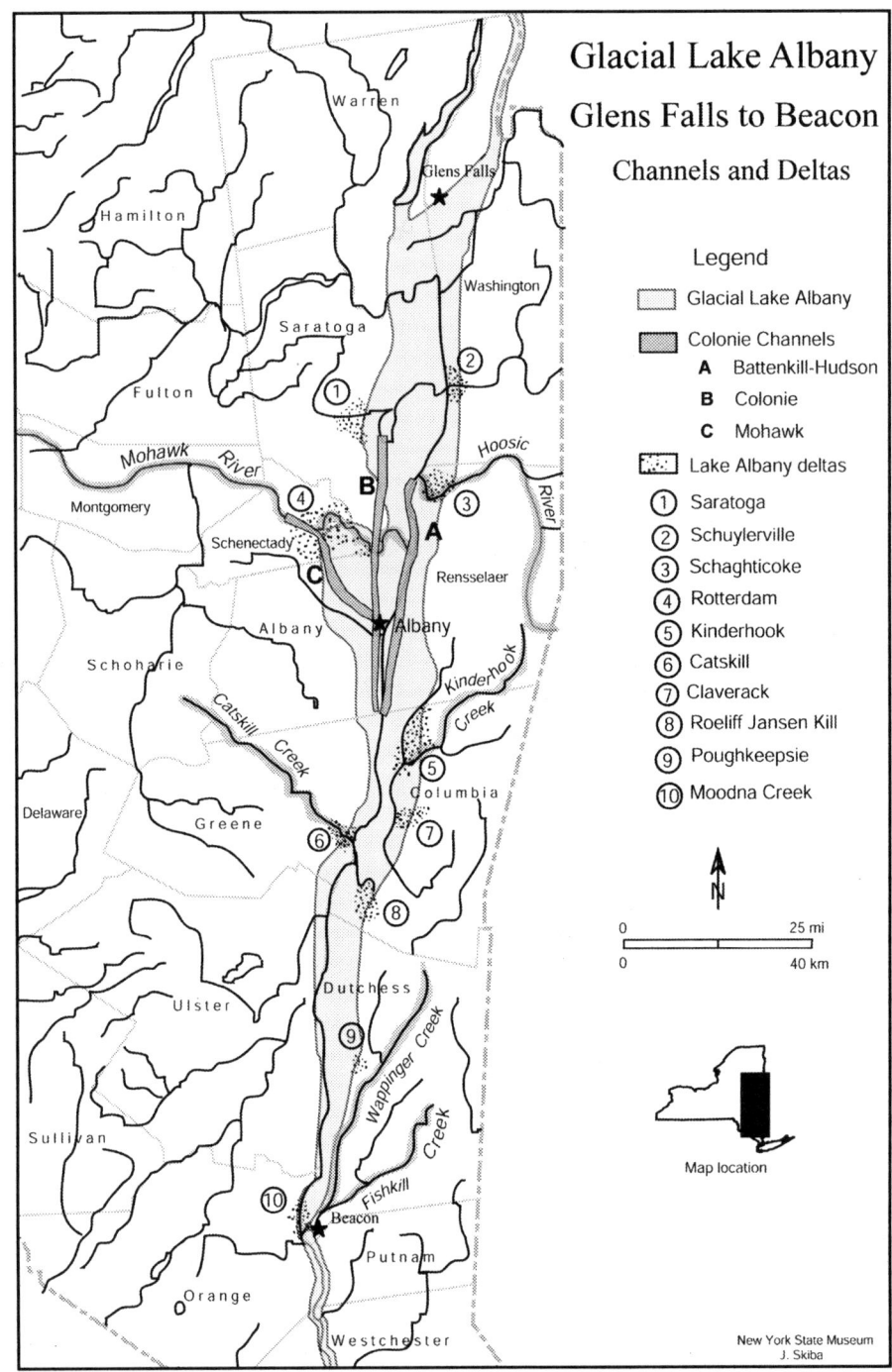

Figure 2.3. Central portion of glacial Lake Albany with location of deltas and the buried Colonie Channels.

erase previous glacial landscapes, most deposits and depositional landscape features in New York are products of late Wisconsinan glaciation.

WISCONSAN GLACIATION

Patterns of ice flow during a glacial advance are largely controlled by preexisting bedrock topography and relief. Glacial erosion is effective on limestone and shale, but less so on sandstone and least effective on regions of crystalline rock. As the Laurentide ice sheet moved into New York State, the Adirondack Highlands obstructed main flow so that the ice spread around the Adirondacks with lobate front southward into the Champlain Valley and south-westward over the Tug Hill Plateau and into the Ontario Lowlands. Eventually the thickening ice sheet topped the Adirondack Highlands and continued its advance into southern New York and Pennsylvania. About cal. 24,300 B.P. (Sirkin 1980), the ice sheet built its Wisconsinan Terminal Moraine along the Atlantic Coast on Long Island (Figure 2.4, nos. 12, 13, 14).

After reaching its maximum extent, the glacier front began a halting northward recession, frequently interrupted by minor readvances. Indicative of ice-marginal accumulation, the east-west orientation of Long Island contains the diagnostic mid-island kame and kettle topography of the Ronkonkoma Moraine that trends to the south shore at Montauk Point. In contrast with the common sandy beaches of the south shore outwash is an irregular north shore terrane controlled by moraines (Figure 2.4, nos. 9, 10, 11), thought to have been constructed during hesitated retreat prior to deglacial uncovering of Long Island Sound. Styles of deglaciation differed among physiographic regions. Recession of the ice margin was accompanied by thinning and regional downwasting thus uncovering upland summits. A few of the highest summits in the Adirondacks briefly supported small cirque glaciers, but waning ice, characterized by thinning and regional downwasting, left areas of remnant ice in valleys of the central and southern Adirondacks. Sediment washed onto dead ice locally resulted in kame and kettle topography, while sinuous esker ridges mark vestiges of subglacial meltwater streams.

The first of several large ice-contact lakes to form at the Laurentide ice sheet margin in New York devel-

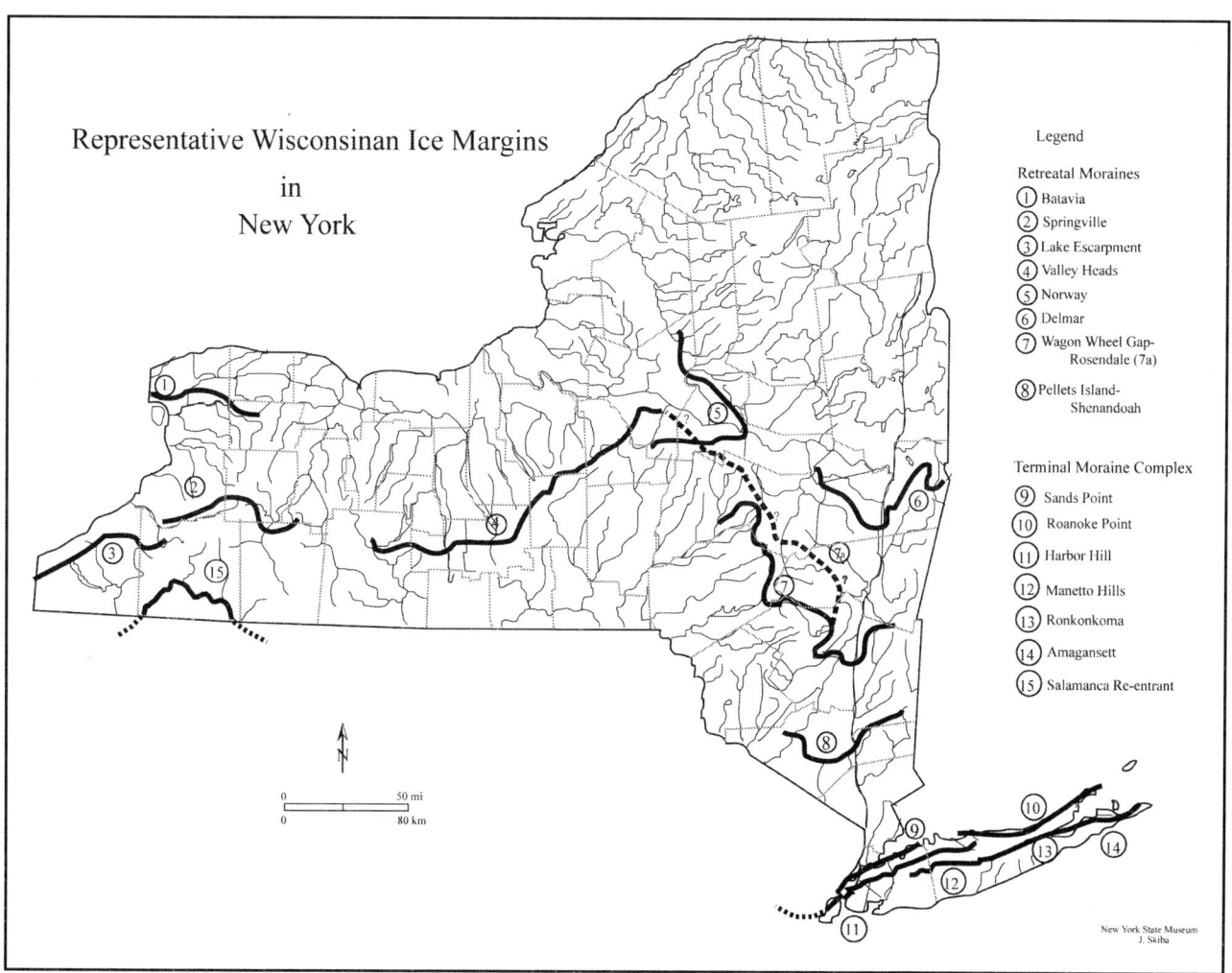

Figure 2.4. Representative Wisconsinan ice margins in New York. Cadwell et al. (1986) illustrate more than 50 ice-margin positions.

oped in the Hudson Valley. Although still open to speculation, the blockage that impounded Lake Albany is thought to have been in the vicinity of New York City, where the Terminal Moraine crosses the Hudson River to Staten Island. The unmistakable evidence of lake sediments extends northward past Glens Falls, through the Champlain Valley. Figure 2.3 illustrates the location of major deltas constructed into glacial Lake Albany during deglaciation (Connally and Sirkin 1973).

In areas of moderate relief, such as the Appalachian Plateau, upland areas were uncovered during glacier recession before adjacent lowlands, thus forming valley ice tongues. Similarly the Taconic and Catskill Mountains were exposed through the thinning ice while ice tongues remained active within the Hudson Valley. Recessional moraines formed following minor glacial readvances or when the ice margin stagnated for a time. General locations of the ice margins at successive positions during general retreat have been traced from the Taconic Mountains, across the Hudson Valley, through the Catskill Mountains and westward into the Appalachian Uplands. Drift deposits in the valleys commonly consist of lake silts greater than 125 m thick and stratified drift whereas on the uplands till is generally less than 5 m.

RECESSION FROM THE VALLEY HEADS MORAINE

As long as the glacier margin remained in the headwaters of the Susquehanna, meltwater escaped freely, spreading outwash gravels in the valleys to the south. However, as glacier recession proceeded north of the bedrock divide, meltwater was impounded in many valleys. About cal. 16,007 B.P., minor fluctuations of the ice margin tended to reach just about the same position without spreading across the divide. Such ice marginal fluctuation built a massive complex of valley-blocking, ice-marginal deposits (Coates 1971). This complex, the Valley Heads Moraine, comprises the present watershed divide between Susquehanna and St. Lawrence drainage basins across the breadth of the Finger Lakes region (Muller and Calkin 1993).

Recession of the ice margin from the Valley Heads Moraine impounded small lakes in major troughs between glacier terminus and moraine dams. Most of these "primitive lakes" initially drained southward across the moraine barrier. Continued recession, although interrupted by minor readvances, uncovered alternative outlets at lower elevations or permitted coalescence of the northward-expanding marginal lakes. Thus ancestral Lake Ithaca in the Cayuga Basin joined Lake Watkins in the Seneca Basin to form Lake Newberry with its outflow south past Horseheads into the Chemung Valley. At maximum extent, Lake Newberry extended into the Keuka Basin as well and received drainage from the Finger Lake troughs as far east as Marcellus, 25 km southwest of Syracuse.

As northward recession of the ice margin exposed lower areas of the uplands, southward drainage of impounded glacial meltwater ended in central and western New York. Instead, lakes contained between retreating ice and deglaciated upland—as in the Finger Lake Basins, the Genesee Valley, and in westernmost New York—developed lower outlets eastward toward the Mohawk Valley. Major meltwater streams carved deep valleys along the north margin of the plateau. Noteworthy among these are the Syracuse Channels (Smoky Hollow, Clark Reservation, Rock Cut, Nottingham, Meadowbrook, and Erie). Withdrawal of the ice border in westernmost New York permitted eastern outflow from glacial Lake Warren (the predecessor to Lake Erie), draining from as far west as northwestern Ohio. As ice withdrawal continued, expansion of impounded waters in the Ontario-Erie Lowlands may have accelerated calving retreat of the glacier front, but also provided conditions that facilitated intermittent readvance.

Ice recession across the Erie-Ontario Lowlands (cal. 16,000–14,000 B.P.) uncovered a landscape dominated by drumlins and areas of kame and kettle topography. It also brought into existence Lake Iroquois, the sizable predecessor of Lake Ontario, with outlet east to the Mohawk Valley across the col near Rome. For many miles both east and west of Rochester, a well-developed barrier beach system was built by wave action along the Lake Iroquois shore.

Continued glacial retreat from the Ontario Basin allowed glacial Lake Iroquois to spread into the St. Lawrence Lowland. Relict strandline features and fine-grained bottom sediments associated with these regional water bodies are found along the eastern shore of Lake Ontario, and along the northern flank of the Adirondack Mountains. When the receding ice margin uncovered alternative outlets on the northern flank of the Adirondacks at Covey Hill, Lake Iroquois outlet to the Mohawk River was replaced by outflow to the St. Lawrence Lowland and directed flow through the Champlain and Hudson Lowlands to the North Atlantic off Long Island (Pair and Rodriquez 1993). Recession of the ice margin northward into Quebec (cal. 13,460 B.P.; Occhietti 2001) opened outflow of St. Lawrence River to the sea. Subsequent rise of sea level due to melting of the world's ice sheets transformed the St. Lawrence Valley and Lake Champlain Valley into the Champlain Sea. The westernmost incursion of marine waters (ca. cal. 13,000 B.P.) reached essentially to the eastern end of the Ontario Basin. Responding more slowly than the climate-induced sea-level rise, and indeed continuing still today, is crustal response to the diminished load due to melting of the ice sheet. This isostatic rebound of the crust left raised beaches, as well as marine clay and silt, yielding mollusk shells and other fossils including bones of a beluga whale found in 1988 near Norfolk in the St. Lawrence Lowland (Pair et al. 1988).

POSTGLACIAL OR HOLOCENE

The beginning of the Holocene is primarily recognized to coincide with the retreat and disappearance of midlatitude continental ice. However, based on stratigraphic and geomorphic evidence, such a boundary must be time transgressive. The Wisconsinan ice margin retreated from its maximum position in southern New York, New Jersey, and Pennsylvania 34,300 B.P., finally retreating into Canada ca. cal. 13,460 B.P. In doing so, the proximity of the receding ice margin favored environments leading to such features as moraines, outwash aprons, deflected stream channels, and impounded lakes. A late pulse of glacial regeneration recognized to have occurred elsewhere, the Younger Dryas, remains elusive yet is to be sought by indirect evidence.

As indicated in previous discussion, progressive northward withdrawal of the ice margin established favorable conditions, thus setting the stage for human occupation. Rising sea level and rebounding land resulted from melting of northern ice remnants. Changing environments are (1) reflected in the transition from glacial outwash to lower terraces, and then from the terraces to the present flood plain (Scully and Arnold 1981); (2) reflected in uninterrupted lake sediments (Mullins 1996); (3) recorded in pollen (Miller 1973) and other fossil evidence contained therein; and (4) illustrated with the periods of alluviation that dominated during the early Holocene and degradation associated with episodes of channel migration during the late Holocene (Knox 1983). With passing of the glacial record, interpretation of the archaeogeological history and the environments encountered by humans remains to be defined in detail, yet field evidence indicates floodplain sites along major avenues of migration and travel were favored.

REFERENCES CITED

Bloom, A.L. (1967). "Fernbank": A rediscovered Pleistocene interglacial deposit near Ithaca, New York. *Geological Society of America*, Abstracts, NE Section, p.15.

Bloom, A.L. (1972). Friends of the Pleistocene. 35th Annual Reunion, Cornell University, Ithaca, N.Y.

Cadwell, D.H. et al. (1986). Surficial Geologic Map of New York. New York State Museum Map and Chart Series No. 40, 5 sheets and authors:
Finger Lakes Sheet. (1986). Edited by E.H. Muller and D.H. Cadwell.
Hudson-Mohawk Sheet. (1987). Edited by D.H. Cadwell and R.J. Dineen.
Niagara Sheet. (1988). Edited by D.H. Cadwell.
Lower Hudson Sheet. (1989). Edited by D.H. Cadwell.
Adirondack Sheet. (1991). Edited by D.H. Cadwell and D.L. Pair.

Coates, D.R., Landry, S.O., and Lipe, W.D. (1971). Mastodon bone age and geomorphic relations in the Susquehanna Valley. *Geological Society of America Bulletin* **82**:2005–2010.

Connally, G.G., and Sirkin, L.A. (1973). Wisconsinan history of the Hudson-Champlain lobe. In *The Wisconsinan Stage*, edited by R.F. Black, R.P. Goldthwait, and H.B. Willman, pp. 47–69. Geological Society of America Memoir No. 136.

Davis, W.M. (1889). Topographic development of the Triassic Formation of the Connecticut Valley. *American Journal of Science* **37**:423–434.

Dineen, R.J., and Hanson, E.L. (1983). *Bedrock Topography and Glacial Deposits of the Colonie Channel Between Saratoga Lake and Coeymans, New York*. New York State Museum Map and Chart Series No. 37, The University of the State of New York, Albany.

Isachsen, Y.W., Landing, E., Lauber, J.M., Rickard, L.V., and Rogers, W.B. (2000). *Geology of New York: A Simplified Account (2nd ed.)*. New York State Museum Education Leaflet No. 28, The University of the State of New York, Albany.

Johnson, D.W. (1916). Plains, planes, and peneplains. *Geographical Review* **1**:443–447.

Knox, J.C. (1983). Responses of river systems to Holocene climates. In *Late Quaternary Environments of the United States, The Holocene, Volume 2*, edited by H.E. Wright, Jr., pp. 26–41. University of Minnesota Press, Minneapolis.

MacClintock, P., and Apfel, E.T. (1944). Correlation of the drifts of the Salamanca Reentrant, New York. *Geological Society of America Bulletin* **55**:1143–1164.

Miller, N.G. (1973). *Late-Glacial and Postglacial Vegetation Change in Southwestern New York State*. New York State Museum Bulletin No. 420, The University of the State of New York, Albany.

Muller, E.H. (1960). Glacial geology of Cattaraugus County, New York. *Guidebook, 23rd Reunion, Friends of Pleistocene*. Geology Department, Syracuse University, Syracuse, N.Y.

Muller, E.H. (1965). Adirondack Mountains. In *A Guidebook; New England, New York State*, edited by J.H. Hartshorn, E.H. Muller, and J.P. Schafer, pp. 47–53. INQUA 7th Conference, Boulder, Colo.

Muller, E.H., and Calkin, P.E. (1993). Timing of Pleistocene events in New York State: *Canadian Journal of Earth Sciences* **30**:1829–1845.

Mullins, H.H. et al. (1996). Seismic stratigraphy of the Finger Lakes: A continental record of Heinrich event H-1 and Laurentide ice sheet instability. In *Subsurface Geologic Investigations of New York Finger Lakes: Implications for Late Quaternary Deglaciation and Environmental Change*, edited by H.T. Mullins and N. Eyles, pp. 1–35. Geological Society of America Special Paper No. 331.

Occhietti, S. (2001). Deglaciation of the Middle Estuary and Charlevoix: An overview. In *Stratigraphy of the Units on Land and Below the St. Lawrence Estuary, and Deglaciation Pattern in the Charlevoix*, edited by S. Occhietti, pp. 11–20. Guidebook 64th Annual Reunion, Northeast Friends of Pleistocene, Quebec City.

Pair, D.L., Muller, E.H., and Anderson, T.W. (1988). Glacial morphostratigraphic relationships and ice recession on the northwest Adirondack slope and adjacent lowlands, New York. *Geological Society of America Abstracts* **20**:60.

Pair, D.L., and Rodriguez, C.G. (1993). Late Quaternary deglaciation of the southwestern St. Lawrence Lowland, New York and Ontario. *Geological Society of America Bulletin* **105**:1151–1164.

Rubin, M., and Alexander, C. (1960). U.S. Geological Survey Radiocarbon Dates V: *Radiocarbon Supplement* **2**:129–185.

Scully, R.W., and Arnold, R.W. (1981). Holocene alluvial stratigraphy in the upper Susquehanna River Basin, New York: *Quaternary Research* **15**:327–344.

Shumaker, R.C. (1957). Till texture variations and Pleistocene deposits of the Union Springs and Scipio Quadrangles, Cayuga County, New York. Unpublished master's thesis, Cornell University, Ithaca, N.Y.

Stuiver, M., Reimer, P.J., Bard, E., Beck, J.W., Burr, G.S., Hughen, K.A., Kromer, B., McCormac, F.G., van der Plicht, J., and Spark, M. (1998). INTCAL98 radiocarbon age calibration 24,000–0 B.P. *Radiocarbon* **40**:1042–1083.

Suess, H. (1954). U.S. Geological Survey Radiocarbon Dates I. *Science* **120**:467–473.

CHAPTER 3

THE LAST DEGLACIATION OF THE NORTHEASTERN UNITED STATES: A COMBINED VARVE, PALEOMAGNETIC, AND CALIBRATED ^{14}C CHRONOLOGY

John C. Ridge

^{14}C-BASED DEGLACIATION MODELS IN THE NORTHEAST

Recent studies have suggested that humans were near the limit of glaciation in the northeastern United States when it was still partly covered by glacial ice (see Marshall 2001 for review). As a result, a better understanding of deglaciation chronology and its associated climate variations have become more important to archaeological investigations. How and when different areas of the northeastern United States became ice-free at the end of the last glaciation places a fundamental constraint on the earliest possible human presence in the region. Habitation and migration patterns in addition to food sources would have been greatly influenced by proximity to ice, postglacial climate oscillations, sea-level change, and the existence of large glacial lakes that sometimes lingered thousands of years after ice recession. It is the goal of this chapter to summarize the late Wisconsinan deglaciation of the northeastern United States and its associated climate oscillations and landscape characteristics. I will leave the issue of when the first humans arrived in the region to archaeologists who are more qualified to make this assessment.

Over the past 50 years Quaternary geologists have used ^{14}C ages from organic sediment in preglacial, glacial, and postglacial deposits to infer chronologies for the last glaciation and postglacial events in the northeastern United States (Schafer and Hartshorn 1965; Connally and Sirkin 1973; M.B. Davis et al. 1980; R.B. Davis and Jacobson 1985; Hughes et al. 1985; Muller and Prest 1985; B. Stone and Borns, 1986; Dyke and Prest 1987; Muller and Calkin 1993). However, few of these ^{14}C ages are precisely tied to glacial events and many of them have ambiguities when used to interpret glacial chronology. The scarcity of unambiguous ^{14}C ages has prompted the reliance on less reliable materials for ^{14}C dating and has contributed to the potential inaccuracy of deglaciation models. Specifically, bulk sediment (or whole-sediment) lacustrine and marine ^{14}C ages have been used verbatim or without sufficient corrections. An additional problem has been the scarcity of detailed reconstructions of end moraines and ice recessional positions that cross the entire region, especially in southern New York State. As a result chronological interpretations have varied greatly and have sometimes overstepped the bounds of existing data.

Problems associated with ^{14}C ages from some organic materials and their required correction factors have been at the heart of chronological uncertainties. Bulk sediment ^{14}C ages from postglacial lake sediment have been used to define limiting ages for ice recession. However, bulk lacustrine sediment ages can have potentially large errors (Shotton 1972; Oeschger et al. 1985; Andrée et al. 1986; Cwynar and Watts 1989; Törnqvist et al. 1992; Wohlfarth et al. 1993, 1995; Lowe et al. 1995; Abbott and Stafford 1996; Wohlfarth 1996; Björck, Bennike et al. 1998; Turney et al. 2000). Both the uptake of dissolved carbon by aquatic plants from water that is not in equilibrium with the atmosphere, and the direct inclusion of sediment containing old carbon has caused ages as much as 10,000 years or more too old in the northeastern United States and adjacent Quebec with errors on the order of 1,000–2,000 years being common (Mott 1975, 1981; Karrow and Anderson, 1975; Miller and Thompson 1979; P.T. Davis and Davis 1980; P.T. Davis et al. 1995; Lini et al. 1995; see Ridge et al. 1999 for a review).

An additional problem is the use of ^{14}C ages from fossils of marine organisms that acquired their carbon from seawater that was not fully equilibrated with the atmosphere. Seawater in glacial and arctic marine environments generally has an anomalous old age that can be the result of (1) an incomplete exchange with

---

Geoarchaeology of Landscapes in the Glaciated Northeast edited by David L. Cremeens and John P. Hart. New York State Museum Bulletin 497. © 2003 by the University of the State of New York, The State Education Department, Albany, New York. All rights reserved.

the atmosphere, (2) the introduction of glacial meltwater carrying dissolved carbon from ancient melting ice or bedrock sources, and (3) poorly ventilated deep-water mixing with surface water (Sutherland 1986; Bard 1988; Rodrigues 1988, 1992; Kasgarian, 1992). Although errors in marine ^{14}C ages have been recognized for the past 30 years (Mangerud 1972; Hjort 1973; Mangerud and Gulliksen 1975; Southon et al. 1990; Bard et al. 1994; Austin et al. 1996; Birks et al. 1996; Bondevik et al. 1999), the application of a marine reservoir correction to specific samples from a glacial marine environment has remained elusive. Reservoir errors may have multiple causes and can vary spatially, bathymetrically, and temporally on a global (Bard 1988; Stocker and Wright 1998) and local (Anderson 1988; Hillaire-Marcel 1988; Rodrigues 1988, 1992; Bondevik et al. 1999) scale. In southern Maine an average Pleistocene marine reservoir correction on the order of 800–1,000 years is supported by the ^{14}C ages of nearby varve sections (revision of Ridge et al. 2001 to refined time scale of this chapter).

Although there was tremendous improvement in the precision and accuracy of ^{14}C ages with the advent of accelerator mass spectrometer measurements, it did not initially lead to a significant refinement of ^{14}C chronologies in the northeastern United States (Borns 1998). Recently there has been a focus on the ^{14}C dating of terrestrial plant macrofossils that eliminate the ambiguities associated with lacustrine bulk sediment and marine samples (Thompson et al. 1996; Dorian 1997; Ridge et al. 1999). This chapter will show a chronology for deglaciation based on selected ^{14}C ages of terrestrial plant macrofossils taken from varved glacial and nonglacial lacustrine sediment. In addition, the New England varve chronology and paleomagnetic declination records will be used to establish regional correlations and formulate a deglaciation chronology for the entire Northeast. The chronology is well anchored along major valleys where glacial lakes accumulated silt and clay varves that allow the application of both varve chronology and paleomagnetism. The ^{14}C calibration of the varve chronology as well as the calibration of ^{14}C ages to approximate calendar years has allowed a more complete formulation of numerical ages for deglaciation at a higher resolution than was previously possible. The use of multiple correlation and relative dating techniques that crosscheck each other, along with ^{14}C ages from terrestrial plant macrofossils, has revised and strengthened the deglacial chronology as compared to previous reconstructions based on limited ice margin reconstructions and less accurate chronologic data.

VARVE CHRONOLOGY IN THE NORTHEAST

In 1920 Ernst Antevs arrived in North America from Sweden as part of a varve expedition led by Gerard De Geer who has long been recognized as the founder of varve chronology. Almost single-handedly Antevs (1922, 1928) created the New England varve chronology. Except for a long sequence in New Jersey (Reeds 1926), Antevs assembled every other glacial varve sequence in the northeastern United States (Figure 3.1).

The New England Varve Chronology

Antevs' work began in the Connecticut Valley of central Connecticut (Figure 3.1) where he worked northward, compiling two nonoverlapping varve sequences (lower and upper Connecticut Valley varves) that are generally referred to as the New England varve chronology (Figure 3.2; Antevs 1922). The lower Connecticut Valley varves, compiled from Hartford to Claremont, were arbitrarily numbered New England varve years (NE yr) 3001–5600, 5709–6277.[1] Taking advantage of correlations to the Hudson (Catskill to Cohoes, New York, NE yr 5501–5800), Merrimack (NE yr 5709–5749, 5771–6352), and Ashuelot (at Keene, New Hampshire, NE yr 5687–5733, 5804–5879) Valleys, Antevs was able to fill gaps in the lower Connecticut Valley varves and extend the sequence. The upper Connecticut Valley varves were compiled at sections in the Connecticut and Passumpsic Valleys from Claremont to St. Johnsbury and were arbitrarily numbered NE yr 6601–7400.

After compiling varve records in Canada (Antevs 1925, 1928) Antevs' work on the New England varve chronology continued (Antevs 1928). Additional work near Hartford and at Newburgh, New York, in the Hudson Valley produced correlations that extended the lower Connecticut Valley sequence to NE yr 2701 (Figures 3.1 and 3.2). Antevs' continued work in the Passumpsic Valley near St. Johnsbury extended the upper Connecticut Valley sequence to NE yr 7750, and he established a correlation to the Winooski Valley (NE yr 7059–7288) in north-central Vermont.

Antevs could not find a match[2] of the lower and upper Connecticut Valley varve sequences. He inferred that there was a gap between the sequences arbitrarily spanning 249 years and here called the Claremont Gap (Figure 3.2). Antevs interpreted the Claremont area as the vicinity of a long stillstand of the receding ice sheet that allowed the lower Connecticut Valley varves to be deposited while areas to the north remained ice covered. Over the past decade my students and I worked with the assumption that there

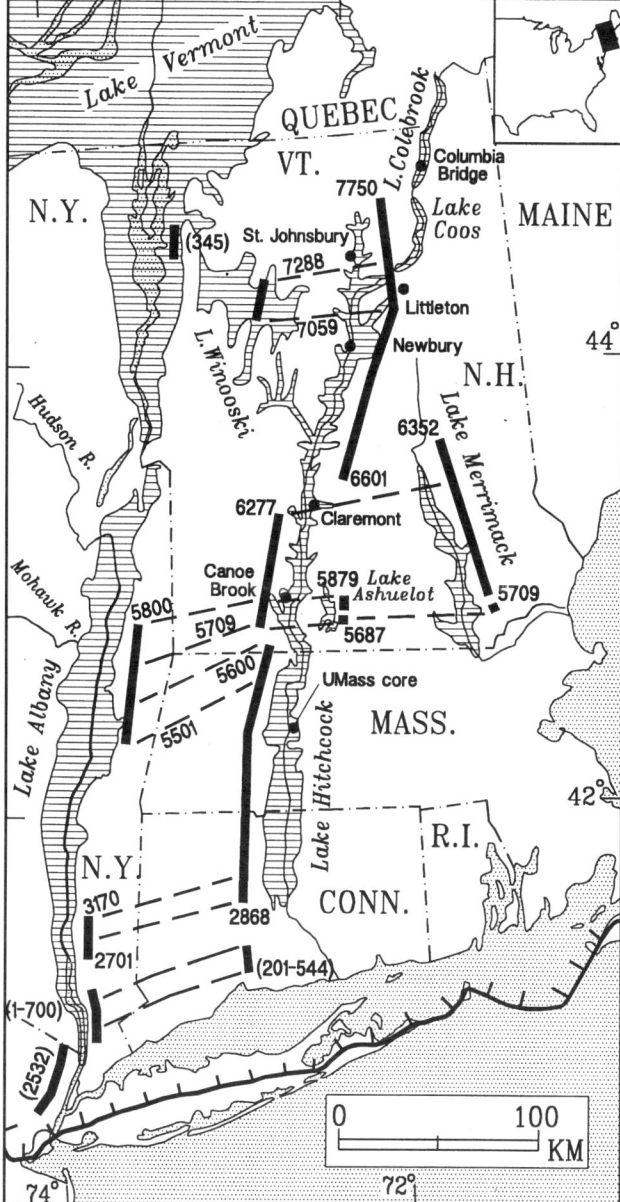

Figure 3.1. Map of western New England and adjacent New York State showing the approximate geographic positions and varve numbers of long measured varve sequences in the northeastern United States. NE yr 2701–6352 and 6601–7750 represent the original New England varve chronology of Antevs (1922, 1928). Varve numbers in parentheses are varve sequences that have not been matched to the New England varve chronology. The ice margin along the southern coast of New England is the maximum extent of late Wisconsinan glaciation.

was no delay in deglaciation because field evidence for a major stillstand event appeared to be absent (Ridge et al. 1996, 1999, 2001). In addition, close inspection of the varve sequences revealed a crude similarity between them with lower Connecticut NE yr 6012 correlative to upper Connecticut NE yr 6601 (Ridge et al. 1996, 1999). The overlap of the two sequences was exactly consistent with continuous rapid ice recession through the Charlestown to Claremont, New Hampshire, area without any delay due to a stillstand of the receding ice front. The overlap was also consistent with paleomagnetic records and ^{14}C ages in both sequences as much as their accuracy and precision allows such a test (Ridge et al. 1999). The weak similarity and apparent lack of a clear match of the two sequences appeared to be the result of exceedingly thick varve deposition (up to 4 m) in the base of the upper Connecticut Valley varves near the mouths of several large rivers. Nonweather-related processes were thought to have masked an annual weather signal in thick upper Connecticut Valley varves and made correlation by varve matching difficult. Over the past 2 years detailed surficial mapping between Charlestown and Claremont has revealed three end moraines associated with small readvances and a 200-yr sequence of persistently thick varves to the south, indicating a delay in deglaciation (Ridge 1999, 2001b). In addition eight new varve sections were discovered in the area that do not match either the lower or upper Connecticut Valley varve sequences and indicate that there is missing varve stratigraphy. Unfortunately none of these new sections is long or complete enough to compile the missing sequence. These new discoveries entirely negate the assumption of continuous rapid deglaciation and also refute the proposed overlap of the lower and upper Connecticut Valley varve sequences. This chapter will later present a revised calibration of the varve chronology that incorporates the Claremont Gap into a calibration of the varves.

During the 1930s the New England varve chronology came under assault. Flint (1929, 1930, 1932, 1933) and later Goldthwait (1938) erroneously interpreted the deglaciation of New England as being the result of regional stagnation of the last ice sheet. Flint's regional stagnation model was incompatible with the systematic onlapping of varves from south to north in the wake of the receding ice sheet (Antevs 1922, 1928) and the interpretations of Lougee (1940). Flint (1930) created serious doubts as to whether the varve chronology was correct but recanted some of his criticisms in later publications (Flint 1932, 1933). It was also during this time that De Geer (1921, 1927, 1929, 1940) proposed annual transatlantic and interhemispheric correlations based on the matching of individual varve records. Most geologists then and today, including Antevs (1935), thought global correlations of this type

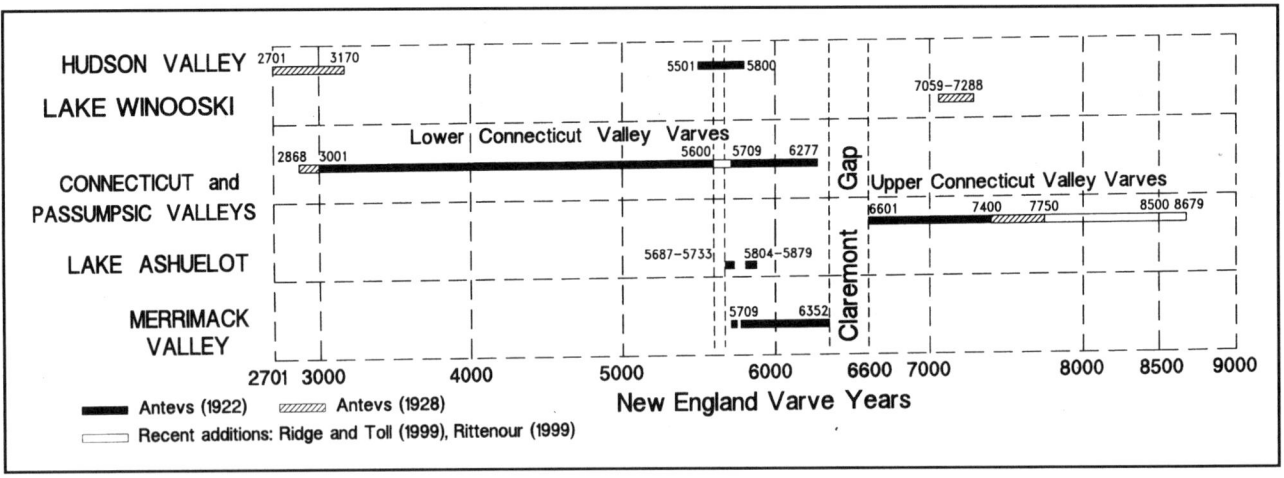

Figure 3.2. Varve sequences of the New England varve chronology as it stands today.

were impossible, leading to further erosion of the perceived accuracy of varve chronology. Later Flint (1956) used erroneous interpretations based on some of the first ^{14}C ages in the Connecticut Valley to refute the New England varve chronology. Despite objections from Antevs (1962), most glacial geologists in the United States came to think of the varve chronology as an outdated and inaccurate technique. Detailed surficial mapping across New England by the U.S. Geological Survey has revealed a systematic south to north recession of the margin of an active ice sheet, a model consistent with Antevs' work (Jahns and Willard 1942; Koteff 1974; Koteff and Pessl 1981). However, by the 1960s the credibility of the varve chronology had been damaged.

Varve Records Outside the New England Varve Chronology

In addition to the New England varve chronology, three other long sequences of varves were compiled in the northeastern United States (Figure 3.1). None of these sequences apparently matches or overlaps the New England varve chronology, and they all have separate numbering systems. None of these sequences has been calibrated, and their importance in the history of deglaciation is that they provide minimum estimates for the duration of glacial lakes and the ages of some glacial events. The oldest of the sequences is in New Jersey and was compiled by Chester Reeds (1926) at the American Museum of Natural History in New York from samples of varves deposited in several different lake stages in the Hackensack Valley (Stanford and Harper 1991). This 2,538-yr sequence was arbitrarily numbered from -1097 to 1434 with the "zero-year" varve being the time of an abrupt lithologic change in the varve sequence. The lithologic change appears to represent a drop in lake level and the beginning of the northward drainage of the lowest level of Lake Hackensack at Sparkill Gap when ice receded into New York State (Stanford and Harper 1991; S. Stanford, personal communication). Antevs (1928) measured additional exposures in the Hackensack Valley that overlap about a quarter of Reed's sequence and match it exactly.

Antevs (1928) compiled the two other long sequences of varves outside of the New England varve chronology. One sequence is a correlation of varves from Lake Albany in the lower Hudson Valley near Haverstraw, New York (yr 1–700, Figure 3.1), and the Quinnipiac Valley (yr 201–544) near New Haven, Connecticut. This sequence appears to be entirely older than the New England varve chronology, but it may overlap the youngest varves in New Jersey. A clear match of the New Jersey and Hudson Valley varves is unlikely, given that the later stages of lacustrine deposition in the Hackensack Valley appears to be nonglacial.[3] The other long sequence was compiled near Essex Junction, Vermont, in the Champlain Valley (Lake Vermont). It is 345 years long and appears to be younger than NE yr 7470. This sequence may overlap the New England varve chronology between NE yr 7500–7900, but a clear match of the sequences is unlikely because the Connecticut Valley varves after NE yr 7470 are nonglacial, whereas the Champlain Valley varves are derived mostly from a glacial source. In addition to the varve sequences described above, Antevs (1928) also compiled many smaller varve sequences (mostly < 100 yr) across New England and into Quebec that have not been matched to any of the longer chronologies.

Improvements and Refinements of the New England Varve Chronology

Although sedimentologic studies of varves in the Northeast have been done (Ashley 1972, 1975; Ashley et al. 1982; Ridge et al. 1990; O'Brien and Pietraszek-Mattner 1998; Ridge 2001a), only in the last decade have further extensions or refinements of the New England varve chronology been accomplished. A large extension of the New England varve chronology (NE yr 2701–6352, 6601–7750) has be accomplished at Newbury (Figure 3.1) where Antevs (1922) originally counted, but did not measure, 750 varves above NE yr 7750. Antevs found the varves too thin for measurement in the field, and he did not analyze the section further. A similar exposure of varves was found a few meters away from Antevs' original outcrop, and the whole section has been studied in detail in duplicate sets of overlapping outcrop cores that were counted and measured using a computer and magnified video images (Ridge and Toll 1999; Ridge et al. 1999). The count has extended the range of the New England varve chronology to NE yr 8679 (+35/-20), 179 years beyond Antevs' original count (Figure 3.2). NE yr 7470–8679 at Newbury are nonglacial varves, the bottom parts of which are very thin (down to 1 mm), and there are some uncertainties in the count which is the reason for the error limits cited above.

Many new sections of varves that exactly match the New England varve chronology have been measured across New England (Lougee 1935; McNish and Johnson 1938; Johnson et al. 1948; Verosub 1979a, 1979b; Thomas 1984; Ridge and Larsen 1990; Ridge et al. 1996, 1999; Levy 1998; Rittenour 1999; Wilson 2000; Larsen et al. 2001), and the New England varve chronology has now been extended to southern Maine (Ridge et al. 2001). Since the 1920s only one error in Antevs' (1922, 1928) original chronology has been found. In the Connecticut Valley at Amherst, Massachusetts, long cores have duplicated much of Antevs' lower Connecticut Valley varves (UMass site on Figure 3.1; NE yr 4682–6027 except for core breaks, Rittenour 1999). In an interval previously covered only by varves measured by Gerard De Geer in the Hudson Valley (NE yr 5600–5709 in Antevs 1922), a sequence of 10 exceedingly thick couplets (NE yr 5669–5678) is absent from a continuous section of the UMass cores. The units missing from the Connecticut Valley record, which De Geer interpreted to be varves in the Hudson Valley, now appear to be thick rhythmic flood events produced by the catastrophic release of water from a glacial lake in the Mohawk Valley. Although NE yr 5669–5678 do not appear to exist, the numbering system for the varve sequence has not yet been changed. For the purposes of this chapter and most other correlation studies, the 10-yr error is not significant enough at this point to warrant the confusion caused by renumbering the varve sequence above NE yr 5668.

Varve Correlation: A Modern View

The importance of varve chronology to formulating the deglaciation history of the Northeast is that it provides regional correlations over thousands of years with unprecedented resolution. The regional correlation of varves from different glacial lakes on an annual scale (see Antevs 1922, 1928; Ridge et al. 1999, 2001) is possible because annual changes in weather and solar output effect rates of glacial melting and therefore the input of sediment to glacial lakes. Across New England these parameters were apparently uniform to the extent that varve records from different lakes match each other over east-west distances of 250 km (Hudson Valley to Merrimack Valley). Correlations among varve records from places where deposition was not dominated by local sedimentation processes are easily recognized because of clear similarities among records of annual varve thickness. In the future another pattern of deposition may be matched across the region in the form of 3- to 22-yr cycles in varve thickness that appear to represent short-term climatic oscillations (Rittenour 1999; Rittenour et al. 2000).

Within the Connecticut Valley varve sequences, it has also been possible to correlate lithologic changes in varves not related to climate. Large flood events, produced by the catastrophic release of water from ice-contact lakes in tributaries, and major drops in lake level caused by the failure of dams are often associated with abrupt changes in varve thickness and lithology that can be recognized basin-wide (Ridge and Toll 1999; Ridge et al. 1999). Events of this type do not produce changes in varve stratigraphy that can be used for interlake correlation, and their recognition is often the result of their appearance in the record of only one lake.

Many varve sequences that record centuries or millennia often show a transition from glacial to nonglacial varves as ice recedes out of a lake basin or becomes too distant to have a strong influence on sedimentation. Nonglacial varves exhibit changes in thickness related to annual variations in sediment delivered to a lake by meteoric streams (runoff). The processes that control meteoric sediment delivery to a lake vary with weather and climate but in a very complex way. Terrestrial erosion and runoff rates, rather than glacial melting, appear to control annual-, decadal-, and centennial-scale sedimentation patterns (Ridge and Toll 1999). Correlation between glacial and nonglacial varves is problematic because these two different environments produce annual thickness patterns that are different responses to solar, climate, and weather variations. For this reason the upper half of the varve record in New Jersey (Reeds 1926) and the New England varve chronology above NE yr 7470, which are both nonglacial, might never be matched annually to contemporaneous glacial varve records. Eventually correlations between different sets of nonglacial varves may provide correlations extending into times greatly postdating ice recession.

DEGLACIATION RELATIVE TO VARVES: BASAL VARVE STRATIGRAPHY

The progressive northward recession of the ice margin in glacial lakes of the Connecticut and Merrimack Valleys (Figure 3.1) was marked by the exposure of till or rock surfaces that were immediately buried by ice-proximal glaciolacustrine sediment. At the time of deglaciation, or within a few decades, varves began to cover the floors of glacial lakes where they were not covered by deltas or were not too steep to receive sediment. These basal varves are generally very thick (up to 4 m) and sandy as compared to other parts of the varve stratigraphy, and they generally become thinner and finer upward with further ice recession. Basal varves in New England become progressively younger northward in the wake of a systematically receding ice sheet, forming an onlapping relationship first documented by Antevs (1922). He was able to define the age of deglaciation in New England varve years at points from south to north in the Connecticut and Merrimack Valleys by matching varve records to the New England varve chronology (Figures 3.3 and 3.4). Antevs (1928) also inferred the approximate age of deglaciation in the Hudson Valley near Newburgh by matching basal varve records at that locality with varves in the Connecticut Valley. Basal varves are exposed at many other localities in the northeastern United States, but correlation to the New England varve chronology through varve matching has been accomplished only in the areas discussed above.

Since the time of Antevs there has been some improvement in our understanding of ice recession with the addition of some basal varve localities and the mapping of glacial readvances. Figures 3.3 and 3.4 show time-distance plots of all known varve sections that have been matched to the New England varve chronology in the Connecticut, Passumpsic, and Merrimack Valleys of Massachusetts, New Hampshire, and Vermont. There is presently no basal varve data in Connecticut that constrains the age of deglaciation relative to the New England varve chronology. Included on Figures 3.3 and 3.4 are all sections measured by Antevs (1922, 1928) to construct the chronology and more recently measured sections (Lougee 1935; Thomas 1984; Ridge and Larsen 1990; Levy 1998; Ridge et al. 1996, 1999, 2001; Ridge and Toll 1999; Rittenour 1999; Rittenour et al. 2000; Wilson 2000; new data).

GLACIAL READVANCES IN THE CONNECTICUT VALLEY

Three readvances of the ice sheet during overall ice recession have been proposed in the Connecticut Valley, the area covered by the New England varve chronology. In southern Massachusetts the age of the Chicopee Readvance (Larsen and Hartshorn 1982) has not been fully bracketed by the varve chronology. Varve sections in the area of the readvance that were not interrupted by advancing ice, and therefore are younger than the readvance, predate NE yr 3800 (Figure 3.3, sites 8, 9, and 17). They provide a minimum varve year age for the readvance of ca. NE yr 3750. A sequence of sandy varves (NE yr 3600–3750) in an uninterrupted varve section just beyond the readvance limit near Springfield (Figure 3.3, site 7; Antevs 1922) may be the result of deposition in close proximity to the ice front at the time of the Chicopee Readvance.

In central Massachusetts deposits interpreted to represent a readvance were first described by Emerson (1898) and later mapped by Larsen and Hartshorn (1982) as the Camp Meeting Cutting Readvance. Near Northampton on the west side of the Connecticut Valley (Figure 3.3, site 20), Antevs (1922) also identified deposits apparently associated with a readvance. The supposed readvance deposits near Northampton rest on NE yr 4634 that forms the top of a sequence of very thick sandy varves. Antevs (1922) also measured a varve section on the east side of the valley 4 km southeast of Amherst (Figure 3.3, site 24) that he interpreted to be interrupted by deformation structures in a 150-varve sequence resting on NE yr 4668. This deformed sequence is in turn overlain by what Antevs (1922) originally interpreted to be till. Readvance of ice to this locality would have occurred after NE yr 4800. Antevs (1928) later retracted his interpretation of a readvance at this site.

At first it seems as if there is ample evidence for a glacial readvance in the Amherst and Northampton area. However, the readvance of ice in the Connecticut Valley south of Antevs' site 22 (Figure 3.3) does not

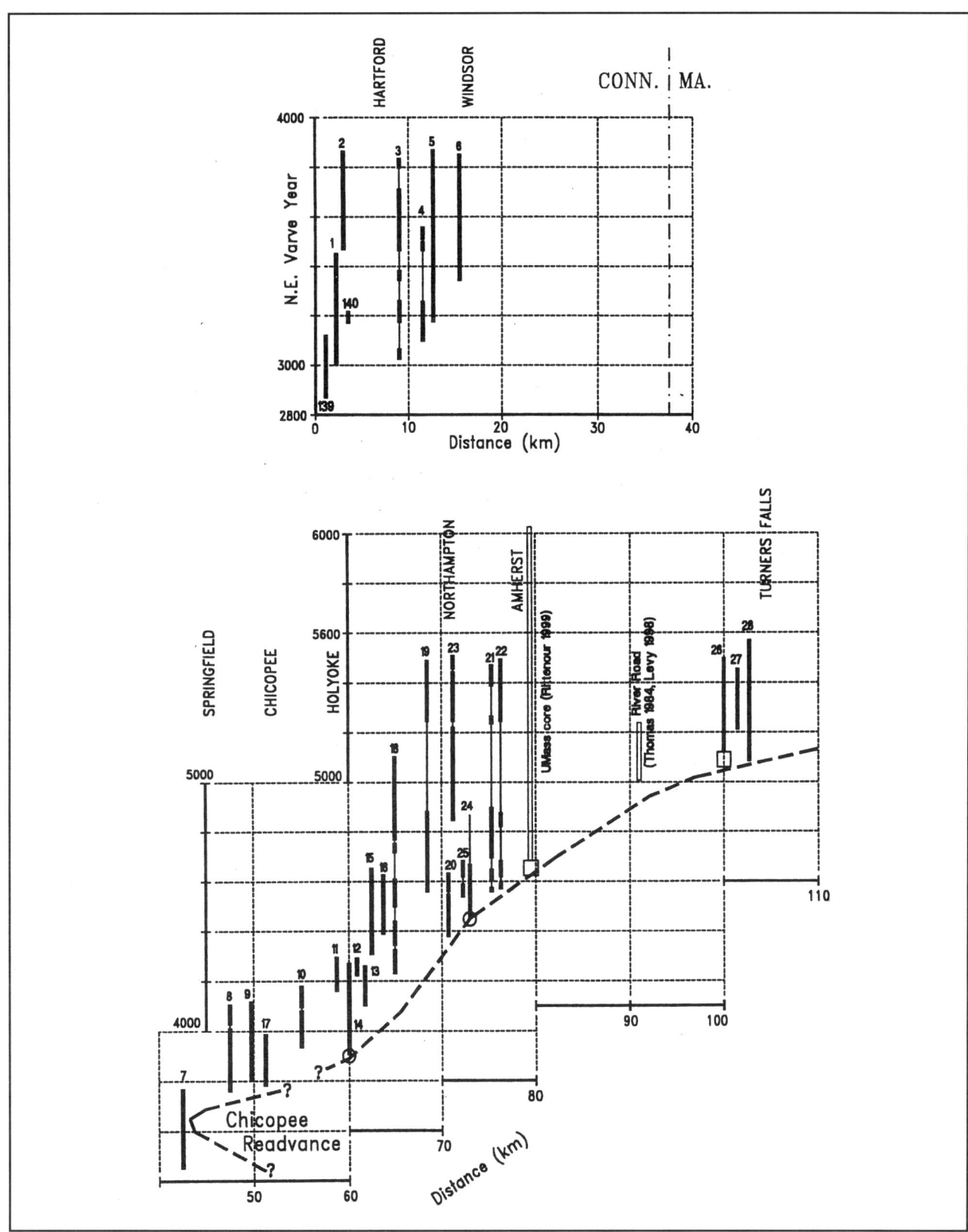

Figure 3.3a, b, and c. Time-distance plot of the time spans of all measured varve sections in the Connecticut and Passumpsic Valleys from Connecticut to northern Vermont and New Hampshire. The distance axis is measured from an arbitrary position near Hartford, Connecticut. The time axis is in New England varve years. The dashed line marking the base of sections resting on till is the approximate age of deglaciation.   *continued*

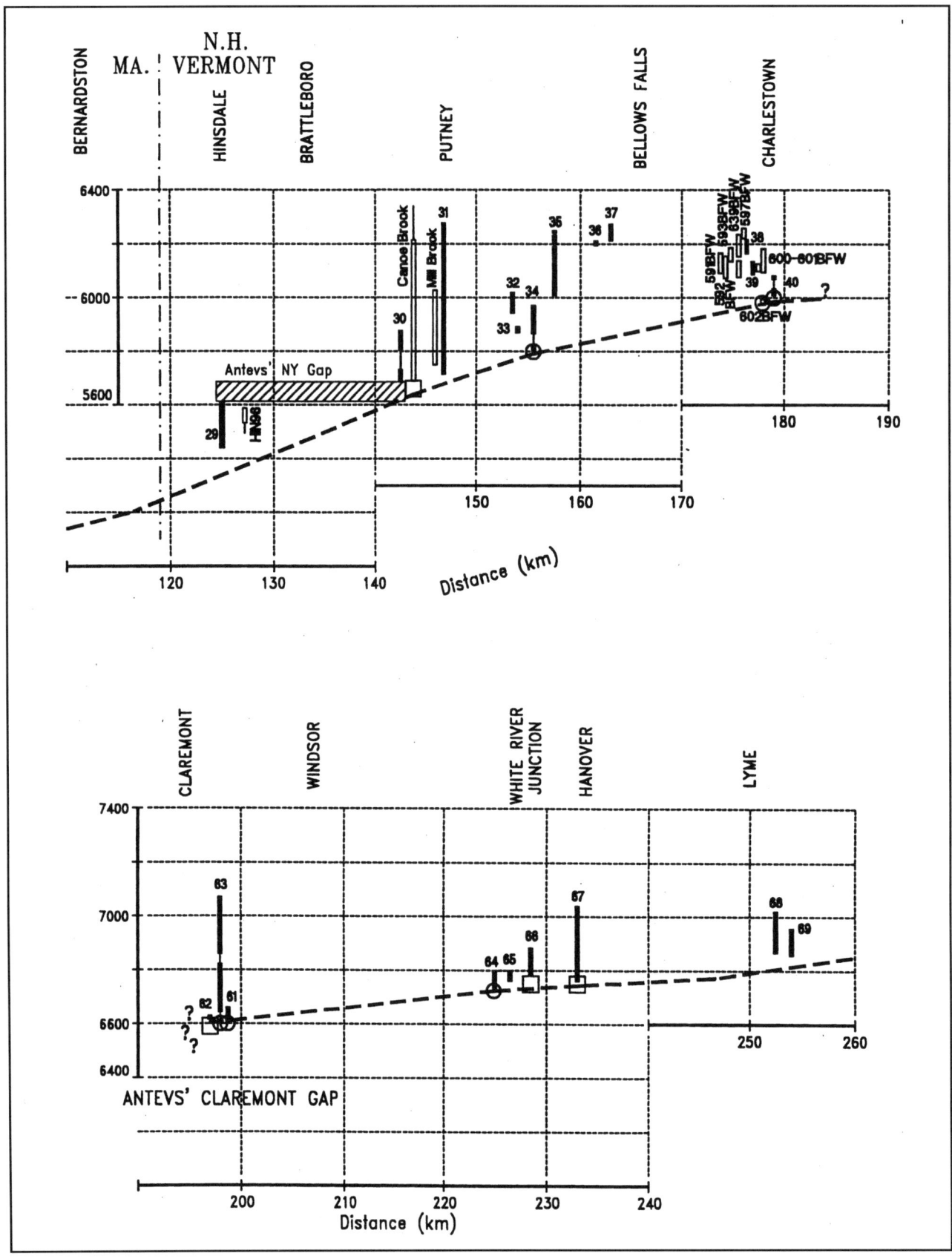

Figure 3.3b *continued*

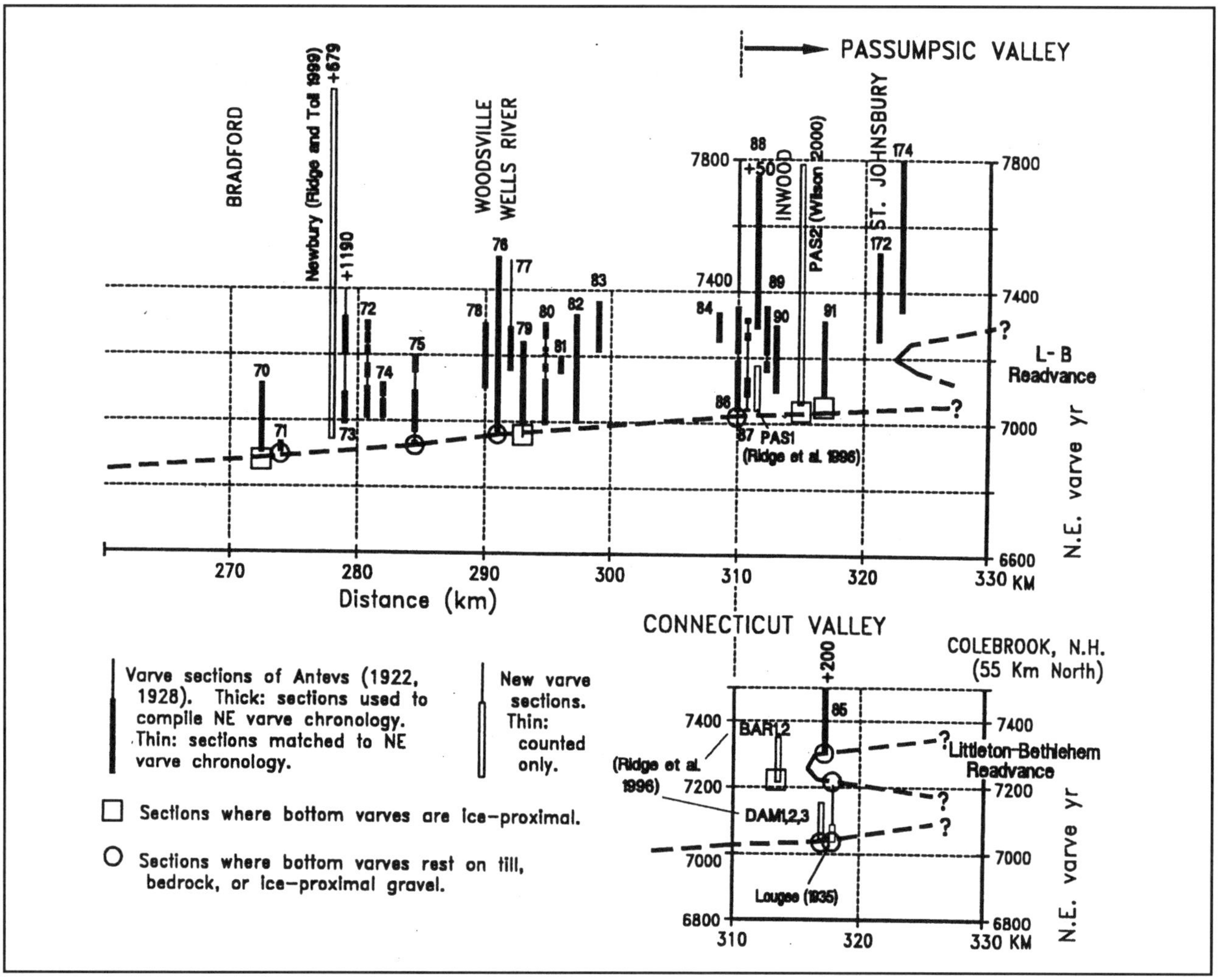

Figure 3.3c

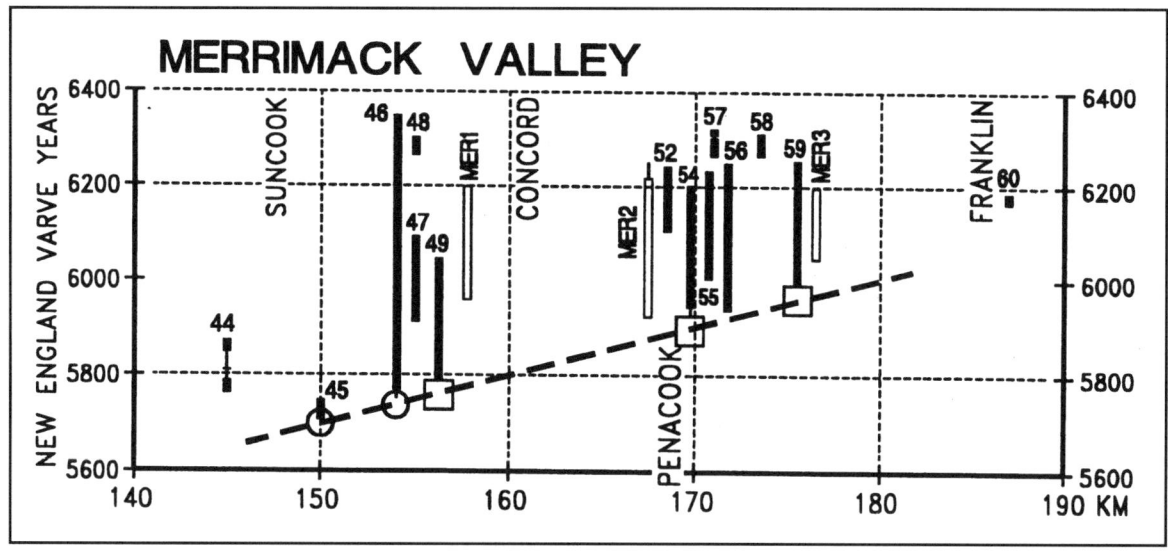

Figure 3.4. Time-distance plot of time spans of all measured varve sections in the Merrimack Valley of New Hampshire. Symbols are the same as for Figure 3.3.

appear possible between NE yr 4650 and 5500 given the continuous varve stratigraphy recorded at other varve sections north of or adjacent to the readvance localities (Figure 3.3, sites 21, 22, and 23). The long cores collected at Amherst (4 km northwest of Antevs' site 24, UMass on Figure 3.3; Rittenour 1999) preserve a continuous sequence of varves (NE yr 4682–6027) that are not interrupted by any features representing the readvance of ice. This evidence precludes any readvance of ice to Amherst or farther south after NE yr 4682. The origin of the supposed readvance features near Amherst and Northampton remains uncertain but a seismic disturbance of the lake floor after NE yr 4600 should not be discarded as a possible explanation.

The third readvance mapped in the Connecticut Valley, the Littleton-Bethlehem Readvance (Figure 3.3), terminated near the Comerford Dam that stretches across the northern Connecticut Valley between Barnet, Vermont, and Monroe, New Hampshire. The readvance was first identified by Antevs (1922) based on varve analysis that showed an apparent delay in deglaciation where a major moraine appeared to cross the area from Bethlehem, New Hampshire (Bethlehem Moraine of Goldthwait 1916 and Crosby 1934a) to St. Johnsbury. The group of features forming the Littleton-Bethlehem moraine system represents the largest and best-preserved ice-front position in northern New England. Crosby (1934b) recorded a multiple till section on the New Hampshire side of the valley at the Comerford Dam. Lougee (1935) found a clayey "till-like" unit overlying partly deformed varves that he matched to the New England varve chronology, thus bracketing the age of the readvance to ca. NE yr 7180–7250 (Figure 3.3). Field mapping, stratigraphic sections, and varve measurements collected over the past decade (Ridge et al. 1996, 1999; Thompson et al. 1996, 1999) are consistent with all of the observations from the first half of this century.

^{14}C AGES AND CALIBRATION OF THE NEW ENGLAND VARVE CHRONOLOGY

Thus far the age of deglaciation has been discussed only in relation to New England varve years, which represent an arbitrarily numbered floating time scale. To use varve analysis to infer the age of deglaciation in New England, it is necessary to apply a numerical dating technique such as ^{14}C dating to the varve chronology. A ^{14}C time scale has been applied to the New England varve chronology (Ridge and Larsen 1990; Ridge et al. 1996, 1999, 2001) based on ^{14}C ages of terrestrial plant macrofossils recovered from varve sections at four localities (Table 3.1). Radiocarbon ages from two localities in the lower Connecticut Valley, Canoe Brook and the UMass site (Figures 3.1 and 3.3), have been used to infer ^{14}C ages for specific years in the lower Connecticut Valley varves. Radiocarbon ages have also been obtained from eight separate years in the upper Connecticut Valley varves at Newbury (Figures 3.1 and 3.3) and in the Passumpsic Valley just south of St. Johnsbury (Figure 3.1; PAS2 on Figure 3.3). Only ^{14}C ages obtained from terrestrial plant macrofossils have been used for assigning ^{14}C ages to the varve chronology.

In previous publications (Ridge et al. 1996, 1999, 2001) three internally consistent ^{14}C ages of samples collected from NE yr 6150 at Canoe Brook were used as the ^{14}C calibration point for the lower and upper Connecticut Valley varves (Table 3.1) assuming the closure of the Claremont Gap. Despite the consistency of the three ^{14}C ages from NE yr 6150, these ages result in a risky calibration point because they occur at the younger limit of a plateau in the ^{14}C time scale when 12,600–12,400 B.P. actually represents a span of 900 calibrated years and therefore varve years. A small error in the ^{14}C ages could lead to a large error in calibration of the varve sequence. Also, as explained earlier, the assumption used to eliminate the Claremont Gap has been invalidated. Radiocarbon ages in the upper Connecticut Valley varves no longer provide a test of the calibration point at NE yr 6150. It is now therefore necessary to separately calibrate the lower and upper Connecticut Valley varve sequences. As a result this chapter will present a revised calibration, eliminating some of the risks and assumptions of previous publications.

A new ^{14}C age obtained from a sample in NE yr 5858 at Canoe Brook (Table 3.1) is here used as the calibration point for the lower Connecticut Valley varves (Figure 3.5). Because the ^{14}C age is older than 12,600 B.P., it avoids uncertainties associated with the ^{14}C plateau at 12,600–12,400 B.P. The accuracy of the ^{14}C age used for the new calibration is crudely tested by the three ^{14}C ages higher in the Canoe Brook section (in NE yr 6150) which appear to be in general agreement with the new calibration point (Figure 3.5). Future refinement of the calibration should focus on obtaining ^{14}C ages from older sections of the lower Connecticut Valley varves. The only other ^{14}C age in the lower Connecticut Valley sequence that is older than 12,600 B.P. (NE yr 6156, Table 3.1) is from a sample of fine peat and gyttja fragments included in the varves. This material is likely in error as a calibration age because it is clearly eroded from some older deposit, and it may contain the remains of aquatic algae and plants. The new calibration point makes the lower Connecticut Valley varve sequence older than in previous publications by about 700 calibrated years, and it opens the Claremont Gap by about 350 calibrated and varve

Table 3.1. ^{14}C Ages from Varves in the New England Varve Chronology

Laboratory number	Age (^{14}C yr BP)	δ^{13}C (‰)	NE varve number	Material dated	Reference
Canoe Brook, Dummerston, Vermont					
GX-25735	12,660 ± 50	-28.9	5858	Woody twigs and *Dryas* leaves	New
GX-14231	12,355 ± 75	-27.2	6150	Bulk sample of silt and clay with nonaquatic leaves and twigs	Ridge and Larsen 1990
GX-14780	12,455 ± 360	-27.6	6150	Handpicked nonaquatic leaves and twigs, mostly *Dryas* and *Salix*	Ridge and Larsen 1990
CAMS-2667	12,350 ± 90	-	6150	*Salix* twig	Norton Miller, personal communication 1993
GX-14781	12,915 ± 175	-27.1	6156	Bulk sample of silt and clay with fragments of peat and gyttja	Ridge and Larsen 1990
Amherst, Massachusetts					
Beta-124780	12,370 ± 120	-27.1	5761-5768	Plant fragments	Rittenour 1999
Newbury, Vermont					
GX-23765	11,530 ± 95	-27.0	7435-7452	Woody twig	Ridge et al. 1999
GX-23766	11,045 ± 70	-27.5	8206	Woody twig	Ridge et al. 1999
GX-23640	10,940 ± 70	-26.8	8357	Woody twig	Ridge et al. 1999
GX-23641	10,080 ± 580	-26.7	8498-8500	Woody twig	Ridge et al. 1999
GX-23767	10,685 ± 70	-26.3	8504	Woody twig	Ridge et al. 1999
GX-23642	10,040 ± 230	-26.5	8542-8544	Chunk of wood	Ridge et al. 1999
GX-23643	10,440 ± 520	-26.8	8652-8662	2 woody twigs	Ridge et al. 1999
East Barnet, Passumpsic River Valley, Vermont					
GX-26456	11,220 ± 50	-27.1	7754	Woody twig	Wilson 2000
Columbia Bridge, Vermont					
WIS-961	11,540 ± 110	-29.0	(>7400)	Wood fragments	Miller and Thompson 1979
WIS-919	11,390 ± 115	-27.5	(>7400)	Wood fragments	Miller and Thompson 1979

years (Figure 3.5).

The calibration of the upper Connecticut Valley varves is the same as in previous publications and is based on the six most consistent ^{14}C ages from the sequence (Table 3.1; Figure 3.5). Despite the use of six ^{14}C ages over a range of 1200 varve years, there is still some uncertainty in the calibration of the upper Connecticut Valley varves on the order of ± 200 years. Three of the four youngest ^{14}C ages used to calibrate the upper Connecticut Valley varves are older than the proposed calibration by about 200 years, whereas the two oldest ^{14}C ages used to calibrate the sequence are about 100 to 200 years younger than the proposed calibration. The calibration presented here essentially represents an average based on all of the ^{14}C ages.

Even with extremely accurate ^{14}C ages, it is still important to remember that terrestrial plant material recovered from varves is always older than the varve in which it is recovered. This age difference between terrestrial plant detritus and the enclosing varve may be insignificant or conversely it could be the source of substantial error in defining the ^{14}C age of a varve. In arctic climates, such as occurred at the end of the last ice age, organic decay is slow and storage of dead plant material on land, resulting in a lag in lacustrine deposition, can be significant (Abbott and Stafford

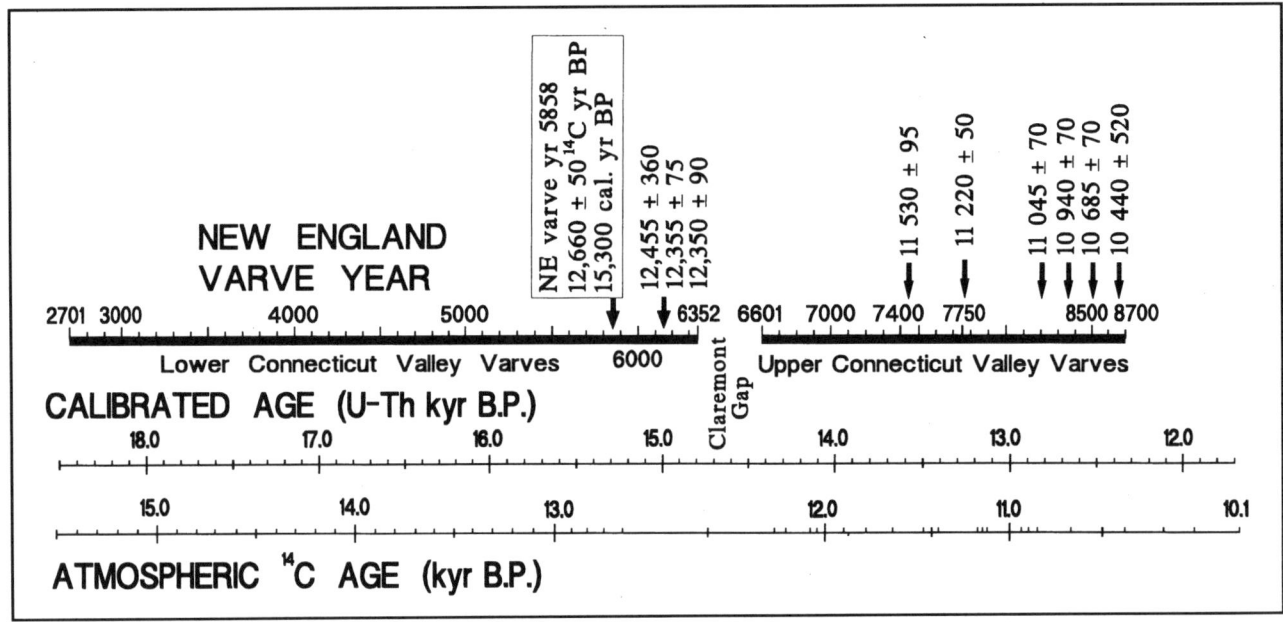

Figure 3.5. Atmospheric ^{14}C and calibrated (U-Th) time scales applied to the New England varve chronology. Some individual ^{14}C ages from the varve chronology (Table 3.1) are plotted for comparison.

1996). For this reason an accurate ^{14}C age from plant remains in a varve represents the maximum possible age of the varve. At this point it seems that the only way to overcome the problem of lag times is to obtain as many ^{14}C ages as possible to develop an internal consistency so that larger lag times associated with some ^{14}C ages can be isolated. Lag times are now thought to produce the scatter in ^{14}C ages of terrestrial plant fossils from varves in Sweden and ages on the younger side of this scatter have been identified as more accurate (Wohlfarth et al. 1995). In other cases the scatter of terrestrial macrofossil ^{14}C ages has yet to be explained (Turney et al. 2000).

The application of a time scale to the varves that approximates calendar years (Figure 3.5) is based on the calibration of ^{14}C ages using the CALIB 4.3 computer program and INTCAL98 calibration data (Stuiver and Reimer 1993; Stuiver et al. 1998). The program converts ^{14}C ages to calibrated ages based on simultaneous ^{14}C and U-Th ages from tropical corals that have well-constrained marine reservoir corrections. To obtain a calibrated (cal.) time scale for the New England varve chronology, the calibrated (U-Th) age of the Canoe Brook ^{14}C calibration point was calculated (NE yr 5858 = 12,660 B.P. = cal. 15,300 B.P.). The calibrated time scale is easily applied to the varve chronology by simply adding or subtracting varve years to the calibrated age of the varve chronology at NE 5858 (Figure 3.5) assuming that calibrated and varve years are the same.

To obtain a complete ^{14}C time scale for the New England varve chronology, the addition of varve years to ^{14}C years will not yield accurate results due to the non-linearity of the ^{14}C time scale created by the secular variation of atmospheric ^{14}C. A ^{14}C time scale has been formulated using the CALIB 4.3 program to determine the calibrated ages of every 100-yr increment of the ^{14}C time scale. The ^{14}C ages were then plotted on an axis adjacent to their corresponding calibrated ages as calculated using the computer program (Figure 3.5). Several ^{14}C plateaus (periods of little or no change) and compressions (periods of rapid change) occur relative to the linear varve- and calibrated-year time scales. A good example of a ^{14}C plateau is the 200-yr ^{14}C time span from 12,600–12,400 B.P. that represents about 900 varve or calibrated years.

PALEOMAGNETIC DECLINATION RECORDS: A REGIONAL CORRELATION TOOL

Outside of the Connecticut, Merrimack, and Hudson Valleys, varves have not been matched to the New England varve chronology. Yet a regional synthesis of deglaciation in the northeastern United States requires correlation of glacial events in central New York, the Hudson-Champlain Valley, and New England. These correlations have been accomplished through paleomagnetic declination stratigraphy recorded by remanent magnetization in laminated muddy lacustrine sediments. The benefit of paleomagnetism is that it is a chronologic correlation tool that extends the ^{14}C and calibrated age chronologies of the New England varve chronology to a much wider region. Correlation based on declination records takes advantage of the secular variation of geomagnetic declination that produces a pattern of eastward and westward declination maxima separated by periods of rapid transition (1°/10–20 yr). The maxima have different amplitudes, and in some cases are unique over the time of deglaciation.

Shown on Figure 3.6 are the remanent declination records obtained from varves and other muddy lacustrine sediment in New England and New York. The Connecticut and Merrimack Valley records were compiled from measurements in varves that have been matched to the New England varve chronology. This work began with the pioneering studies of sedimentary paleomagnetism by McNish and Johnson (1938) and Johnson et al. (1948). Verosub (1979a, 1979b) later verified these early studies in most of the lower Connecticut Valley varves (NE yr 3000–5350) using modern laboratory techniques. Over the past decade nearly all of the remaining lower and upper Connecticut varves, including the extension of the chronology to NE yr 8500, have been verified (Ridge et al. 1996, 1999).

In central New York to northern Vermont, lacustrine sediment used to construct a declination record has not been matched to the New England varve chronology. The time scale of this record is simply a relative age plot that is nonlinear (Figure 3.6). Relative ages for the western Mohawk Valley samples are based on the positions of units within a complex stratigraphic sequence (Ridge et al. 1990, 1991; Ridge and Franzi 1992; Ridge 1997). In other regions the succession of deglaciation and glacial lakes in New York has been used to determine the relative ages of paleomagnetic samples (Brennan et al. 1984; Pair et al. 1994; Ridge et al. 1999).

The correlation of the New England and New York declination records (Figure 3.6) is based on the matching of declination maxima of similar amplitude. Also the general assumption that the two records must in some way overlap in time is justified given the time covered by the New England record (15,200–10,500 B.P.; cal. 18,200–12,500 B.P.). This is the time frame generally recognized as the approximate age of deglaciation in central to northern New York (Connally and Sirkin 1973; Muller and Prest 1985; Muller and Calkin 1993). Assisting in the correlation of the declination records is the occurrence of a unique declination maximum on both records. An extreme western declination maximum of 320–295° (40–65° West) appears in both records and is unique during the period of late Wisconsinan deglaciation in both regions. To test this correlation, independent ^{14}C ages associated with both records can be compared from times that appear to be paleomagnetically correlative. Radiocarbon ages from Canoe Brook indicate an age of 12,600–12,500 B.P. (Figure 3.6) as the extreme western declination approaches its maximum in varves of the Connecticut Valley. In the Ontario Basin the western maximum occurs in sediment deposited at about the time of the initiation of Lake Iroquois (Brennan et al. 1984). The early sediment of Lake Iroquois has an age of 12,500–12,100 B.P. based on ^{14}C ages of spruce twigs from Lake Iroquois deposits (Muller and Prest 1985; Muller and Calkin 1993). The declination records also match at approximately equivalent ^{14}C ages (ca. 11,000 B.P.) when the Champlain Sea entered the Champlain and St. Lawrence Valley and a declination of 0° is recorded during a declination shift from west to east on both records.

CHRONOLOGY OF LATE WISCONSINAN GLACIATION: NEW YORK TO MAINE

The application of ^{14}C and calibrated-year time scales to the New England varve chronology is the key to formulating the deglaciation model of the northeastern United States presented in this chapter. The age of deglaciation along the axes of the Connecticut and Merrimack Valleys is defined where basal varve sections have been matched to the New England varve chronology. The correlation of the declination records in New England and New York makes it possible to define the ages of glacial events in New York that can be tied to the declination stratigraphy (Figure 3.6). Using these correlations as a foundation, two maps of the northeastern United States have been prepared showing the age of deglaciation in ^{14}C (Figure 3.7; Table 3.2) and calibrated (Figure 3.8) years. Again, the nonlinearity of the ^{14}C time scale becomes apparent giving the impression that ice recession occurred more rapidly in ^{14}C years than in calibrated years at various times.

Beyond the limits of the New England varve chronology and paleomagnetic records, the age of deglaciation can be determined only with site-specific ^{14}C ages that can then be converted to calibrated years. This can be done for ^{14}C ages back to ca. 20,300 B.P. (cal. 24,000 B.P.) which is the limit of the data sets available to the CALIB 4.3 calibration program (Stuiver and Reimer

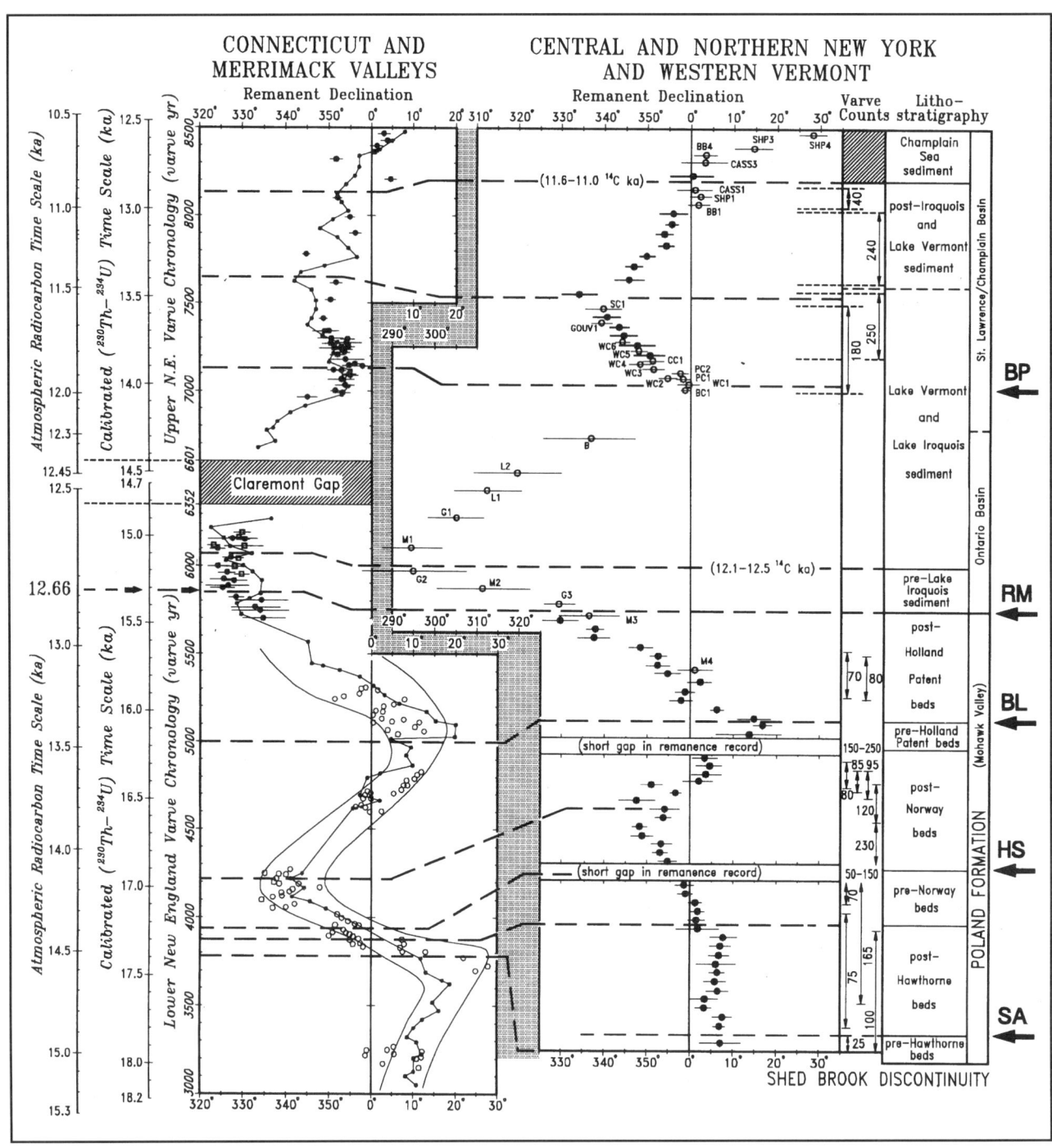

Figure 3.6. Correlation of late Wisconsinan paleomagnetic declination records from New England and New York State. Error bars on some data points are a_{95} (cone of confidence) values. Data sources in the Connecticut Valley are Johnson et al. (1948; dots with tie line), Verosub (1979a; open circles and envelope), and Ridge et al. (1999; solid circles). Data from the Merrimack Valley are from Ridge et al. (2001; open squares). Data from New York are from Ridge et al. (1990; dark circles, Mohawk Valley), Brennan et al. (1984; open circles, Ontario Basin and Mohawk Valley), Pair et al. (1994; open circles, St. Lawrence Basin), and Ridge et al. (1999; solid circles, Champlain Basin). The abbreviations of glacial readvances in New York (Table 3.2) are given at the right side of the diagram.

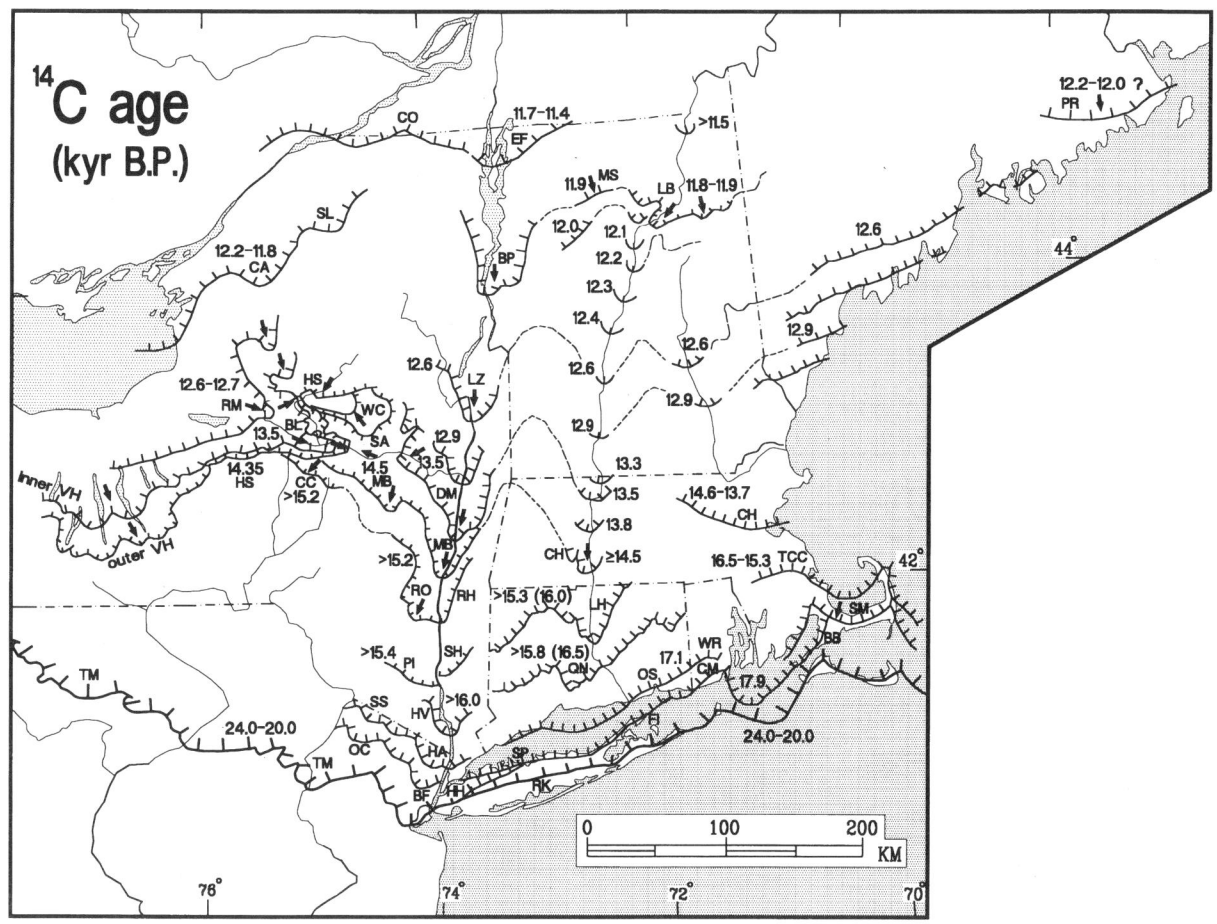

Figure 3.7. The deglaciation chronology of the northeastern United States in atmospheric ^{14}C years B.P. Arrows indicate ice-front positions that are the limits of glacial readvances. The abbreviations of moraines, ice margins, and glacial readvances are given on table 3.2.

Table 3.2. Abbreviations of Moraines, Ice Margins, and Glacial Readvances on Figures 3.6, 3.7, 3.8, and 3.9

BB	Buzzards Bay Moraine	MS	Middlesex Readvance
BF	Bloomfield ice margin	OC	Ogdensburg–Culvers Gap Moraines
BL	Barneveld–Little Falls Readvance	OS	Old Saybrook Moraine
BP	Bridport Readvance	PI	Pellets Island Moraine
CA	Carthage-Harrisonville ice margin	PR	Pineo Ridge Moraine
CC	Cassville-Cooperstown Moraine	QN	Quinnipiac ice margin
CH	Chicopee Readvance	RH	Red Hook Moraine
CM	Charlestown Moraine	RK	Ronkonkama Moraine
CO	Covey Hill ice margin	RM	Rome Readvance
DM	Delmar Readvance	RO	Rosendale Readvance
EF	Enosburg Falls ice margin	SA	Salisbury Readvance
FI	Fishers Island Moraine	SH	Shenandoah Moraine
HA	Lake Hackensack ice margin	SL	Star Lake Moraine
HH	Harbor Hill Moraine	SM	Sandwich Moraine
HS	Hinckley-St. Johnsville Readvance	SP	Sands Point Moraine
HV	Haverstraw ice margin	SS	Sussex Moraine
LB	Littleton-Bethlehem Readvance	TCC	Lakes Taunton and Cape Cod ice margin
LC	Lake Charles ice margin	TM	Terminal Moraine
LH	Lake Hitchcock dam ice margin	VH	Valley Heads Moraines
LZ	Luzerne Readvance	WC	West Canada Readvance
MB	Middleburg Readvance	WR	Wolf Rock Moraine

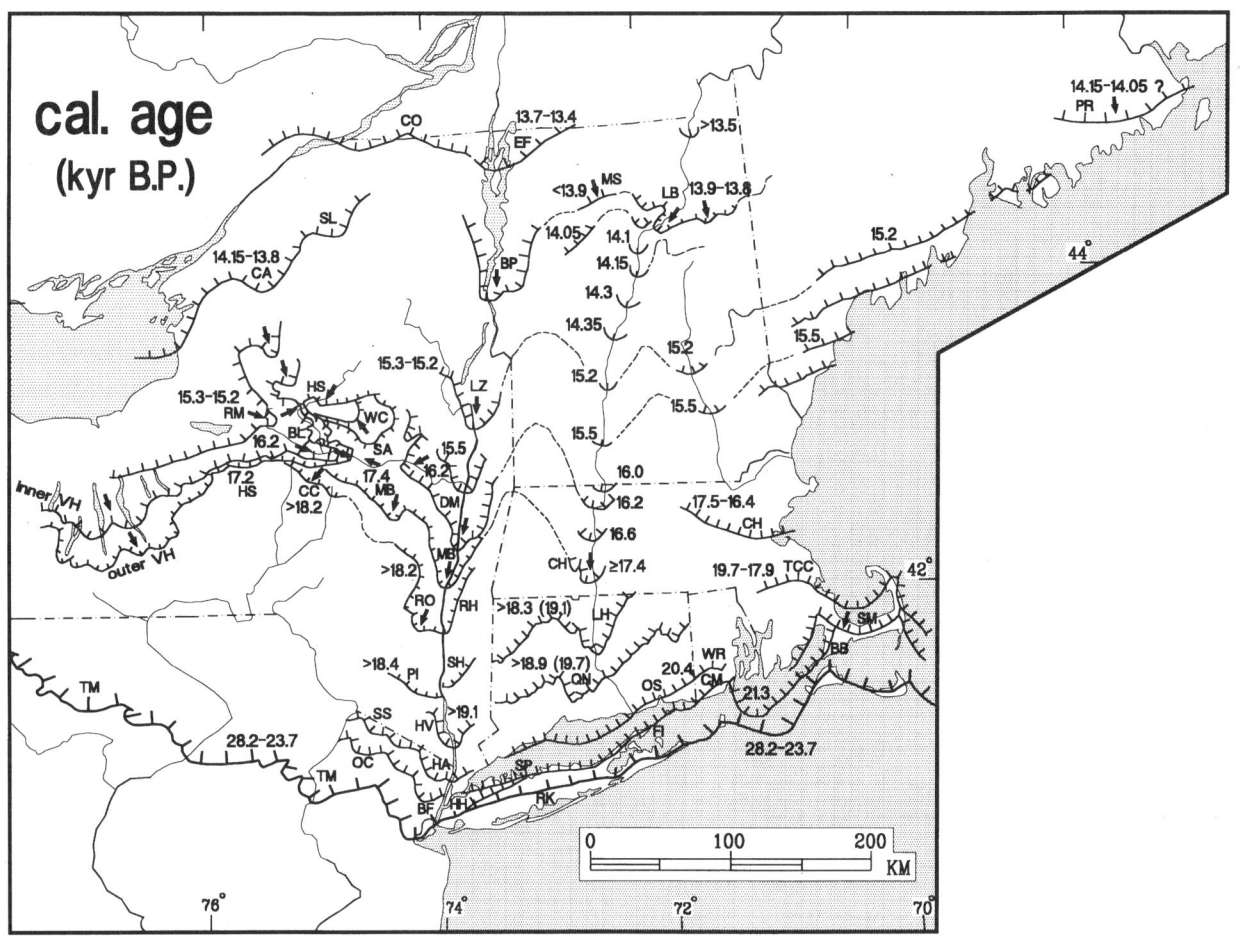

Figure 3.8. The deglaciation chronology of the northeastern United States in calibrated (U-Th) years B.P. Arrows indicate ice-front positions that are the limits of glacial readvances. The abbreviations of moraines, ice margins, and glacial readvances are given on Table 3.2.

1993; Stuiver et al. 1998). In the following discussion the ^{14}C and calibrated age difference of 3,700 years at 20,300 B.P. (cal. 24,000 cal. B.P.) with an additional ±500 years is used as an estimate of a possible correction for ^{14}C ages greater than 20,300 B.P. Unfortunately there are relatively few unambiguous ^{14}C ages directly tied to glacial events in the Northeast, and even fewer from prior to 15,400 B.P. (cal. 18,400 B.P.), which is the beginning of the New England varve chronology. For these reasons it is difficult to make definitive statements about the ages of events prior to ca. 15,400 B.P. For some events that do not have ^{14}C age control, limiting ages in terms of an event's latest possible ^{14}C age can be defined from varve counts outside of the New England varve chronology. In southeastern New England moraines that represent readvances or stillstand events have been correlated by Boothroyd et al. (1998) to oxygen isotope records from the Greenland Ice Sheet Project 2 (GISP2) ice core (Cuffey et al. 1995; Stuiver et al. 1995). Ages from the Greenland ice core are then used to infer ages of deglaciation in New England. Some ice margin reconstructions for which there is limited age control are still shown on Figures 3.7 and 3.8 to depict the overall pattern of ice recession.

The Late Wisconsinan Maximum (24,000–20,000 B.P.; cal. 28,200–23,700 B.P.)

During the later part of the last glaciation (late Wisconsinan), ice reached its farthest advance across almost all of New York and into Pennsylvania and New Jersey (Figures 3.7 and 3.8) where the limit of advance is recorded by the "Terminal Moraine" (Lewis 1884; Crowl and Sevon 1980; Cotter et al. 1986; Stanford and Harper 1991; Muller and Prest 1993; Stanford 1993). The terminal position of the ice sheet is also represented by end moraines along the southern coast of New England from Long Island eastward onto the continental shelf southeast of Cape Cod (Oldale 1982; Sirkin 1982, 1986; B. Stone and Borns 1986; Uchupi et al. 2001). Along the perimeter of the ice sheet in the northeast there are very few ^{14}C ages from terrestrial plant fossils to indicate the time that ice reached its maximum. This chapter relies heavily on the analysis of ^{14}C ages by others that include a heavy dose of bulk sediment and marine fossil ages. The few ^{14}C ages obtained on wood samples seem to indicate that ice reached its maximum extent in central New York no earlier than 24,000 B.P. (ca. cal. 27,200–28,200 B.P.; Muller and Calkin 1993). Radiocarbon ages of bulk sediment samples from New Jersey indicate that the ice advance reached its limit at ca. 22,800 B.P. (cal. 25,800–26,800 B.P.; Harmon 1968), whereas early recession occurred no later than 18,600–19,300 B.P. (cal. 22,100–22,900 B.P.; Cotter et al. 1986; D.H. Cadwell, personal communication 1993), 19,340 ± 695, GX-4279). Radiocarbon ages from wood, bulk sediment, and marine fossils in southern New England and on Long Island indicate that ice reached its maximum extent at ca. 24,000–20,000 B.P. (cal. 28,200–23,700 B.P.; B. Stone and Borns 1986).

It is not yet clear from ^{14}C ages if the last ice sheet reached its maximum extent simultaneously everywhere in the Northeast with possible age differences of as much as a few thousand years. The limit of ice advance appears to be represented by a synchronous feature, the "Terminal Moraine" (Figures 3.7 and 3.8), across Pennsylvania (Crowl and Sevon 1980) and New Jersey (Stanford 1993). On eastern Long Island the terminal position is marked by the Ronkonkama Moraine, and this moraine is apparently truncated to the west by the younger Harbor Hill Moraine (Sirkin 1982, 1986) that appears to be continuous with the "Terminal Moraine" in New Jersey (Stanford and Harper 1991; Stanford 1993). If the interpretation of moraines on Long Island is correct, it implies that most of the terminal position of the last ice sheet south of New England is slightly older than the terminal position of the ice sheet through New Jersey and Pennsylvania. At this point, based on limited ^{14}C ages from terrestrial plant fossils, a reasonable estimate for the age of the maximum extent of glaciation across the region is 24,000–20,000 B.P. (cal. 28,200–23,700 B.P.).

Early Ice Recession (20,000–15,300 B.P.; cal. 23,700–18,300 B.P.)

Two problems exist in defining the chronology of deglaciation prior to 15,400 B.P. (cal. 18,400 B.P.). First, there are few ^{14}C ages from this time except for bulk sediment sample ages. Second, there are few detailed reconstructions of ice margins from this time across south-central New York. Most reconstructions of recessional positions are in New England where they have not been dated using ^{14}C ages; instead, ages have been inferred based on the correlation of moraines in Rhode Island with cold pulses on oxygen isotope records from the GISP2 ice core (Boothroyd et al. 1998). This procedure seems to fit the regional time frame but makes the assumption that cold pulses in Greenland correspond to times of glacial readvance or stillstand in New England. Where this idea has been tested with the Valley Heads Readvances in New York and the Littleton-Bethlehem Readvance in northern New Hampshire (Ridge and Toll 1999), this assumption appears to be correct (Figure 3.9).

Two large recessional positions in southern New England have been reconstructed. They include the Sand Point to Fishers Island Moraines on Long Island (Sirkin 1982, 1986), which appear to truncate the Harbor Hill Moraine on western Long Island and have been traced across the ocean floor to the Charlestown Moraine in Rhode Island (Figures 3.7 and 3.8; Schafer and Hartshorn 1965; Boothroyd et al. 1998). These features may be correlative to the Buzzards Bay Moraine on Cape Cod (Oldale 1982). A correlation to the GISP2 ice core (Cuffey et al. 1995; Stuiver et al. 1995) gives an age of 17,900 B.P. (cal. 21,300 B.P.) for these features (Boothroyd et al. 1998; Figure 3.9). Approximately contemporaneous features in northern New Jersey may be moraines near Ogdensburg and Culvers Gap (Minard 1961; Sirkin and Minard 1972) and the Bloomfield ice margin (Stanford and Harper 1991; Stanford 1993).

A second large recessional position is marked by submarine moraines along the southern coast of Connecticut (Figure 3.7 and 3.8) that come on land to form the Old Saybrook Moraine in Connecticut (Goldsmith, 1982; J. Stone, DiGiamoco-Cohen, et al. 1998; J. Stone, Schaefer et al. 1998) and the Wolf Rocks Moraine in Rhode Island (Boothroyd et al. 1998). This ice front may be the same age as the Sandwich Moraine on Cape Cod, which truncates the eastern extension of the Buzzards Bay Moraine (Oldale 1982). The projected age of these features of 17,100 B.P. (cal. 20,400 B.P.) is based on correlation to the GISP2 ice core (Boothroyd et al. 1998; Figure 3.9). In northern New Jersey approximately contemporaneous features may be the Sussex Moraine and an ice margin formed in northern glacial Lake Hackensack that extends into New York (Stanford and Harper 1991; Stanford 1993). The lithologic break at the 0-yr varve of the Hackensack Valley (Antevs 1928) may represent northward recession of ice into New York to a point where Sparkill Gap became the northward draining spillway for a nongla-

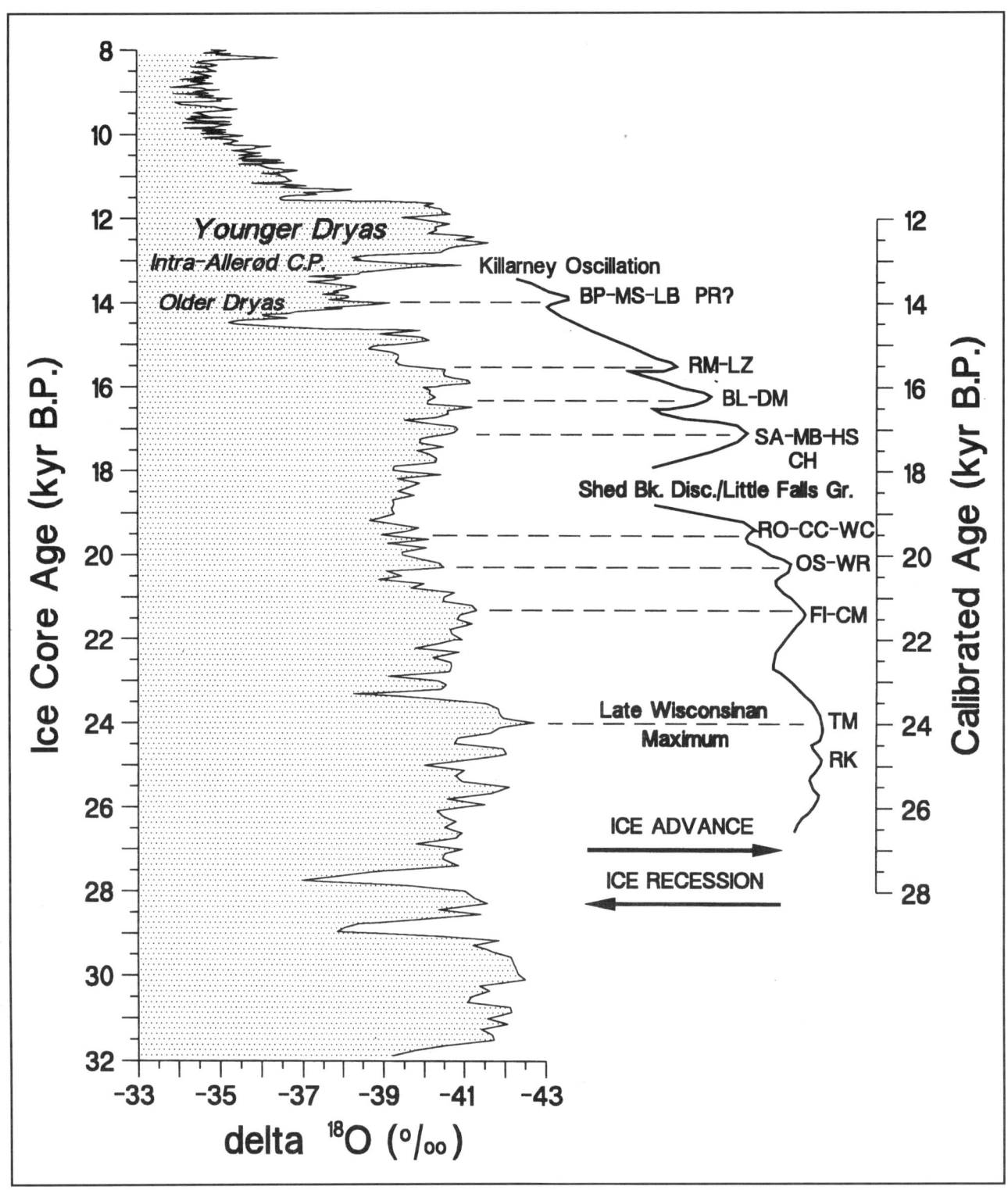

Figure 3.9. The Greenland Ice Sheet Project 2 (GISP2) ice-core record of oxygen isotopes (Cuffey et al. 1995; Stuiver et al. 1995) plotted with the general inferred history of deglaciation in New York and New England. The abbreviations of moraines, ice margins, and glacial readvances are given on Table 3.2. Time scale for the oxygen isotope record is based on annual layer counts in the GISP2 ice core. The time scale for the deglaciation chronology is the calibrated (U-Th) time scale of events in New York and New England discussed in text.

cial lake in the Hackensack Valley. If this is true, ice recession from the Bloomfield ice margin to just over the New York border took more than 1,097 years based on varve counts in the Hackensack Valley (Antevs 1928).

In the Hudson Valley Antevs measured varves in two areas, near Haverstraw and in the Newburgh-Beacon area (Figures 3.7 and 3.8), where varve records place constraints on the age of deglaciation. Seven hundred varves measured by Antevs (1928) near Haverstraw do not appear to overlap the New England varve chronology. Therefore, the oldest Haverstraw varve recorded by Antevs is at least 700 years older than NE yr 2701. The oldest measured varve at Haverstraw, which is not the bottom of varves at this locality, must then have an age of no less than 16,000 B.P. (cal. 19,100 B.P.). This is also the latest possible time when deglaciation could have occurred at a Haverstraw ice margin. Near Newburgh Antevs (1928) measured the oldest varve connected to the New England varve chronology, but he was not able to measure to the bottom of Hudson Valley varves in this area. The occurrence of thick varves (8–20 cm) and bedrock not far beneath Antevs' measured varves indicate that the oldest measured varve might not be too much younger than the age of deglaciation. The oldest measured varve at Newburgh (NE yr 2701) has an age of 15,400 B.P. (cal. 18,400 B.P.), which is also the latest possible age for deglaciation at Newburgh. This is approximately the position of the Pellets Island and Shenandoah Moraines (Connally and Sirkin 1970, 1973, 1986; Figures 3.7 and 3.8).

On Cape Cod and in southeastern Massachusetts (Figures 3.7 and 3.8) ice simultaneously impounded glacial Lakes Taunton and Cape Cod (Larson 1982; B. Stone and Peper 1982). The age of this ice margin can be bracketed only from the ages of surrounding ice margins. Farther west in Connecticut two ice-front positions have been reconstructed that mark the limit of ice when the ice front crossed the Quinnipiac Valley near New Haven and when the dam for glacial Lake Hitchcock was constructed at Rocky Hill (J. Stone, DiGiamoco-Cohen et al. 1998; J. Stone, Schafer et al. 1998). These ice-front positions appear to postdate the Haverstraw ice margin and predate the ice margin at Newburgh in the Hudson Valley. They may be approximately the same age as the ice margin for Lakes Taunton and Cape Cod although the Quinnipiac ice margin appears geographically to represent a closer correlation. Varves measured by Antevs (1928) in the Quinnipiac Valley do not overlap the New England varve chronology. The oldest Quinnipiac varve measured by Antevs, which is not the bottom of the Quinnipiac varve section, is at least 500 years older than NE yr 2701. This corresponds to an age of no later than 15,800 B.P. (cal. 18,900 B.P.) which also represents the youngest possible age for deglaciation at the Quinnipiac ice margin. The estimated age for this ice margin of 16,500 B.P. (cal. 19,700 B.P.; J. Stone, DiGiamoco-Cohen et al. 1998) is in agreement with the varve-based chronology and is consistent with the inferred ages of surrounding features (Figures 3.7 and 3.8).

The ice-front position for the Lake Hitchcock dam at Rocky Hill is older than the oldest varve so far recorded for Lake Hitchcock in surface exposures near Hartford (NE yr 2868; Antevs 1928) giving it an age of no younger than 15,300 B.P. (cal. 18,300 yr B.P.). The oldest varve recorded by Antevs from Lake Hitchcock was measured at a site where there are more varves in the subsurface. The estimated age of 16,000 B.P. (cal. 19,100 B.P.) for the construction of the Lake Hitchcock dam given by J. Stone, DiGiamoco-Cohen et al. (1998); J. Stone, Schafer et al. (1998) seems compatible with the estimated ages of surrounding features based on ice core correlations (Boothroyd et al. 1998) and the ^{14}C calibration of varves from the Connecticut Valley.

Deglaciation (15,300–14,600 B.P.; cal. 18,300–17,500 B.P.)

In the mid-Hudson Valley is the limit of the Rosendale Readvance (Figures 3.7 and 3.8) near Kingston, New York (Connally 1968; Connally and Sirkin 1973, 1986). The eastern extension of the readvance limit appears to correspond to the Red Hook Moraine (Connally and Sirkin 1986), whereas west of the Hudson Valley the limit of the readvance follows the flank of the Catskill Mountains (Dineen 1986). This is the youngest readvance in the Hudson Valley that is not correlative to post-14,600 B.P. readvances in the western Mohawk Valley, which are paleomagnetically correlated to the New England varve chronology. Farther west on the south side of the Mohawk Valley, the Cassville-Cooperstown Moraine (Krall 1977) also represents a readvance position (Fleisher 1986). Based on similar elevations for ice-front positions along opposite sides of the Mohawk Valley, the Cassville-Cooperstown Moraine has been correlated with the West Canada Readvance in the western Mohawk Valley (Ridge et al. 1991; Ridge and Franzi 1992). In this chapter the Rosendale Readvance is considered an approximate equivalent of the Cassville-Cooperstown Moraine and the West Canada Readvance, because all of these features and events represent the youngest ice-front positions in their respective valleys that are older than the Middleburg and Salisbury Readvances of the Mohawk Valley (see discussion below).

The West Canada Readvance occurred in glacial

Lake Newport at a time when the upper Newport beds were deposited and preserved a paleomagnetic declination of 20–30° East (Ridge et al. 1990, 1991). There is no equivalent period of extreme eastern declination at the beginning of the declination record for the New England varve chronology (NE yr 3000–3700 on Figure 3.6) that could represent the same period. The West Canada Readvance, glacial Lake Newport, and the Cassville-Cooperstown Moraine all predate a major break in the lacustrine stratigraphy of the western Mohawk Valley represented by the Shed Brook Discontinuity and the Little Falls Gravel. The Little Falls Gravel and Shed Brook Discontinuity (Ridge 1997) represent a time of ice recession at the eastern end of the Mohawk Valley that allowed subaerial erosion and fluvial drainage eastward in the Mohawk Valley and may be equivalent to the Erie Interstade in the eastern Great Lakes region (Ridge 1997). A limiting age of no younger than 15,200 B.P. (cal. 18,200 B.P.) for the West Canada Readvance, and by correlation the Cassville-Cooperstown Moraine and Rosendale Readvance, is based on the paleomagnetic record for the contemporaneous upper Newport Beds that predate NE yr 3000. This interpretation is consistent with the fact that all of these events occurred prior to the formation of the Little Falls Gravel and Shed Brook Discontinuity, which predate 14,600 B.P. (Figure 3.6; cal. 17,500 B.P.; Ridge 1997).

Deglaciation (14,600–12,000 B.P.; cal. 17,500–14,100 B.P.)

The age of deglaciation in the Connecticut Valley from southern Massachusetts to northern New Hampshire and Vermont, along a 30-km stretch of the Merrimack Valley, and in Lake Winooski of north-central Vermont (Larsen 1987) can be inferred from the calibrated ages (Figures 3.7 and 3.8; 14,600–12,000 B.P.; cal. 17,500–14,100 B.P.) of basal varves matched to the New England varve chronology (Antevs, 1922, 1928; Ridge et al. 1999, 2001; Larsen et al. 2001). The age of the Chicopee Readvance is limited to no later than 14,500 B.P. (cal. 17,400 B.P.), but it may be represented by thick sandy varves beyond the readvance limit that were deposited at this time (NE yr 3600–3750). If this is the age of the Chicopee Readvance, then it is equivalent to the Middleburg and Salisbury Readvances in the Hudson and Mohawk Valleys (discussed below).

In eastern Massachusetts an ice margin in glacial Lake Charles is shown on Figures 3.7 and 3.8 to indicate the general pattern of ice recession, but its age cannot be inferred any more precisely than 14,600–13,700 B.P. (cal. 17,500–16,400 B.P.). The nearly continuous tracing of ice-front positions from the Merrimack Valley across eastern New Hampshire (Koteff et al. 1993) to southern Maine (Smith 1980, 1982, 1985; Thompson and Borns 1985; Smith and Hunter 1989) has allowed the construction of terrestrial ice-front positions to a region where the last ice sheet receded in marine water. The correlation of ice-front positions and calibration of the New England varve chronology has allowed the application of terrestrial ^{14}C ages to ice recession where previously only marine ^{14}C ages had been obtained (Ridge et al. 2001). Marine ^{14}C ages in southern Maine appear to be about 800 to 1,000 years older on average than the terrestrial ^{14}C ages for single ice-front positions as revised from Ridge et al. (2001) and presented here.

The ages of four readvances in central New York (Figures 3.7 and 3.8; Salisbury, St. Johnsville-Hinckley, Little Falls–Barneveld, and Rome Readvances) represented by diamicton units in the Poland Formation and later deposits of the eastern Ontario Basin and western Mohawk Valley (Muller et al. 1986; Ridge et al. 1990; Ridge and Franzi 1992; Muller and Calkin 1993; Ridge 1997) are shown on Figure 3.6 relative to the New York declination record. The correlation of the New York paleomagnetic stratigraphy with the New England varve chronology has allowed the application of numerical ages to these events in central New York. The Hinckley-St. Johnsville (14,350 B.P.; cal. 17,200 B.P.) and Barneveld–Little Falls (13,500 B.P.; cal. 16,200 B.P.) Readvances of the Ontario Lobe covered the area of earlier readvances from the east and truncate ice marginal landforms of the Cassville-Cooperstown Moraine and West Canada and Salisbury Readvances making a very complex deglacial history (Figures 3.7 and 3.8). Based on varve counts between till units that represent the Salisbury (14,500 B.P.; cal. 17,400 B.P.) and Hinckley-St. Johnsville Readvances (Hawthorne and Norway Diamictons of Ridge et al. 1990) about 165 years separated these events. These two readvances may represent a response to the same climatic stimulus but are slightly out of phase possibly due to varying dynamic responses to rising water levels in deep lakes (> 200 m) along their margins. The Hinckley-St. Johnsville and Barneveld–Little Falls Readvance limits appear to be traceable southwest of the Mohawk Valley to outer and inner moraine complexes of the Valley Heads Moraines south of Lake Ontario. The Rome Readvance (Muller and Calkin 1993; Ninemile Readvance of Ridge and Franzi 1992) has been assigned to the Port Huron Stade of the eastern Great Lakes region (Muller et al. 1986) and paleomagnetic correlations to New England (Figure 3.6) indicate a consistent age of 12,700–12,600 B.P. (cal. 15,300–15,200 B.P.).

In the Hudson and eastern Mohawk Valleys, ages of readvances (Figures 3.7 and 3.8) have been inferred by correlation to ice margins and lake levels in the western Mohawk Valley. Based on similar ice margin elevations along opposite sides of the Mohawk Valley, the Salisbury Readvance (Ridge and Franzi 1992) is correlated to the Middleburg Readvance limit (Fleisher 1986) which has been traced eastward to the Hudson Valley (Dineen 1986). In the Hudson and eastern Mohawk Valley, the Delmar Readvance (Dineen 1986) seems to have impounded a lake consistent with water levels recorded by deltas at the margin of the Barneveld–Little Falls Readvance in the western Mohawk Valley (Lake Gravesville of Ridge and Franzi 1992). The Delmar Readvance is therefore given an age similar to the Barneveld–Little Falls Readvance of 13,500 B.P. (cal. 16,200 B.P.).

Nonannual flood units in the Hudson Valley varve stratigraphy that were misidentified as varves by Gerard De Geer (in Antevs 1922, NE yr 5669–5678; see Rittenour 1999) appear to represent the catastrophic release of water into the Hudson Valley from a low-level Mohawk Valley glacial lake (Lake Amsterdam of LaFleur 1979, 1983). The flood units have an age of 12,900 B.P. (cal. 15,500 B.P.) and represent the final blockage of the eastern Mohawk Valley by Hudson Valley ice. Shortly after this (12,700–12,600 B.P.; cal. 15,300–15,200 B.P.) was the Luzerne Readvance in the upper Hudson Valley (Connally and Sirkin 1971, 1973) which here has been assigned an age based on what are roughly equivalent ice margin positions in the Connecticut Valley, represented by small end moraines and readvances near Claremont (Ridge 2001b). The Luzerne Readvance has been assigned to the Port Huron Stade (Muller and Calkin 1993) and appears to be correlative to the Rome Readvance to the west. Continued ice recession in the northern Connecticut Valley from 12,400–12,000 B.P. (cal. 14,600–14,100 B.P.) is recorded by onlapping basal varves that have been matched to the upper Connecticut Valley varves of the New England varve chronology.

Final Ice Recession and the Champlain Sea (12,000–11,000 B.P.; cal. 14,100–13,000 B.P.)

The final phase of deglaciation in the Northeast began with the readvance of ice in northern New England. A large moraine system was deposited at the margin of the Littleton-Bethlehem Readvance from Bethlehem, New Hampshire, to St. Johnsbury, Vermont, at 11,900–11,800 B.P. (cal. 13,900–13,800 B.P.). In north-central Vermont the Middlesex Readvance buried wood with a ^{14}C age of 11,900 ± 50 B.P. (GX-26457; Larsen 2001). The Littleton-Bethlehem Readvance occurs at the time of the Older Dryas event (Ridge and Toll 1999; Thompson et al. 1999) in ice core records from Greenland and appears to add to existing evidence of synchronous cooling in the North Atlantic region (Dansgaard et al. 1989, 1993; Stuiver et al. 1995; Ingólfsson et al. 1997; Björck, Walker et al. 1998).

In the Champlain Valley the Bridport Readvance (Connally and Sirkin 1973) seems to have an advance limit that may join the Middlesex and Littleton-Bethlehem Readvance limits farther east. It also occurs in a position where its age relative to varves (younger than NE yr 6601–7000 and older than Essex Junction varve sequence of Antevs 1928) and paleomagnetic samples (Figure 3.6) indicates that it is close in age to readvance events to the east. On the northwest side of the Adirondacks, the Carthage-Harrisonville ice margin (Pair and Rodrigues 1993) and the Star Lake Moraine (Clark and Davis 1988) may be approximately the same age as the Bridport Readvance. Radiocarbon ages that have been used to estimate the age of these features as 12,500–12,600 B.P. are from bulk sediment samples taken from lake bottoms and may be in error. The bulk sediment ages are the same as or older than ^{14}C ages from spruce twigs marking the initiation of Lake Iroquois in the Ontario Basin, but Lake Iroquois had to be initiated prior to moraine deposition northwest of the Adirondacks. A more likely younger age for the moraines in the northwestern Adirondacks is 12,200–11,800 B.P. (cal.14,150–13,800 B.P.; Figures 3.7 and 3.8), which is approximately the same as is inferred for the Bridport Readvance.

In eastern Maine ice readvanced to form the Pineo Ridge delta and moraine complex which has been given an age of ca. 13,000 B.P. based on uncorrected marine ^{14}C ages (Stuiver and Borns 1975; Thompson and Borns 1985; Kaplan 1999). If the revised marine reservoir correction of 800–1,000 years in southern Maine (revision of Ridge et al. 2001) is applied to the marine ^{14}C age of the Pineo Ridge Readvance, it would have an age of 12,200–12,000 B.P. (cal. 14,150–14,050 B.P.). At this point such a projection is speculative and based solely on the comparison of terrestrial and marine ^{14}C ages in southern Maine. If the reservoir adjustment is correct, the Pineo Ridge Readvance would be approximately the same age as the Littleton-Bethlehem Readvance in northern New Hampshire and Vermont and may also be an Older Dryas feature (Figure 3.9).

The final recession of ice from the Connecticut Valley of Vermont was accomplished before 11,500 B.P. (cal. 13,500 B.P.), which is the age of terrestrial plant macrofossils in varves deposited after deglaciation at Columbia Bridge near the Canadian border (Figure 3.1; Miller and Thompson 1979). The recession of ice into Canada occurred at about the time that water was released from Lake Iroquois at Covey Hill into Lake

Vermont in the Champlain Valley (Parent and Occhietti 1988; Pair and Rodrigues 1993). An age of 11,700–11,400 B.P. (cal. 13,700–13,400 B.P.) has been assigned to this ice margin based on its relationship to varves at Enosburg Falls in the Missisquoi Valley of northern Vermont (Ridge et al. 1999). After 11,400 B.P. (cal. 13,400 B.P.) glacial ice had completely receded from the northeastern United States except for lingering ice masses in the uplands of northwestern Maine. Soon after the recession of ice into Canada, lakes in the St. Lawrence and Champlain Valleys were lowered to the level of the ocean as marine water of the Champlain Sea invaded these basins. Paleomagnetic records of sediment recording the transition from lacustrine to marine deposits in these basins show a declination of 0° at the time of a transition from western to eastern declination (Pair et al. 1994; Ridge et al. 1999). Magnetic correlation of this event to the New England varve chronology (Figure 3.6; Ridge et al. 1999) indicates an age range of 11,100–10,600 B.P. (cal. 13,100–12,700 B.P.). This age estimate is crude given the quality of the paleomagnetic data and a compression of the ^{14}C time scale, but it overlaps age estimates for the Champlain Sea invasion that have attempted to eliminate marine ^{14}C reservoir errors (Anderson, 1988; Rodrigues 1988, 1992).

DISCUSSION: CLIMATE AND LANDSCAPE DURING DEGLACIATION

Deglaciation is often conceptualized as being the result of climatic warming, but it is important to remember that environments south of receding ice in the northeastern United States were mostly very cold and tundra-like until the last phases of deglaciation. Evidence of a cold climate is relict periglacial ground ice features (pingo scars and ice wedge casts) formed on glacial deposits in southern to central New England until ca. 12,500 B.P. (cal. 14,700 B.P.; Schafer 1968; Larsen 1979; Black 1983; J. Stone and Ashley 1992; B. Stone et al. 1992; Rittenour 1999). The time of 12,500 B.P. (cal. 14,700 B.P.) is an important one because it represents about the half-way point in the uncovering of the landscape in the northeastern United States. Although it took about 7,500 years (cal. 9,000 years) for the first half of deglaciation to occur, almost all of the remaining landscape was deglaciated in no more than 1,100 years (cal. 1,200 years). From this perspective the period before 12,500 B.P. was relatively cold, whereas after this time came the greatest changes in late Wisconsinan climate and ice recession rates. In addition, the general pattern of deglaciation in the Northeast appears to match the pattern of oxygen isotope variations seen in the GISP2 (Cuffey et al. 1995; Stuiver et al. 1995) and Greenland Ice Sheet Project (GRIP) (Dansgaard et al. 1989, 1993) ice cores (Figure 3.9). The oxygen isotope records from Greenland ice cores are a reasonable proxy for temperature around the North Atlantic region with the northeastern United States being no exception (Björck, Walker et al. 1998b).

Summary of Glacial and Climatic Oscillations (20,000–10,000 B.P.; cal. 24,000–11,500 B.P.)

Initially ice recession was slow, averaging about 25 m/cal. yr in New England and up to 40 m/cal. yr in southern New York from 20,000–16,000 B.P. (cal. 24,000–19,100 B.P.). While warming started, conditions were not likely to have been much different from preceding times of glacial advance or full glacial conditions. There were probably minor oscillations of the ice front, as any minor cooling event would have easily retarded recession that was driven by melting. In other areas of the eastern United States, reduced melting rates, rather than increased accumulation rates, have been identified as the dominant mechanism controlling glacial readvances during overall ice recession (Lowell et al. 1998). Where the ice margin receded on land, calving did not influence ablation except over short distances where recession occurred in deep glacial lakes.

Apparently rapid ice recession across central New York from 16,000–14,600 yr B.P. (cal. 19,100–17,500 B.P., Figure 3.9), where later readvances again covered the land surface, may indicate a period of increased warmth. The full extent of ice recession in New York at this time is not known, but the Mohawk Valley became ice free and was not blocked by ice in the Hudson Valley to the east, allowing fluvial drainage and subaerial erosion to occur (Ridge 1997). The minimum rate of recession in the Hudson Valley would have been 65 m/cal. yr, whereas the withdrawal rate of ice into the Ontario Basin may have been 125 m/cal. yr or more.

A return to cold conditions seems to be represented by major readvances in New York and possibly the Connecticut Valley at 14,600–12,500 B.P. (cal. 17,500–14,700 B.P.). The number, magnitude, and rapidity of readvances in New York suggest periods of cold conditions that were rapidly punctuated by warmer intervals. Ice recession rates in New York at this time are somewhat ambiguous as a climate indicator due to the highly dynamic nature of ice-front oscillation in deep glacial lakes. However, in New England, where ice does not appear to have oscillated as much, recession initially had a rate of 30–45 m/cal. yr and then accelerated to 90 m/cal. yr as ice receded into southern New Hampshire and Vermont. Ice recession on the continental shelf of New England probably accelerated as calving occurred along a marine ice margin in the Gulf of Maine. Oxygen isotope records from Greenland indicate that this period was a time of oscillations in temperature (Figure 3.9).

From 12,500–12,000 B.P. (cal. 14,700–14,050 B.P.) ice recession accelerated across the Northeast as a result of sharply warming climate. This time has been recognized globally as a period of warming and rapid ice recession (Björck et al. 1996; Björck, Walker et al. 1998) and rapid sea-level rise (Fairbanks 1989; Bard et al. 1990, 1993; Clark et al. 1996; Adkins et al. 1998; Lambeck and Chappell 2001). Over Greenland the warmest conditions during late Wisconsinan ice recession occurred at this time (Figure 3.9; Dansgaard et al. 1989, 1993; Cuffey et al. 1995; Stuiver et al. 1995). Ice recession in the Champlain Valley and through central to northern New Hampshire and Vermont occurred at a rate of about 230 m/cal. yr (Ridge et al. 1999). Similar, or perhaps more rapid, ice recession occurred on the coast of Maine and in central New York as ice receded in deep-water marine and lacustrine environments. Calving may have accelerated ice recession along these aquatic ice fronts, but in other sections of the Northeast ice fronts in lakes were too narrow and mostly too shallow to control glacier dynamics. Accelerated ice recession along terrestrial ice fronts was the result of warming climate that greatly increased melting.

A final readvance along an ice front from the Champlain Valley to the coast of Maine appears to have occurred at ca. 12,200–11,800 B.P. (cal. 14,150–13,800 B.P.). This event appears to correspond to the Older Dryas event (Thompson et al. 1999; Ridge and Toll 1999) of northern Europe and Greenland ice core records (Dansgaard et al. 1989, 1993; Björck et al. 1995; Björck, Walker, et al. 1998; Stuiver et al. 1995; Ingólfsson et al. 1997). Cold conditions abruptly ended and rapid ice recession returned from 11,800–11,100 B.P. (cal. 13,800–13,100 B.P.), although it was not as warm as prior to the Older Dryas event. Cold climatic conditions briefly returned to northern New England during the Killarney Oscillation at 11,100 B.P. (cal. 13,100 B.P.), which has been recognized in lake cores from the maritime provinces of Canada and other regions of eastern North America (Levesque et al. 1997; Cwynar and Levesque 1995; Yu and Eicher 1998). This event appears to be the equivalent of the Intra-Allerød Cold Period of Europe and Greenland ice core records (Figure 3.9; Levesque et al. 1997; Björck, Walker et al. 1998). Except for possible lingering ice in northwest Maine, glaciers had receded from the northeastern United States before either of these events occurred.

More extreme cooling occurred during the Younger Dryas Event (10,700–10,000 B.P.; cal. 12,800–11,500 B.P.), which has been recorded in lake cores in New England, Canada, and Europe (Mott et al. 1986; Peteet et al. 1990; Mayle et al. 1993; S. Wilson et al. 1993; Mayle and Cwynar 1995; Cwynar and Levesque 1995; Thompson et al. 1996; Levesque et al. 1994, 1997; Björck, Walker et al. 1998) and in Greenland ice core records (Figure 3.9; Dansgaard, White et al. 1989; Dansgaard, Johnson et al. 1993; Stuiver et al. 1995). Evidence of elevated terrestrial erosion rates in northern New England during both the Killarney Oscillation and the Younger Dryas Event have been suggested from the varve stratigraphy of the upper Connecticut Valley (Ridge and Toll 1999).

Landscape Characteristics: Proglacial and Deglaciated Terrains

In addition to temperature and vegetation patterns dictated by climate, other aspects of the landscape require our attention if we are to make smart decisions about where to look for the earliest human habitation sites and migration routes. Two important landscape characteristics related to glaciation are sea-level position and the formation of large glacial lakes across the region. Lowered sea level during the last glaciation would have been an important factor to human migration during the early stages of deglaciation. It would have created a perimeter of dry land on the continental shelf of New Jersey and southern New England beyond the edge of the ice sheet, where marine water covers the shelf today (Uchupi et al. 2001). During the later stages of deglaciation from the coast of New Hampshire to eastern Maine, isostatic depression of the land surface allowed marine flooding and prevented human migration in a coastal area that is above sea level today (Thompson and Borns 1985; Koteff et al. 1993).

Glacial lakes are important to understanding possible human habitation and migration into newly deglaciated areas of the northeastern United States because they can occupy a large percentage of the land surface in some areas and they form in long north-south corridors that are generally recognized as migration routes. This is especially true in eastern Massachusetts where much of the land surface was at least briefly covered by glacial lakes during deglaciation (B. Stone and Peper, 1982; B. Stone et al. 1992). Seasonally frozen lake surfaces would have provided easy access to some regions and lakes that persisted for centuries to millennia after deglaciation undoubtedly had an impact on food sources. Nowhere is this more apparent than in the northern Connecticut Valley, where glacial lakes evolved into nonglacial lakes that persisted for at least 1,800 years after deglaciation (until at least

10,400 B.P.; cal. 12,300 B.P. at Newbury; Ridge et al. 1999; Ridge and Toll 1999). The upper Connecticut Valley lakes survived not only the early postglacial tundra climate but eventually became surrounded by spruce forests. Mussels (*Pyganodon fragilis*, Smith and Ridge 2001) inhabited the upper Connecticut Valley lakes by NE yr 8265 (at Newbury; 10,700 B.P.; cal. 12,800 B.P.), and fish trails have been found in varves as early as NE yr 7359 at Newbury and NE yr 7370 in the Passumpsic Valley (PAS2 on Figure 3.3; 11,700 B.P.; cal. 13,700 B.P.). Efforts to locate evidence of the earliest human arrival might best be spent along the shorelines of upper Connecticut Valley lakes rather than on younger fluvial terrace surfaces. Drained glacial lakes also have an influence on the landscape. The poorly drained clayey and silty flat surfaces of lake floors often became wetlands after lake drainage that provided food sources for early mammals and waterfowl in addition to the earliest humans entering the region.

CONCLUSION

My purpose has been to provide a detailed chronology of ice recession at the close of the last glaciation in the northeastern United States. Ice recession in the Northeast occurred ca. 20,000–11,000 B.P. (cal. 23,700–13,000 B.P.) and limited the extent to which any organisms could inhabit the region (Figures 3.7 and 3.8). The glacier itself was a barrier, but patterns of ice recession and oscillation driven by abrupt swings in climate also need to be considered (Figure 3.9). From the standpoint of temperature-controlled climate, there may have been times more favorable to human migration into the region depending on lifestyle requirements. If warming of climate from full glacial conditions was required for early human habitation or migration into previously glaciated areas, there appear to be three windows of opportunity during late Wisconsinan deglaciation: 16,000–14,600, 12,500–12,000, and 11,800–11,100 B.P. (cal. 19,000–17,450, 14,700–14,100, and 13,800–13,100 B.P.). Conversely, if human lifestyles at the time were adapted to landscapes and vegetation of cold tundra-like conditions, then 14,600–12,500, 12,000–11,800, and 11,100–10,000 B.P. (cal. 17,500–14,700, 14,100–13,800, and 13,100–11,500 B.P.) would have been favorable times for entry into previously glaciated landscapes.

Certain climatic conditions may have been favored over others providing windows of opportunity for habitation or migration, but another aspect of climate patterns needs to be considered. Rapid changes in climate as marked by sudden changes in glacier dynamics and ice-front oscillations that characterize an unstable climate mode would have forced people either to adapt rapidly to new conditions or to find refuge through rapid migration to more favorable areas. Periods of unstable rapidly oscillating cold and warmer periods such as occurred 14,600–12,500 B.P. (cal. 17,500–14,700 B.P.) are times when prolonged human settlement without rapid adaptation or migration may not have been possible (Figure 3.9). Given the generally cold conditions, the unstable mode of climate during deglaciation becomes a very important factor when one considers possible human habitation and migration at the edge of human tolerances.

In addition to climatic windows and patterns of climate change, the earliest human inhabitants almost certainly took advantage of favorable landscape features such as glacial and postglacial lakes to aid their migration and search for food. It is important when searching for possible archaeological sites to consider not only where the glacier front was at any given time, but also what climatic conditions were controlling glacier dynamics and what landscape conditions were left in the wake of receding ice.

END NOTES

1. Varve year numbers of the New England varve chronology will be identified with an abbreviation (NE yr) to indicate specific years in the sequence; for example, NE yr 5647 indicates varve year 5647 in the New England varve chronology. All varve sequences in North America and Sweden get higher in number as they get younger and may start with negative numbers.

2. In this chapter the terms *match* and *overlap* have different meanings. Matching varve sequences have similar annual thickness records that allow a definitive annual correlation. Overlapping varve sequences cover the same time period regardless of whether their annual thickness records match.

3. The term *glacial varve* as used in this chapter is defined as a varve that is deposited in a lake where the delivery of sediment by glacial meltwater controls the variation in annual layer thickness. A "nonglacial varve" is deposited in a lake where delivery of sediment from meteoric (runoff) sources dominates or is the only source.

REFERENCES CITED

Abbott, M.B., and Stafford, T.W. (1996). Radiocarbon geochemistry of modern and ancient arctic lake systems, Baffin Island, Canada. *Quaternary Research* **45**:300–311.

Adkins, J.F., Cheng, H., Boyle, E.A., Druffel, E.R.M., and Edwards, R.L. (1998). Deep-sea coral evidence for rapid change in ventilation of the deep North Atlantic 15,400 years ago. *Science* **280**:725–728.

Anderson, T.W. (1988). Late Quaternary pollen stratigraphy of the Ottawa Valley–Lake Ontario region and its application in dating the Champlain Sea. In *The Late Quaternary Development of the Champlain Sea Basin*, edited by N.R. Gadd, pp. 205–224. Special Paper No. 35, Geological Association of Canada, St. John's, Newfoundland.

Andrée, M., Oeschger, H., Siegenthaler, U., Riesen, T., Möll, M., Ammann, B., and Tobolski, K. (1986). ^{14}C dating of plant macrofossils in lake sediment. *Radiocarbon* **28**:411–416.

Antevs, E. (1922). *The Recession of the Last Ice Sheet in New England*. Research Series No. 11. American Geographical Society, New York.

Antevs, E. (1925). *Retreat of the Last Ice Sheet in Eastern Canada*. Memoir No. 146, Geological Survey of Canada, Ottawa.

Antevs, E. (1928). *The Last Glaciation with Special Reference to the Ice Sheet in North America*. Research Series No. 17, American Geographical Society, New York.

Antevs, E. (1935). Telecorrelations of varve curves. *Geologiska Foreningens i Stockholm Förhandlingar* **57**:47–58.

Antevs, E. (1962). Transatlantic climatic agreement versus ^{14}C dates. *Journal of Geology* **70**:194–205.

Ashley, G.M. (1972). *Rhythmic Sedimentation in Glacial Lake Hitchcock, Massachusetts-Connecticut*. Geology Department Contribution No. 10, University of Massachusetts, Amherst.

Ashley, G.M. (1975). Rhythmic sedimentation in glacial Lake Hitchcock, Massachusetts-Connecticut. In *Glaciofluvial and Glaciolacustrine Sedimentation*, edited by A.V. Jopling and B.C. McDonald, pp. 304–320. Special Publication No. 23, Society of Economic Paleontologists and Mineralogists, Tulsa, Okla.

Ashley, G.M., Thomas, G., Retelle, M., and Hartshorn, J. (1982). Sedimentation in a proglacial lake: Glacial Lake Hitchcock. In *Guidebook for Field Trips in Connecticut and South Central Massachusetts*, edited by R. Joesten and S.S. Quarrier, pp. 89–102. Guidebook No. 5, State Geological and Natural History Survey of Connecticut, Hartford.

Austin, W.E., Bard, E., Hunt, J.B., Kroon, R., and Peacock, J.D. (1996). The Icelandic Vedde ash and the Younger Dryas marine reservoir factor. *Radiocarbon* **37**:53–62.

Bard, E. (1988). Correction of accelerator mass spectrometry ^{14}C ages measured in planktonic foraminifera: Paleoceanographic implications. *Paleoceanography* **3**:635–645.

Bard, E., Arnold, M., Fairbanks, R.G., and Hamelin, B. (1993). ^{230}Th-^{234}U and ^{14}C ages obtained by mass spectrometry on corals. *Radiocarbon* **35**:191–199.

Bard, E.B., Arnold, M., Mangerud, J., Paterne, M., Labeyrie, L., Duprat, J., Mélières, M-A., Sønstegaard, E., and Duplessy, J.C. (1994). The North Atlantic atmosphere-sea surface ^{14}C gradient during the Younger Dryas climatic event. *Earth and Planetary Science Letters* **126**:275–287.

Bard, E., Hamelin, B., Fairbanks, R.G., and Zindler, A. (1990). Calibration of the ^{14}C time scale over the past 30,000 years using mass spectrometric U-Th ages from Barbados corals. *Nature* **345**:405–410.

Birks, H.H., Gulliksen, S., Haflidason, H., Mangerud, J., and Possnert, G. (1996). New radiocarbon dates for the Vedde Ash and the Saksunarvatn Ash from western Norway. *Quaternary Research* **45**:119–127.

Björck, S., Bennike, O., Possnert, G., Wohlfarth, B., and Digerfeldt, G. (1998). A high-resolution ^{14}C dated sediment sequence from southwest Sweden: Age comparisons between different components of the sediment. *Journal of Quaternary Science* **13**:85–89.

Björck, S., Kromer, B., Johnsen, S., Bennike, O., Hammarlund, D., Lemdahl, G., Possnert, G., Rasmussen, T.L., Wohlfarth, B., Hammer, C.U., and Spurk, M. (1996). Synchronized terrestrial-atmospheric deglacial records around the North Atlantic. *Science* **274**:1155–1160.

Björck, S., Walker, M.J.C., Cwynar, L.C., Johnsen, S., Knudsen, K.-L., Lowe, J.J., Wohlfarth, B., and INTIMATE members. (1998). An event stratigraphy for the last Termination in the North Atlantic region based on the Greenland ice-core record: A proposal by the INTIMATE group. *Journal of Quaternary Science* **13**:283–292.

Björck, S., Wohlfarth, B., and Possnert, B. (1995). ^{14}C AMS measurements from the late Weichselian part of the Swedish time scale. *Quaternary International* **27**:11–18.

Black, R.F. (1983). Pseudo-ice-wedge casts of Connecticut, northeastern United States. *Quaternary Research* **20**:74–89.

Bondevik, S., Birks, H.H., Gulliksen, S., and Mangerud, J. (1999). Late Weichselian marine ^{14}C reservoir ages at the western coast of Norway. *Quaternary Research* **52**:104–114.

Boothroyd, J., Freedman, J.H., Brenner, H.B., and Stone, J.R. (1998). The glacial geology of southern Rhode Island. In *Guidebook to Field Trips in Rhode Island and Adjacent Regions of Connecticut and Massachusetts*, edited by Murray, D.P., pp. C-5:1–25. 90th Annual Meeting, New England Intercollegiate Geological Conference, Kingston, R.I.

Borns, H.W., Jr. (1998). The progress of radiocarbon dating. *Geological Society of America Abstracts with Programs* **30(1)**:6.

Brennan, W.J., Hamilton, M., Kilbury, R., Reeves, R.L., and Covert, L. (1984). Late Quaternary secular variation of geomagnetic declination in western New York. *Earth and Planetary Science Letters* **70**:363–372.

Clark, P.U., Alley, R.B., Keigwin, L.D., Licciardi, J.M., Johnsen, S.J., and Wang, H. (1996). Origin of the first global meltwater pulse following the last glacial maximum. *Paleoceanography* **11**:563–577.

Clark, P.U., and Davis, P.T. (1988). Deglacial chronology of the northwestern Adirondack Mountains. *Geological Society of America Abstracts with Programs* **20**:12.

Connally, G.G. (1968). The Rosendale readvance in the lower Wallkill Valley, New York. In *National Association of Geology Teachers Guidebook, Eastern Section*, pp. 22–28, New Paltz, N.Y.

Connally, G.G., and Sirkin, L.A. (1970). Late glacial history of the upper Wallkill Valley, New York. *Geological Society of America Bulletin* **81**:3297–3306.

Connally, G.G., and Sirkin, L.A. (1971). The Luzerne readvance near Glens Falls, New York. *Geological Society of America Bulletin* **82**:989–1008.

Connally, G.G., and Sirkin, L.A. (1973). Wisconsinan history of the Hudson-Champlain lobe. In *The Wisconsinan Stage*, edited by R.F. Black, R.P. Goldthwait, and H.B. Willman, pp. 47–69. Geological Society of America Memoir No. 136.

Connally, G.G., and Sirkin, L.A. (1986). Woodfordian ice margins, recessional events, pollen stratigraphy of the mid-Hudson Valley. In *The Wisconsinan Stage of the First Geological District, Eastern New York*, edited by D.H. Cadwell, pp. 50–72. New York State Museum Bulletin 455. The University of the State of New York, Albany.

Cotter J.F.P., Ridge, J.C., Evenson, E.B., Sevon, W.D., Sirkin, L., and Stuckenrath, R. (1986).The Wisconsinan history of the Great Valley, Pennsylvania, and New Jersey, and the age of the "Terminal Moraine." In *The Wisconsinan Stage of the First Geological District, Eastern New York*, edited by D.H. Cadwell, pp. 22–49. New York State Museum Bulletin 455. The University of the State of New York, Albany.

Crosby, I.B. (1934a). Extension of the Bethlehem, New Hampshire, moraine. *Journal of Geology* **42**:411–421.

Crosby, I.B. (1934b). Geology of Fifteen Mile Falls development. *Civil Engineering* **4**:21–24.

Crowl, G.H., and Sevon, W.D. (1980). Glacial border deposits of late Wisconsinan age in northeastern Pennsylvania. *General Geology Report* No. G-71, Pennsylvania Geological Survey, Harrisburg.

Cuffey, K.M., Clow, G.D., Alley, R.B., Stuiver, M., Waddington, E.D., and Saltus, R.W. (1995). Large arctic temperature change at the Wisconsin-Holocene glacial transition. *Nature* **270**:455–458.

Cwynar, L.C., and Levesque, A.J. (1995). Chironomid evidence for late-glacial climatic reversals in Maine. *Quaternary Research* **43**:405–413.

Cwynar, L.C., and Watts, W.A. (1989). Accelerator mass spectrometer ages for late-glacial events in Ballybetagh, Ireland. *Quaternary Research* **31**:377–380.

Dansgaard, W., Johnsen, S.J., Clausen, H.B., Dahl-Jensen, D., Gundestrup, N.S., Hammer, C.U., Hvidberg, C.S., Steffensen, J.P., Sveinbjörnsdottir, A.E., Jouzel, J., and Bond, G. (1993). Evidence for general instability of past climate from a 250-kyr ice-core record. *Nature* **364**:218–220.

Dansgaard, W., White, J.W.C., and Johnsen, S.J. (1989). The abrupt termination of the Younger Dryas climate event. *Nature* **339**:532–534.

Davis, M.B., Spear, R.W., and Shane, L.C. (1980). Holocene climate of New England. *Quaternary Research* **14**:240–250.

Davis, P.T., and Davis, R.B. (1980). Interpretation of minimum-limiting radiocarbon dates for deglaciation of Mount Katahdin area, Maine. *Geology* **8**:396–400.

Davis, P.T., Dethier, D.P., and Nickmann, R. (1995). Deglaciation chronology and late Quaternary pollen records from Woodford Bog, Bennington County, Vermont. *Geological Society of America Abstracts with Programs* **27**(1):38.

Davis, R.B., and Jacobson, G.L., Jr. (1985). Late glacial and early Holocene landscapes in northern New England and adjacent areas of Canada. *Quaternary Research* **23**:341–368.

De Geer, G. (1921). Correlation of late glacial clay varves in North America with the Swedish time scale. *Geologiska Foreningens i Stockholm Förhandlingar* **43**:70–73.

De Geer, G. (1927). Late glacial clay varves in Argentina measured by D. Carl Caldenius, dated and connected with the solar curve through the Swedish time scale. *Geografiska Annaler* **9**:1–8.

De Geer, G. (1929). Gotiglacial clay-varves in southern Chile measured by Dr. Carl Caldenius, identified with synchronous varves in Sweden, Finland, and U.S.A. *Geografiska Annaler* **11**:247–256.

De Geer, G. (1940). Geochronologia Suecica principles. *Kungl. Svenska Vetenskapsakademiens Handlingar*, Tredje Series, Band 18, No. 6.

Dineen, R.J. (1986). Deglaciation of the Hudson Valley between Hyde Park and Albany, New York. In *The Wisconsinan Stage of the First Geological District, Eastern New York*, edited by D.H. Cadwell, pp. 89–108. New York State Museum Bulletin 455. The University of the State of New York, Albany.

Dorion, C.C. (1997). An updated high resolution chronology of deglaciation and accompanying marine transgression in Maine. Unpublished master's thesis, Department of Geology, University of Maine, Orono.

Dyke, A.S., and Prest, V.K. (1987). Late Wisconsinan and Holocene retreat of the Laurentide ice sheet. *Géographie physique et Quaternaire* **41**:237–263.

Fairbanks, R.G. (1989). A 17,000-year glacio-eustatic sea level record: Influence of glacial melting rates on the Younger Dryas event and deep-ocean circulation. *Nature* **342**:637–642.

Fleisher, P.J. (1986). Glacial geology and late Wisconsinan stratigraphy, upper Susquehanna drainage basin, New York. In *The Wisconsinan Stage of the First Geological District, Eastern New York*, edited by D.H. Cadwell, pp. 121–142. New York State Museum Bulletin 455. The University of the State of New York, Albany.

Flint, R.F. (1929). The stagnation and dissipation of the last ice sheet. *Geographical Review* **19**:256–289.

Flint, R.F. (1930). *The Glacial Geology of Connecticut*. Connecticut Geological Survey Bulletin No. 47, Hartford.

Flint, R.F. (1932). Deglaciation of the Connecticut Valley. *American Journal of Science* **24**:152–156.

Flint, R.F. (1933). Late-Pleistocene sequence in the Connecticut Valley. *Geological Society of America Bulletin* **44**:965–988.

Flint, R.F. (1956). New radiocarbon dates and late-Pleistocene stratigraphy. *American Journal of Science* **254**:265–287.

Goldsmith, R. (1982). Recessional moraines and ice retreat in southeastern Connecticut. In *Late Wisconsinan Glaciation of New England*, edited by G.J. Larson and B.D. Stone, pp. 61–76. Kendall/Hunt Publishing, Dubuque, Iowa.

Goldthwait, J.W. (1916). Glaciation in the White Mountains of New Hampshire. *Geological Society of America Bulletin* 27:263–294.

Goldthwait, J.W. (1938). The uncovering of New Hampshire by the last ice sheet. *American Journal of Science* 36:345–372.

Harmon, K.P. (1968). Late Pleistocene forest succession in northern New Jersey. Unpublished doctoral dissertation, Department of Geology, Rutgers University, New Brunswick, N.J.

Hillaire-Marcel, C. (1988). Isotope composition (^{18}O, ^{13}C, ^{14}C) of biogenic carbonates in Champlain Sea sediments. In *The Late Quaternary Development of the Champlain Sea Basin*, edited by N.R. Gadd, pp. 177–194. Special Paper No. 35, Geological Association of Canada, St. John's, Newfoundland.

Hjort, C. (1973). A sea correction for East Greenland: *Geologiska Föreningens i Stockholm Förhandlingar* 95:132–134.

Hughes, T., Borns, H.W., Jr., Fastook, J.L., Hyland, M.R., Kite, J.S., and Lowell, T.V. (1985). Models of glacial reconstruction and deglaciation applied to Maritime Canada and New England. In *Late Pleistocene History of Northeastern New England and Adjacent*, edited by H.W. Borns, Jr., P., LaSalle, and W.B. Thompson, pp. 139–150. Special Paper No. 197, Geological Society of America, Boulder, Colo.

Ingólfsson, Ó., Björck, S., Haflidason, H., and Rundgren, M. (1997). Glacial and climatic events in Iceland reflecting regional North Atlantic climatic shifts during the Pleistocene-Holocene transition. *Quaternary Science Reviews* 16:1135–1144.

Jahns, R.H., and Willard, M. (1942). The Pleistocene and recent deposits in the Connecticut Valley, Massachusetts. *American Journal of Science* 240:161–191, 265–287.

Johnson, E.A., Murphy, T., and Torreson, O.W. (1948). Prehistory of the earth's magnetic field. *Terrestrial Magnetism and Atmospheric Electricity* (now *Journal of Geophysical Research*) 53:349–372.

Kaplan, M.R. (1999). Retreat of a tidewater margin of the Laurentide ice sheet in eastern coastal Maine between ca. 14,000 and 13,000 ^{14}C yr B.P. *Geological Society of America Bulletin* 111:620–632.

Karrow, P.F., and Anderson, T.W. (1975). Palynological studies of lake sediment profiles from SW New Brunswick: Discussion. *Canadian Journal of Earth Science* 12:1808–1812.

Kasgarian, M. (1992). *Radiocarbon in coastal waters off the eastern United States*. Doctoral dissertation, Department of Geology and Geophysics, Yale University, University Microfilms, Ann Arbor, Mich.

Koteff, C. (1974). The morphologic sequence concept and deglaciation of southern New England. In *Glacial Geomorphology*, edited by D.R. Coates, pp. 121–144. Publications in Geomorphology, State University of New York, Binghamton.

Koteff, C., and Pessl, F., Jr. (1981). Systematic ice retreat in New England. *Professional Paper* No. 1179. U.S. Geological Survey, Reston, Va.

Koteff, C., Robinson, G.R., Goldsmith, R., and Thompson, W.B. (1993). Delayed postglacial uplift and synglacial sea levels in coastal central New England. *Quaternary Research* 40:46–54.

Krall, D.B. (1977). Late Wisconsinan ice recession in east-central New York. *Geological Society of America* 88:1697–1710.

LaFleur, R.G. (1979). Deglacial events in the eastern Mohawk—northern Hudson lowland. In *Guidebook for Field Trips, Joint Annual Meeting*, edited by G.M. Friedman, pp. 326–350. Joint Annual Meeting, New York State Geological Association and New England Intercollegiate Geological Conference, Rensselaer Polytechnic Institute, Troy, N.Y.

LaFleur, R.G. (1983). Mohawk Valley episodic discharges—the geomorphic and glacial sedimentary record. In *Eastern Section Guidebook for Field Trips*, edited by G.M. Friedman, pp. 45–68. National Association of Geology Teachers, Eastern Section, Spring Meeting, Rensselaer Polytechnic Institute, Troy, N.Y.

Lambeck, K., and Chappell, J. (2001). Sea level change through the last glacial cycle. *Science* 292:679–686.

Larsen, F.D. (1979). Retreat of ice left eastern upland of Massachusetts subject to deep freeze. *Professional Paper* No. 1150. U.S. Geological Survey, Reston, Va.

Larsen, F.D. (1987). History of glacial lakes in the Dog River Valley, central Vermont. *In Guidebook for Field Trips in Vermont, Volume 2*, edited by D.S. Westerman, pp. 213–236. 79th Annual Meeting of the New England Intercollegiate Geological Conference, Montpelier, Vt.

Larsen, F.D. (2001). The Middlesex readvance of the late-Wisconsinan ice sheet in central Vermont at 11,900 ^{14}C years B.P. *Geological Society of America Abstracts with Programs* 33(1):A-15.

Larsen, F.D., and Hartshorn, J.H. (1982). Deglaciation of the southern portion of the Connecticut Valley of Massachusetts. In *Late Wisconsinan Glaciation of New England*, edited by G.J. Larson and B.D. Stone, pp. 115–128. Kendall/Hunt Publishing, Dubuque, Iowa.

Larsen, F.D., Ridge, J.C., and Wright, S.F. (2001). Correlation of varves of glacial lake Winooski, north-central Vermont. *Geological Society of America Abstracts with Programs* 33(1):A-66.

Larson, G.J. (1982). Nonsynchronous retreat of ice lobes from southeastern Massachusetts. In *Late Wisconsinan Glaciation of New England*, edited by G.J. Larson and B.D. Stone, pp. 101–114. Kendall/Hunt Publishing, Dubuque, Iowa.

Levesque, A.J., Cwynar, L.C., and Walker, I.R. (1994). A multiproxy investigation of late-glacial climate and vegetation change at Pine Ridge Pond, southwest New Brunswick, Canada. *Quaternary Research* 42:316–327.

Levesque, A.J., Mayle, F.E., Walker, I.R., and Cwynar, L.C. (1997). The amphi-Atlantic oscillation: A proposed late-glacial climatic event. *Quaternary Science Reviews* 12:629–643.

Levy, L.B. (1998). Interpreting the carbonate concretions of glacial Lake Hitchcock. Unpublished bachelor's honors thesis, Department of Geology, Mount Holyoke College, South Hadley, Mass.

Lewis, H.C. (1884). Report on the terminal moraine in Pennsylvania and western New York. *Report Z*. Pennsylvania Geological Survey, Harrisburg.

Lini, A., Bierman, P.R., Lin, L., and Davis, P.T. (1995). Stable carbon isotopes in postglacial lake sediments: A technique for timing the onset of primary productivity and verifying AMS 14-C dates. *Geological Society of America Abstracts with Programs* **27**(6):58.

Lougee, R.J. (1935). Time measurements of an ice readvance at Littleton, N.H. *Proceedings of the National Academy of Sciences* **21**:36–41.

Lougee, R.J. (1940). Deglaciation of New England. *Journal of Geomorphology* **3**:189–217.

Lowe, J.J., Coope, G.R., Harkness, D.D., Sheldrick, C., and Walker, M.J.C. (1995). Direct comparison of UK temperatures and Greenland snow accumulation rates, 15-12,000 calendar years ago. *Journal of Quaternary Science* **10**:175–180.

Lowell, T.V., Hayward, R.K., and Denton, G.H. (1998). Role of climate oscillations in determining ice-margin position: Hypothesis, examples, and implications. In *Glacial Processes Past and Present*, edited by D.M. Mickelson and J.W. Attig, pp. 193–203. Special Paper No. 337, Geological Society of America, Boulder, Colo.

Mangerud, J. (1972). Radiocarbon dating of marine shells including a discussion of apparent age of recent shells from Norway. *Boreas* **1**:143–172.

Mangerud, J., and Gulliksen, S. (1975). Apparent radiocarbon ages of recent marine shells from Norway, Spitsbergen, and Arctic Canada. *Quaternary Research* **5**:263–274.

Marshall, E. (2001). Pre-Clovis sites fight for acceptance. *Science* **291**:1730–1732.

Mayle, F.E., and Cwynar, L.C. (1995). Impact of the Younger Dryas cooling event upon lowland vegetation of maritime Canada. *Ecological Monographs* **65**:129–154.

Mayle, F.E., Levesque, A.J., and Cwynar, L.C. (1993). Accelerator-mass-spectrometer ages for the Younger Dryas event in Atlantic Canada. *Quaternary Research* **39**:355–360.

McNish, A.G., and Johnson, E.A. (1938). Magnetization of unmetamorphosed varves and marine sediments. *Terrestrial Magnetism and Atmospheric Electricity* (now *Journal of Geophysical Research*) **43**:401–407.

Miller, N.G., and Thompson, G.G. (1979). Boreal and western North American plants in the Pleistocene of Vermont. *Journal of the Arnold Arboretum* **60**:167–218.

Minard, J.P. (1961). End moraines on Kittatinny Mountain, Sussex County, New Jersey. *Professional Paper* No. 424-C, pp. C-61–C-64, U.S. Geological Survey.

Mott, R.J. (1975). Palynological studies of lake sediment profiles from southwestern New Brunswick. *Canadian Journal of Earth Science* **12**:273–288.

Mott, R.J. (1981). Appendix 4, palynology of southeastern Quebec. In *Surficial Geology of the Lac Mégantic Area, Québec*, by W.W. Shilts, pp. 99–102. Memoir No. 397, Geological Survey of Canada, Ottawa.

Mott, R.J., Grant, D.R., Stea, R., and Occhietti, S. (1986). Late-glacial climatic oscillation in Atlantic Canada equivalent to the Allerød/Younger Dryas event. *Nature* **323**:247–250.

Muller, E.H., and Calkin, P.E. (1993). Timing of Pleistocene glacial events in New York State. *Canadian Journal of Earth Science* **30**:1829–1845.

Muller, E.H., and Prest, V.K. (1985). Glacial lakes in the Ontario Basin. In *Quaternary Evolution of the Great Lakes*, edited by P.F. Karrow and P.E. Calkin, pp. 212–229. Special Paper No. 30, Geological Association of Canada, St. John's, Newfoundland.

Muller, E.H., Franzi, D.A., and Ridge, J.C. (1986). Pleistocene geology of the western Mohawk Valley, New York. In *The Wisconsinan Stage of the First Geological District, Eastern New York*, edited by D.H. Cadwell, pp. 143–157. New York State Museum Bulletin 455. The University of the State of New York, Albany.

O'Brien, N.R., and Pietraszek-Mattner, S. (1998). Origin of the fabric of laminated fine-grained glaciolacustrine deposits. *Journal of Sedimentary Research* **68**:832–840

Oeschger, H., Andrée, M., Moell, M., Riesen, T., Siegenthaler, U., Ammann, B., Tobolski, K., Bonani, B., Hofmann, H.J., Morenzoni, E., Nessi, M., Suter, M., and Wölfli, W. (1985). Radiocarbon chronology of Lobsigensee: Comparison of materials and methods. In *Swiss Lake and Mire Environments During the Last 15,000 Years*, edited by G. Lang, pp. 135–139. Dissertationes Botanicae No. 87.

Oldale, R.N. (1982). Pleistocene stratigraphy of Nantucket, Martha's Vineyard, the Elizabeth Islands, and Cape Cod, Massachusetts. In *Late Wisconsinan Glaciation of New England*, edited by G.J. Larson and B.D. Stone, pp. 1–34. Kendall/Hunt Publishing, Dubuque, Iowa.

Pair, D.L., Muller, E.H., and Plumley, P.W. (1994). Correlation of late Pleistocene glaciolacustrine and marine deposits by means of geomagnetic secular variation, with examples from northern New York and southern Ontario. *Quaternary Research* **42**:277–287.

Pair, D.L., and Rodrigues, C.G. (1993). Late Quaternary deglaciation of the southwestern St. Lawrence Lowland, New York and Ontario. *Geological Society of America Bulletin* **105**:1151–1164.

Parent, M., and Occhietti, S. (1988). Late Wisconsinan deglaciation and Champlain Sea invasion in the St. Lawrence Valley, Quebec. *Géographie physique et Quaternaire* **42**:215–246.

Peteet, D.M., Vogel, J.S., Nelson, D.E., Southon, J.R., Nickmann, R.J., and Heusser, L.E. (1990). Younger Dryas climatic reversal in northeastern USA? AMS ages for an old problem. *Quaternary Research* **33**:219–230.

Reeds, C.A. (1926). The varved clays at Little Ferry, New Jersey. *American Museum Novitates* No. 209, American Museum of Natural History, New York.

Ridge, J.C. (1997). Shed Brook Discontinuity and Little Falls Gravel: Evidence for the Erie interstade in central New York. *Geological Society of America Bulletin* **109**:652–665.

Ridge, J.C. (1999). Surficial geologic map of the Bellows Falls Quadrangle, Cheshire and Sullivan Counties, N.H., and Windham and Windsor Counties, Vt. *New Hampshire Geological Survey Open-File Report*, Concord, N.H., scale 1:24,000, 3 sheets.

Ridge, J.C. (2001a). Speculation on glacial varve deposition in the northeastern US. *Geological Society of America Abstracts with Programs* **33**(1):A-14.

Ridge, J.C. (2001b). Surficial geologic map of part of the Springfield Quadrangle, Sullivan County, N.H., and Windsor County, Vt. *New Hampshire Geological Survey Open-File Report*, Concord, N.H., scale 1:24,000, 3 sheets.

Ridge, J.C., Besonen, M.R., Brochu, M., Brown, S.L., Callahan, J.W., Cook, G.J., Nicholson, R.S., and Toll, N.J. (1999). Varve, paleomagnetic, and ^{14}C chronologies for late Pleistocene events in New Hampshire and Vermont (U.S.A.). *Géographie physique et Quaternaire* **53**:79–107.

Ridge, J.C., Brennan, W.J., and Muller, E.H. (1990). The use of paleomagnetic declination to test correlations of late Wisconsinan glaciolacustrine sediments in central New York. *Geological Society of America Bulletin* **102**:26–44.

Ridge, J.C., Canwell, B.A., Kelly, M.A., and Kelley, S.Z. (2001). Atmospheric ^{14}C chronology for late Wisconsinan deglaciation and sea-level change in eastern New England using varve and paleomagnetic records. In *Deglacial History and Relative Sea-Level Changes, Northern New England and Adjacent Canada*, edited by T.K. Weddle and M.J. Retelle, pp. 173–191. Special Paper No. 351, Geological Society of America, Boulder, Colo.

Ridge, J.C., and Franzi, D.A. (1992). Late Wisconsinan glacial lakes of the western Mohawk Valley region of central New York. In *New York State Geological Association Field Trip Guidebook, 64th Annual Meeting*, edited by R.H. April, pp. 97–120. Colgate University, Hamilton.

Ridge, J.C., Franzi, D.A., and Muller, E.H. (1991). Late Wisconsinan, pre-Valley Heads glaciation in the western Mohawk Valley, central New York, and its regional implications. *Geological Society of America Bulletin* **103**:1032–1048.

Ridge, J.C., and Larsen, F.D. (1990). Reevaluation of Antevs' New England varve chronology and new radiocarbon dates of sediments from glacial Lake Hitchcock. *Geological Society of America Bulletin* **102**:889–899.

Ridge, J.C., Thompson, W.B., Brochu, M., Brown, S., and Fowler, B. (1996). Glacial geology of the upper Connecticut Valley in the vicinity of the lower Ammonoosuc and Passumpsic Valleys of New Hampshire and Vermont. In *Guidebook to Field Trips in Northern New Hampshire and Adjacent Regions of Maine and Vermont*, edited by M.R. Van Baalen, pp. 309–340. 88th Annual Meeting of the New England Intercollegiate Geological Conference, Harvard University, Cambridge, Mass.

Ridge, J.C., and Toll, N.J. (1999). Are late-glacial climate oscillations recorded in varves of the upper Connecticut Valley, northeastern United States? *Geologiska Foreningens i Stockholm Förhandlingar* **121** (3):187–193.

Rittenour, T.M. (1999). Drainage history of glacial Lake Hitchcock, northeastern USA. Unpublished master's thesis, Department of Geology and Geography, University of Massachusetts, Amherst.

Rittenour, T.M., Brigham-Grette, J., and Mann, M.E. (2000). El Niño-like climate teleconnections in New England during the late Pleistocene. *Science* **288**:1039–1042.

Rodrigues, C.G. (1988). Late Quaternary invertebrate faunal associations and chronology of the western Champlain Sea Basin. In *The Late Quaternary Development of the Champlain Sea Basin*, edited by N.R. Gadd, pp. 155–176. Special Paper No. 35, Geological Association of Canada, St. John's, Newfoundland.

Rodrigues, C.G. (1992). Successions of invertebrate microfossils and the late Quaternary deglaciation of the central St. Lawrence Lowland, Canada and United States. *Quaternary Science Reviews* **11**:503–534.

Schafer, J.P. (1968). Periglacial features and pre-Wisconsin weathered rock in the Oxford-Waterbury-Thomaston area, western Connecticut. In *Guidebook for Fieldtrips in Connecticut*, edited by P.M. Orville, pp. 1–5. 60th Annual Meeting, New England Intercollegiate Geological Conference, Connecticut Geological and Natural History Survey Guidebook No. 2, New Haven.

Schafer, J.P., and Hartshorn, J.H. (1965). The Quaternary of New England. In *The Quaternary of the United States*, edited by H.E. Wright, Jr., and D.G. Frey, pp. 113–128. Princeton University Press, Princeton, N.J.

Shotton, F.W. (1972). An example of hard water error in radiocarbon dating of vegetable matter. *Nature* **240**:460–461.

Sirkin, L.A. (1982). Wisconsinan glaciation of Long Island, New York to Block Island. In *Late Wisconsinan Glaciation of New England*, edited by G.J. Larson and B.D. Stone, pp. 35–59. Kendall/Hunt Publishing, Dubuque, Iowa.

Sirkin, L.A. (1986). Palynology and stratigraphy of Cretaceous and Pleistocene sediments on Long Island, New York: a basis for correlation with New Jersey coastal plain sediments. *U.S. Geological Survey Bulletin*, Report: B 1559, Reston, Va.

Sirkin, L.A., and Minard, J.P. (1972). Late Pleistocene glaciation and pollen stratigraphy in northwestern New Jersey. *Professional Paper* No. 800-D, pp. D-51–D-56. U.S. Geological Survey, Reston, Va.

Smith, G.W. (1980). *End moraines and Glaciofluvial Deposits, Cumberland and York Counties, Maine* (map): Maine Geological Survey, Augusta, scale 1:125,000, 1 sheet.

Smith, G.W. (1982). End moraines and the pattern of last ice retreat from central and south coastal Maine. In *Late Wisconsinan Glaciation of New England*, edited by G.J. Larson and B.D. Stone, pp. 195–209. Kendall/Hunt Publishing, Dubuque, Iowa.

Smith, G.W. (1985). Chronology of late Wisconsinan deglaciation of coastal Maine. In *Late Pleistocene History of Northeastern New England and Adjacent Quebec*, edited by H. W. Borns, Jr., P. LaSalle, and W.B. Thompson, pp. 29–44. Special Paper No. 197, Geological Society of America, Boulder, Colorado.

Smith, G.W., and Hunter, L.E. (1989). Late Wisconsinan deglaciation of coastal Maine. In *Studies in Maine Geology, Volume 6: Quaternary Geology*, edited by R.D. Tucker and R.G. Marvinney, pp. 13–32. Maine Geological Survey, Augusta.

Smith, G.W. and Ridge, J.C. (2001). *Pyganodon fragilis* (Lamarck, 1819) from late-glacial varves in northern New England. *Freshwater Mussel Conservation Society Abstract with Programs*, Pittsburgh.

Southon, J.R., Nelson, D.E., and Vogel, J.S. (1990). A record of past ocean-atmosphere radiocarbon differences from the northeast Pacific. *Paleoceanography* **5**:197–206.

Stanford, S.D. (1993). Late Wisconsinan glacial geology of the New Jersey Highlands. *Northeastern Geology* **15**:210–223.

Stanford, S.D., and Harper, D.P. (1991). Glacial lakes of the lower Passaic, Hackensack, and lower Hudson Valleys, New Jersey and New York. *Northeastern Geology* **13**:271–286.

Stocker, T.F., and Wright, D.G. (1998). The effect of a succession of ocean ventilation changes on ^{14}C. *Radiocarbon* **40**(1):359–366.

Stone, B.D., and Borns, H., Jr. (1986). Pleistocene glacial and interglacial stratigraphy of New England, Long Island, and adjacent Georges Bank and Gulf of Maine. In *Quaternary Glaciations in the Northern Hemisphere*, edited by V. Sibrava, D.Q. Bowen, and G.M. Richmond, pp. 39–52. Pergamon, New York.

Stone, B.D., Lapham, W.L., and Larsen, F.D. (1992). Glaciation of the Worcester plateau, Ware-Barre area, and evidence for the succeeding late Woodfordian periglacial climate. In *Guidebook for Field Trips in the Connecticut Valley Region of Massachusetts and Adjacent States*, edited by P. Robinson and J.B. Brady, pp. 467–487. 84th Annual Meeting of the New England Intercollegiate Geological Conference, Amherst, Mass.

Stone, B.D., and Peper, J.D. (1982). Topographic control of the deglaciation of eastern Massachusetts: Iice lobation and the marine incursion. In *Late Wisconsinan Glaciation of New England*, edited by G.J. Larson and B.D. Stone, pp. 145–166. Kendall/Hunt Publishing, Dubuque, Iowa.

Stone, J.R., and Ashley, G.M. (1992). Ice-wedge casts, pingo scars, and the drainage of Lake Hitchcock. In *Guidebook for Field Trips in the Connecticut Valley Region of Massachusetts and Adjacent States*, edited by P. Robinson, and J.B. Brady, pp. 305–331. 84th Annual Meeting of the New England Intercollegiate Geological Conference, Amherst, Mass.

Stone, J.R., DiGiacomo-Cohen, M., Lewis, R.S., and Goldsmith, R. (1998). Recessional moraines and the associated deglacial record of southeastern Connecticut. In *Guidebook to Field Trips in Rhode Island and Adjacent Regions of Connecticut and Massachusetts*, edited by D.P. Murray, pp. B-5:1–20. 90th Annual Meeting, New England Intercollegiate Geological Conference, Kingston, R.I.

Stone, J.R., Schafer, J.P., London, E.H., Lewis, R.L., DiGiacomo-Cohen, M.L., and Thompson, W.B. (1998). Quaternary geologic map of Connecticut and Long Island Sound Basin. *Open-File Report* No. 98-371, U.S. Geological Survey, Reston, Va.

Stuiver, M., and Borns, H.W., Jr. (1975). Late Quaternary marine invasion in Maine: Its chronology and associated crustal movement. *Geological Society of America Bulletin* **86**:99–104.

Stuiver, M., Grootes, P.M., and Braziunas, T.F. (1995). The GISP2 δ^{18}O climate record of the past 16,500 years and the role of the sun, ocean, and volcanoes: *Quaternary Research* **44**:341–354.

Stuiver, M., and Reimer, P.J. (1993). Extended ^{14}C data base and revised CALIB 3.0 ^{14}C age calibration program. *Radiocarbon* **35**:215–230.

Stuiver, M., Reimer, P.J., Bard, E., Beck, J.W., Burr, G.S., Hughen, K.A., Kromer, B., McCormac, F.G., van der Plicht, J., and Spark, M. (1998). INTCAL98 radiocarbon age calibration 24,000–0 cal. B.P. *Radiocarbon* **40**:1041–1083.

Sutherland, D.G. (1986). A review of Scottish marine shell radiocarbon dates, their standardization and interpretation. *Scottish Journal of Geology* **22**:145–164.

Thomas, G.M. (1984). A comparison of the paleomagnetic character of some varves and tills from the Connecticut Valley. Unpublished master's thesis, Department of Geology and Geography, University of Massachusetts, Amherst.

Thompson, W.B., and Borns, H.W., Jr. (editors). (1985). *Surficial geologic map of Maine* (map): Maine Geological Survey, Augusta, scale 1:500,000, 1 sheet.

Thompson, W.B., Fowler, B.K., and Dorian, C.C. (1999). Deglaciation of the northwestern White Mountains, New Hampshire. *Géographie physique et Quaternaire* **53**:59–78.

Thompson, W.B., Fowler, B.K., Flanagan, S.M., and Dorion, C.C. (1996). Recession of the Late Wisconsinan ice sheet from the northwestern White Mountains, N.H. In *Guidebook to Field Trips in Northern New Hampshire and Adjacent Regions of Maine and Vermont*, edited by M.R. Van Baalen, pp. 203–234. 88th Annual Meeting of the New England Intercollegiate Geological Conference, Harvard University, Cambridge, Mass.

Törnqvist, T.E., de Jong, A.F.M., and van der Borg, K. (1992). Accurate dating of organic deposits by AMS ^{14}C measurement of macrofossils. *Radiocarbon* **34**:566–577.

Turney, C.S.M., Coope, G.R., Harkness, D.D., Lowe, J.J., and Walker, M.J.C. (2000). Implications for the dating of Wisconsinan (Weichselian) late-glacial events of systematic radiocarbon age differences between terrestrial plant macrofossils from a site in SW Ireland. *Quaternary Research* **53**:114–121.

Uchupi, E., Driscoll, N., Ballard, R.D., and Bolmer, S.T. (2001). Drainage of late Wisconsinan glacial lakes and the morphology and late Quaternary stratigraphy of the New Jersey–southern New England continental shelf. *Marine Geology* **172**:117–145.

Verosub, K.L. (1979a). Paleomagnetism of varved sediments from western New England: Secular variation. *Geophysical Research Letters* **6**:245–248.

Verosub, K.L. (1979b). Paleomagnetism of varved sediments from western New England: Variability of the recorder. *Geophysical Research Letters* **6**:241–244.

Wilson, B.R. (2000). A chronology and environmental interpretation of glacial to nonglacial lacustrine varves in the Passumpsic Valley, Barnet, Vermont. Unpublished bachelor's thesis, Department of Geology, Tufts University, Medford, Mass.

Wilson, S.E., Walker, I.R., Mott, R.J., and Smol, J.P. (1993). Climatic and limnological changes associated with the Younger Dryas in Atlantic Canada. *Climate Dynamics* **8**:177–187.

Wohlfarth, B. (1996). The chronology of the last termination: A review of radiocarbon-dated, high-resolution terrestrial stratigraphies. *Quaternary Science Reviews* **15**:267–284.

Wohlfarth, B., Björck, S., Possnert, G., Lemdahl, G., Brunnberg, L., Ising, J., Olsson, S., and Svennson, N.-O. (1993). AMS dating Swedish varved clays of the last glacial/interglacial transition and the potential/difficulties of calibrating Late Weichselian 'absolute' chronologies. *Boreas* **22**:113–128.

Wohlfarth, B., Björck, S., and Possnert, G. (1995). The Swedish time scale: A potential calibration tool for the radiocarbon time scale during the late Weichselian. *Radiocarbon* **37**:347–359.

Yu, Z., and Eicher, U. (1998). Abrupt climate oscillations during the last deglaciation in central North America. *Science* **282**:2235–2238.

SECTION II
UPLAND SETTINGS

CHAPTER 4

GEOARCHAEOLOGY OF SOILS ON STABLE GEOMORPHIC SURFACES: MATURE SOIL MODEL FOR THE GLACIATED NORTHEAST

David L. Cremeens

Cultural resources are associated with a contemporary or former ground surface. People in most cases did not occupy freshly deposited surfaces of unaltered sediments; they lived on ground surfaces defined by vegetated soils. Even as technology increased and people were able to excavate deeper and deeper, they still performed these activities from a particular ground surface, and the record of their occupation is associated with that ground surface. One of the objectives of the geoarchaeologist is to determine the information in a ground surface record that allows correct determination and evaluation of this association or, more properly, the archaeological context. The pedological environment is the immediate ground surface environment resulting from a period of geomorphic stability and the consequent period of soil formation (pedogenesis). The pedological environment is not only the environment following occupation; it is largely the environment of occupation. Cultural deposition, a large family of processes that transforms artifacts from systemic context to archaeological context, is necessarily associated with a landscape and therefore with a ground or soil surface (Schiffer 1987). In this chapter I define and present the pedological environment in terms conducive to evaluation of archaeological context. A model is presented, based on specific subsoil horizons, the time-diagnostic rates of horizon formation, and the resultant soil horizon arrangement of a mature soil that defines vertical zonation in the archaeological record as a function of the evolving pedological record. The model is used to distinguish between pedogenic processes that affect artifact distribution and depositional processes involving sedimentation and artifact/site burial, a distinction that is important because soils result from a hiatus in deposition of sediments and the development of a terrestial ecosystem on a relatively stable landscape. This chapter is an expansion of earlier ideas of archaeological context in a soil horizon sequence presented in Cremeens and Hart (1995).

ARCHAEOLOGICAL CONTEXT AND SPATIAL CONCEPTS IN GEOARCHAEOLOGY

Archaeological context can be defined as the spatial and temporal location of the material remains of past cultural systems, the processes that have affected the deposition of those remains, and their alteration after deposition. It is the nonbehavioral states of artifacts, as opposed to, in the systemic context, the behavioral states of artifacts (Schiffer 1988). To fully comprehend archaeological remains, it is crucial to understand both how and why artifacts were deposited (Schiffer's c-transforms) and what has occurred to the artifacts since deposition (Schiffer's n-transforms). Once a site of occupation is abandoned, ground surface (pedological and geomorphic) processes determine whether or not the archaeological record is preserved, modified, or destroyed (Bettis 1988; Waters 1992). Pedological and geomorphic processes belong to the larger group of natural and cultural processes known as site formation processes (Schiffer 1987). Site-formation processes introduce variability and patterning into the archaeological record in the form of modified, destroyed, or added materials, and modified, destroyed, or added patterns. If a landscape is geomorphically stable, a site may remain at the surface without becoming buried by sedimentation or destroyed by erosion. However, pedological processes are operative on stable landscapes. The open-system, dynamic nature of soil at the ground surface results in changes in soil morphology with time, including thickness and chemical and physical properties of individual soil horizons (Johnson and Watson-Stegner 1990).

To resolve the archaeological context of cultural remains, at least four spatial concepts have to be addressed: (arbitrary) excavation levels; provenience; natural strata; and soil horizons. Excavation levels are operationally defined in the field as the actual fieldwork and a record of how the site was excavated and how

---

Geoarchaeology of Landscapes in the Glaciated Northeast edited by David L. Cremeens and John P. Hart. New York State Museum Bulletin 497. © 2003 by the University of the State of New York, The State Education Department, Albany, New York. All rights reserved.

artifacts were recovered. Provenience refers to the archaeological find-spot (Schiffer 1987). Natural strata imply deposition and the law of superposition, a concept that most archaeologists are familiar with. Natural strata are preoccupation, syn-occupation, or postoccupation. Soil horizons are in situ and postdepositional with regard to natural strata and define a ground surface with which the cultural remains are associated. The vertical and horizontal distributions of artifacts have to be evaluated in light of these spatial concepts, but some confusion arises from misunderstanding of what these concepts mean. Many archaeologists confuse natural strata with soil horizons. Artifacts in the subsoil (B horizon) are often explained as a depositional phenomena; that is, the artifacts in the B horizon were deposited in that "soil" and then subsequently buried by the A-horizon "soil." Significant interpretations are often made as to the age of artifacts, and the resultant delineation of discrete occupation surfaces, based on vertical position below datable layers. These interpretations are based on the assumption that the artifacts were deposited below, hence before the dated layers, a strictly sedimentary perspective. A plausible, alternative explanation is that the artifacts were deposited on or above the datable layer, and then moved down over time (Frolking and Lepper 2001). This alternative scenario would place the age of the artifacts as contemporary with, or even younger, than the datable layer.

Figure 4.1, modified from Rapp and Hill (1999), shows the interpretative process in evaluating the archaeological context of a site. Geoarchaeologists come into the picture at the point of final context and have to work backward to arrive at an initial context from which behavioral context is inferred. Sedimentary processes and pedological processes are not always easily distinguished, especially on younger landscapes. However, with increasing time and as a geomorphic surface becomes stable, pedological processes become distinguishable from and dominant to sedimentary processes. The result is that, on stable geomorphic surfaces, pedological processes dominate the "n-transforms" and archaeological context increasingly becomes a function of the soil environment that develops on the surface.

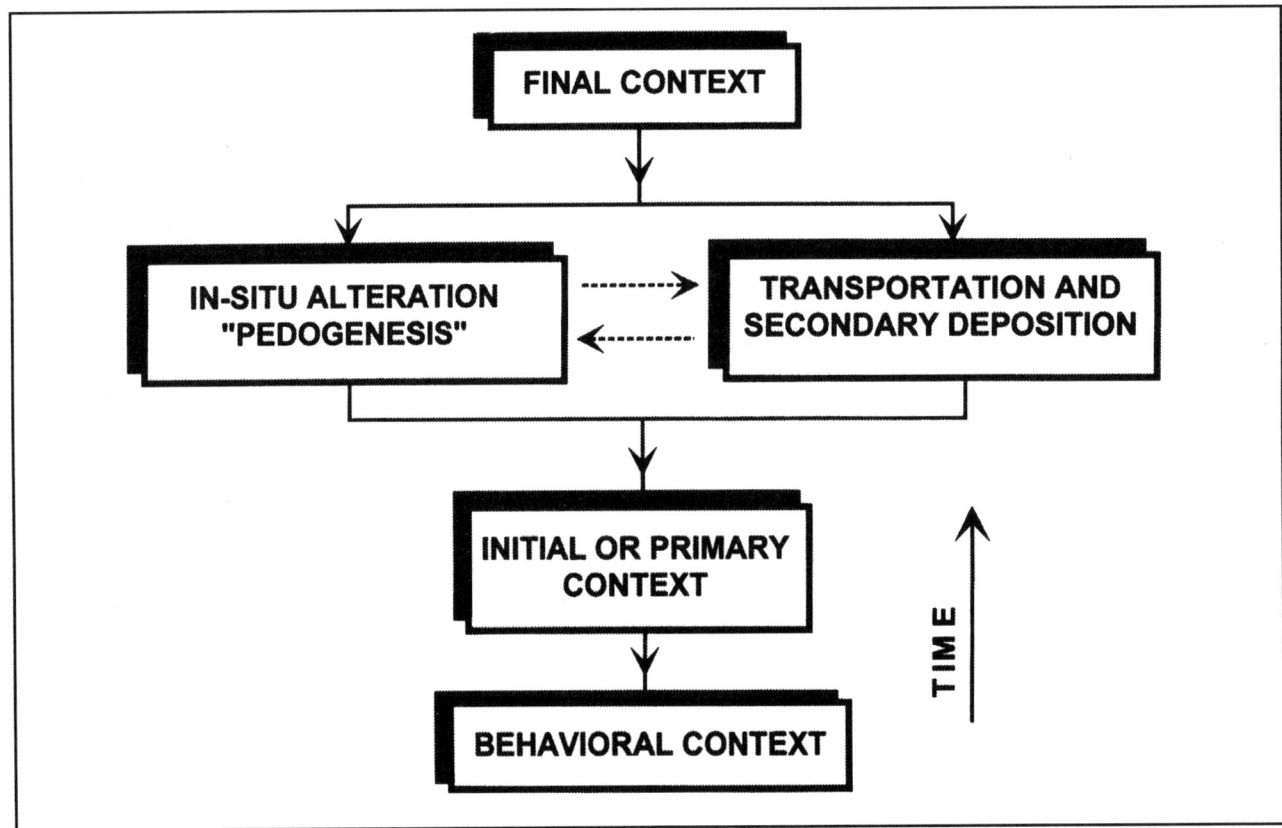

Figure 4.1. Interpretive process in evaluating archaeological context. Modified from Rapp and Hill (1999).

STABLE GEOMORPHIC SURFACES AND SOILS

Stable geomorphic surfaces are those on which no erosion or deposition is occurring and on which a terrestrial ecosystem and an associated soil are developing. Stable geomorphic surfaces can originate as a depositional surface or an erosional surface, and they can be level or steeply sloping. To describe a geomorphic surface as stable gives no indication as to its geological origin, or current geometry (topography). It indicates only that the surface has had no erosion or deposition within a time frame long enough for a soil to develop. In this chapter the term *upland* will be used to refer to areas topographically above and geomorphically older than the immediate valley floor. Thus uplands can include older fluvial terraces, lacustrine terraces, and various types of proglacial deposits that can occur adjacent to or on a valley floor but are obviously higher and older than the current flood plain. Whether or not the uplands are stable is denoted by the presence of a mature soil.

The presence of a soil indicates a period of time, a hiatus, during which landscape conditions are relatively stable (Holliday 1990). The soil results from the development and persistence, with time, of a terrestrial ecosystem associated with the specific climatic conditions. In this way, the upper portion of the regolith (soil parent material) for the particular geomorphic surface is transformed into soil. A special form of biological weathering that occurs at the exposed ground surface is termed *pedogenesis*. The changes occur in response to the soil-forming factors—a set of earth surface conditions that determine pedogenesis, and via a set of pedological processes—the mechanisms by which the changes occur. The factors of soil-formation approach to studying pedogenesis had its beginning in 19th-century Russia and early 20th-century United States and set the stage for the concept of soils as organized natural bodies that vary in response to changes in the factors.

Perhaps the most visibly recognizable or distinguishable subdivision of a soil is a soil horizon. Soil horizons are zones within the soil that generally parallel the earth's surface and that differ from the underlying material as a result of the interactions, through time, of geological parent materials, climate, living organisms, and topography (the soil-forming factors mentioned above). Soil horizons are more discrete and internally less variable than the soil as a whole (Slater et al. 1994) and are the products of soil formation (pedogenesis) that develop in situ and subsequent to the formation of the geological material (parent material) on which the soil occurs (National American Commission on Stratigraphic Nomenclature 1983). In some cases horizons are entities with a common genetic context. In other cases complex and compound horizons result from the polygenetic interaction of a variety of processes and material sources (Morrison 1978; Slater et al. 1994).

Genetic Soil Horizons

Pedologists express a qualitative judgment about the kinds of changes that are believed to have taken place in the particular soil by the use of genetic soil horizons. The horizons are designated with an alphanumeric code that consists of a capital letter or letters modified by lower-case letter suffixes, Arabic numeral prefixes and suffixes, and primes (Soil Survey Division Staff 1993). The capital letters O, A, E, B, C, and R represent the master horizons, which give the overall character of the horizon relative to position in the profile and dominant properties. These capital letters form the base symbols to which the other characters (lower-case letters, Arabic numerals, primes) are often added to indicate subordinate properties and to complete the designation. The resultant symbol is a representation of the morphological characteristics of the particular horizon as determined in the field. For example, the symbol 2Btg3 indicates the third subdivision of a B horizon formed in the second layer or stratum of parent material, with evidence of translocated clay (t) and dull gray colors associated with water saturation (g).

Two surface horizons are recognized: O and A. O horizons consist primarily of undecomposed or partially decomposed organic litter that has been deposited on the surface, or organic material that was deposited under saturated conditions and has decomposed to varying degrees (Soil Survey Staff 1999). A horizons are mineral horizons, formed at the surface, or below an O horizon, and characterized by an accumulation of humified organic matter intimately mixed with the mineral fraction. They exhibit obliteration of all or much of the original rock structure (sediment stratification). A horizons commonly display properties resulting from cultivation, pasturing, or similar kinds of surface disturbance.

E horizons are near surface mineral horizons in which the main characteristics are the loss of silicate clay, iron, or aluminum (or some combination of all these), leaving a relatively light colored concentration of uncoated sand and silt particles. An E horizon is commonly above a B horizon, and much or all of the original rock structure is obliterated.

B horizons are subsurface horizons that have formed below an O, A, or E horizon. They are dominated by obliteration of all or much of the original rock structure and exhibit one or more of the following: an illuvial (washed in or translocated) accumulation of silicate clay, iron, aluminum, humus, carbonates, gypsum, or silica, alone or in combination; evidence of removal of carbonates; residual concentration of iron and aluminum compounds; and/or coatings of iron and aluminum compounds, and other minor characteristics (Soil Survey Staff 1999).

C horizons are subsolum (below the A and B horizons) horizons or layers that are relatively unaffected by the pedogenic processes and lack the properties of overlying O, A, E, and B horizons. Original rock structure, such as sedimentary stratification, is present in C horizons. The material of C horizons may be either like or unlike the material from which the overlying horizons formed. R horizons indicate hard, unweathered bedrock.

Transitional horizons are dominated by properties of one master horizon but have subordinate properties of another master horizon (Soil Survey Division Staff 1993). Transitional horizons are designated by two capital-letter symbols (e.g., AB, BA, BE, BC). The first letter indicates the dominant properties. Combination horizons have recognizable properties of two kinds of master horizons as indicated by capital letters separated by a virgule (e.g., A/O, E/B, B/E) (Soil Survey Staff 1999). The first letter indicates material with greater volume.

Lower-case letter suffixes are used to modify the concept of master horizons by providing more information on more specific characteristics of the particular horizon. The various rules for using the lower-case letters are provided by the Soil Survey Division Staff (1993) and the Soil Survey Staff (1999) and reviewed by Holliday (1990). Arabic numeral suffixes indicate vertical subdivision of horizons that are designated by a single combination of letters. The successive horizon subdivisions are numbered in sequence until the letter designation changes (e.g., Bt1-Bt2-Bt3-Btg1-Btg2). Horizons are vertically subdivided for minor morphological reasons, reasons that do not change the letter designation. The numbering of vertical subdivisions within a horizon is not interrupted at a lithological discontinuity (discussed below) if the same letter combination is used in both materials.

Lithological discontinuities, defined as significant changes in the parent materials (regolith) from which the horizon formed, are indicated by Arabic numeral prefixes. The first stratum or layer of material, nearest the surface, is not identified with a prefix (i.e., the "1" is understood). Underlying contrasting materials are then numbered consecutively, starting with 2, to indicate a change in material, not a type of material. For example, if two similar materials (e.g., two strata of colluvium) are separated by a third dissimilar material (e.g., alluvium), the sequence would still be numbered consecutively (e.g., A-BE-2Bt1-3Bt2-3BC).

During pedogenesis, the subsoil B horizon undergoes morphological changes and becomes the most persistent and time-diagnostic soil horizon. B horizons are more resistant to erosion than horizons above. In fact, many paleosols are identified on the basis of B-horizon properties alone. The top or middle of a B horizon in such a paleosol can be an erosion surface. As B horizons evolve with time, they can become increasingly resistant to penetration by flora and fauna, including humans. This resistance results in a concentration of biological activity above the B horizon and a concentration of coarse fragments including artifacts. As the B horizon becomes increasingly difficult for people to penetrate, it has an influence on systemic context. The morphological changes a B horizon undergoes obviously depend on the conditions or soil-forming factors at a particular location. However, all B horizons develop intrahorizon zonation with time. The top of the B develops into a transition zone (BE or BA horizon) that usually has a smaller size and stronger grade of soil structure (Soil Survey Division Staff 1993) and a differing color than the remainder of the B horizon. The middle of the B is where the maximum expression of the B horizon properties are usually expressed. The lower B is characterized by a weaker-grade and coarser-sized soil structure and less development of the characteristic color and other properties that define the B horizon. This becomes a transitional BC horizon that grades into relatively unaltered parent material (C horizon).

Diagnostic Subsurface Horizons

The specific, quantitative expression of soil horizons, reflecting specific sets of pedogenic processes under specific conditions of the soil-forming factors, are implied in the definitions of the diagnostic horizons. Diagnostic horizons are used as part of the differentiae in various soil classification systems worldwide, and in particular in the United States (Soil Survey Staff 1999). In the U.S. Department of Agriculture system, *Soil Taxonomy*, diagnostic horizons occur as surface horizons (epipedons—synonymous with topsoil) and as subsurface horizons (both subsurface and subsoil). The horizons are defined by quantitative properties that by inference indicate a specific genetic pathway. Four main diagnostic subsurface horizons are commonly found in the northeastern United States and southeastern Canada: cambic horizons (indicated by a Bw symbol), argillic horizons (Bt), fragipans (Bx), and spodic horizons (Bhs). All of these horizons occur in glaciated as well as unglaciated regions. Cambic horizons are characterized by the development of soil structure and/or brighter, redder soil color and/or leaching of certain soil minerals relative to the C horizon (Soil Survey Staff 1999). Implied in the concept of a cambic horizon is that of immature subsoil development, because of insufficient time for further development or due to some other soil-forming factor not conducive to further development. In general, cambic horizons require a minimum of a few hundreds of years to form (Bilzi and Ciolkosz 1977; Birkeland 1999).

Argillic horizons, designated as Bt, are subsurface horizons where translocated clay has accumulated to a significant extent (Soil Survey Staff 1999). The implication of an argillic horizon is that of a subsurface horizon more mature than a cambic horizon and formed by an incremental accumulation of clay in a relatively base-rich parent material, in a mostly forested environment. In the eastern United States research has shown that approximately 2,000 years of pedogenesis are required for a B horizon to make the transition from a cambic horizon to an argillic horizon (Bilzi and Ciolkosz 1977; Ciolkosz et al. 1994; Cremeens 1995; Birkeland 1999). More mature argillic horizons can take thousands to tens of thousands of years to form, and are characterized by extensive and often abrupt clay accumulation that increasingly forms a barrier to biological penetration.

Fragipans are diagnostic subsurface horizons (designated as Bx) that are characterized by seemingly cemented, brittle consistence (strength) and that are restrictive to the entry of water and roots into the soil matrix (Soil Survey Staff 1999). Fragipans often occur in the same profile as, and below, cambic, argillic, or spodic horizons (defined below). Although fragipans are the traditional "hardpans" of the eastern woodlands, their genesis is obscure (Soil Survey Staff 1999; Ciolkosz et al. 1992). Like argillic horizons, fragipans require a minimum of a few thousand years to form (Ciolkosz et al. 1992; Cremeens et al. 1998). The root barrier function of fragipans results in numerous tree tips in upland areas dominated by fragipan containing soils. Forests in northern Pennsylvania and southern New York are notable for their tree tip microtopography. Once a fragipan develops to where it becomes a root barrier, most other biological vectors will not be able to penetrate it.

Spodic horizons, designated with Bhs and Bs, are characterized by an accumulation of translocated organic carbon with aluminum and/or iron. Spodic horizons can evidently form quite rapidly, in a matter of a few hundred years under ideal conditions (Callum 1995; Soil Survey Staff 1999). However, the time frame of spodic horizon formation has been little studied. Most studies have focused on the mechanism of formation. In the glaciated Northeast, spodic horizons generally occur in coarse-textured, pro- and postglacial meltwater deposits under coniferous and hardwood vegetation. Spodic horizons are not impenetrable to biological vectors in the same sense as a well-developed argillic horizon or a fragipan. However, a special type of cemented horizon consisting of spodic materials, called ortstein, occurs throughout the Northeast. Ortstein is identified with a Bhsm designation, the "m" referring to cemented conditions. With a relatively horizontal orientation, ortstein becomes root restricting (Soil Survey Staff 1999). This is more common in wetter landscapes. Vertical orientation of ortstein also occurs as irregular columns, more commonly in better-drained conditions. The vertically oriented ortstein tends to be less root restrictive than the horizontally oriented variety. Regardless of the orientation, the cemented nature of ortstein still presents a pedogenically developed barrier to penetration by many biological factors.

The genetic horizons used in delineating the diagnostic horizons and the diagnostic horizons are not mutually exclusive but generally correspond. For example, not all Bt horizons are argillic horizons; they may not meet the clay content criteria. But all argillic horizons are Bt horizons; they have evidence of translocated clay. In the following discussion the argillic Bt horizon is used for illustrative purposes, but the mechanisms are applicable to other soils with diagnostic subsurface horizons.

Soil Profiles

A soil profile (Figure 4.2) is a vertical section of a soil through all its horizons including and extending into the C horizon (Soil Science Society of America 1997). Beyond that simple definition, however, soil profiles provide critical information for the interpretation of the immediate ground surface with which the profile is associated. Geomorphic, pedological, and cultural processes all leave a record that can, for the most part, be interpreted from a soil profile. This record is critical to interpreting the cultural resources associated with a site both from a syn-occupation perspective, and from a post-occupation perspective. It is a pedological paradigm that soil horizons in a profile occur in a logical vertical arrangement due to surface and subsurface processes (Cremeens and Hart 1995). This arrangement provides the key to interpreting a soil profile. Fortunately soil profiles are analogous to geopetal features in rock (Cremeens and Hart 1995)—any rock feature that indicates the relation of top to bottom at the time of the rock's formation (American Geological Institute 1976). The soil profile provides information as to the ground surface associated with that soil, even if that soil is buried or part of a complex profile.

The pedological environment that evolves on a stable geomorphic surface is expressed in soils that develop increasing horizon expression and differentiation with time (Figure 4.3). In the traditional model of pedogenesis, profile development is represented as a continuous function of time. With time, profile development pro-

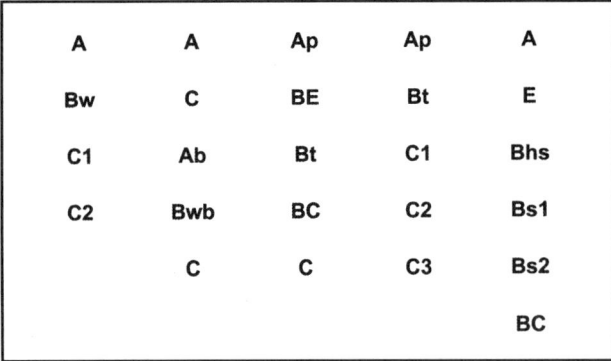

Figure 4.2. Examples of soil profiles as arrangements of soil horizons.

ceeds through a continuous, orderly path of increasingly differentiated horizon expression, especially the B horizon and the upper B subsurface horizons (E and BE or BA). This model assumes little or no change in the other soil-forming factors, a fairly safe assumption during the Holocene on stable upland geomorphic surfaces in the glaciated Northeast.

This model does not assume that the ground surface is static. The pedological environment is an open-system; physically, chemically, and biologically dynamic. Johnson and Watson-Stegner (1990) refer to pedoturbation as a normal ground-surface set of processes, which can function to retard and/or mix (partially or wholly destroy) horizons that have developed according to the traditional model above. As part of the "organism's" soil-forming factor, plant and animal activity (floralturbation and faunalturbation or, collectively, bioturbation) are very much a part of the pedological environment on stable geomorphic surfaces. Bioturbation can be abrupt and on a rather large scale, such as the tree tips described above, or more incremental and subtle, such as biomantle formation (Johnson and Watson-Stegner 1990; Van Nest 1993). Biomantle formation results when small organisms, earthworms, insects, and so forth, move finer particles, while larger particles such as artifacts and rock fragments will be buried by the fines and will settle due to density to the bottom of the zone of activity, generally the bottom of the A horizon.

Johnson and Watson-Stegner (1990) defined the orderly formation of a profile horizon sequence via the traditional model as proanisotropic or progressive pedogenesis (Figure 4.4). Profile retarding or mixing processes including sedimentation, erosion, and pedoturbation are defined as regressive pedogenesis. On stable geomorphic surfaces, progressive pedogenesis is dominant although pedoturbation processes are active. Immature B horizons are relatively easy to penetrate by biological agents, and mixing occurs throughout the solum portions of the profile (A and B horizon combined). As the B horizon evolves, it becomes increasingly difficult to penetrate by much of the flora and fauna, especially when it evolves into fragipans, ortstein, or clay-rich argillic horizons. The resistance to penetration tends to focus or concentrate the bioturbation processes above the main body of the B horizon. Occasional tree tip or large-animal burrowing (Butler 1995) can penetrate these B horizons, but the majority of bioturbation energy becomes concentrated above the B horizon.

Upland Soils of the Northeastern United States and Southeastern Canada

In the northeastern United States and southeastern Canada, multiple glaciations and the associated variations in climate have essentially created new parent material across most of the region and set the pedological clock of soil formation at a known time (Ciolkosz et al. 1989). The known time is the later stages of the late Wisconsin (Woodfordian) episode of the Pleistocene: essentially 18,000 years ago at the "Terminal Moraine" in Pennsylvania and New York to less than 11,000 years ago in northern New England and the Atlantic Provinces of Canada. Since the deposition of glacial materials, the variation of climate, vegetation, and other soil-forming factors, although not static, has been subordinate. Thus more is known about the genesis of soils in glaciated areas than of soils in nonglaciated areas. As

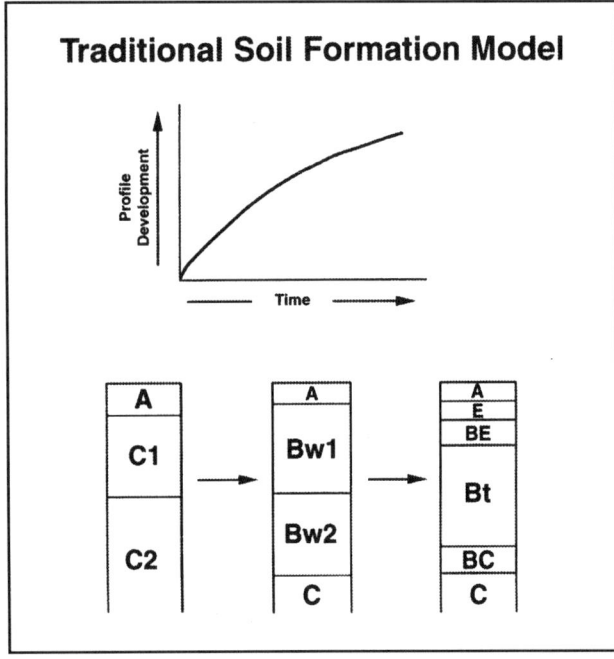

Figure 4.3. Traditional soil formation model.

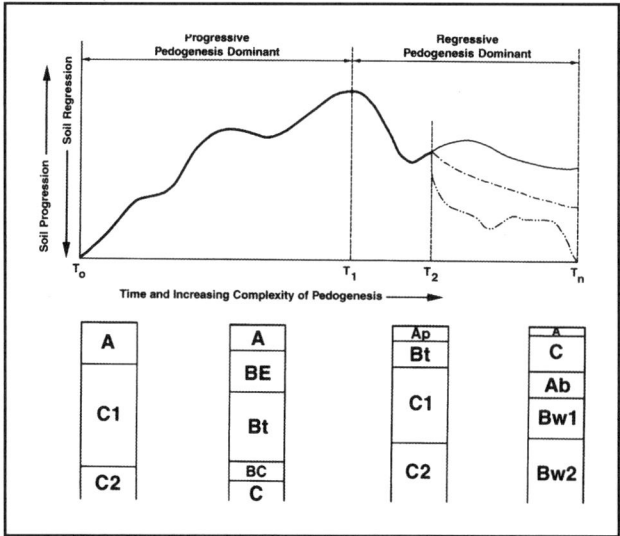

Figure 4.4. Progressive-regressive pedogenesis. Modified from Johnson and Watson-Stegner (1990).

discussed above, the term *upland* refers to topographically higher, and geomorphically older, landscapes of various origins and may include older terraces.

An idealized developmental sequence for soils in the northeastern United States was presented by Ciolkosz et al. (1989). Soils show a transition with time from weakly developed entisols (A-C profile) to moderately developed inceptisols (A-Bw-C or A-Bw-Bx-C profiles) where cambic Bw and/or fragipan Bx subsoil horizons form. Further development leads to alfisols (with A-E-Bt-C and/or A-E-Bt-Bx-C profiles) where clay-rich argillic Bt subsoil horizons form.

Inceptisols with fragipans are extensive on glaciated upland surfaces throughout southern New England, southern New York, and northeastern and northwestern Pennsylvania. In northeastern Pennsylvania, fragipan Bx horizons occur on 55 percent of the land surface, with 45 percent of the land surface having less than 8 percent slope (Ciolkosz et al. 1999). Here soils developed in acid loamy tills derived from sandstone and shale bedrock. The dominant soils on gently sloping uplands are fragiaquepts (wet inceptisols with fragipans). In northwestern Pennsylvania where 71 percent of the land surfaces have less than 8 percent slope, fragipan Bx horizons occur on 63 percent of land surfaces and argillic Bt horizons occur on 66 percent of land surfaces. This area is mapped extensively with fragiudalfs (moist alfisols with argillic horizons and fragipans) and fragiaqualfs (wet alfisols with argillic horizons and fragipans). Argillic Bt horizons are also extensive throughout glaciated central New York due to the presence of easily leached carbonates (Ciolkosz et al. 1999). These areas are largely mapped with hapludalfs (moist alfisols with argillic horizons) and haplaqualfs (wet alfisols with argillic horizons).

Spodosols are characterized with a spodic Bhs subsoil horizon and can also contain fragipan Bx horizons and/or argillic Bt horizons and can contain the cemented ortstein spodic horizon (Bhsm). Spodosols are found primarily in northern portions of the glaciated Northeast, including northern New York, northern Vermont and New Hampshire, and most of Maine. Spodosols formed in acidic, sandy parent materials associated with stratified drift, proglacial and meltwater deposits, under coniferous or deciduous vegetation, and under a frigid soil temperature regime (Ciolkosz et al. 1989). Stanley and Ciolkosz (1981) noticed a trend of increasing Bhs spodic horizon development from unglaciated areas in the southern Appalachian Mountains into northern New York and northern New England.

In southeastern Canada, spodosols, podzols as mapped on the Food and Agriculture Organization (FAO) World Soil Map, seldom contain fragipans. Ontario and the Atlantic Provinces are largely mapped with spodosols (orthic podzols) containing spodic Bhs horizons and alfisols (albic luvisols) containing argillic Bt horizons. However, some areas in the Atlantic Provinces and in large portions of Quebec are mapped with inceptisols (cambisols) containing cambic Bw horizons. The remainder of Quebec is mapped with spodosols.

The following section is illustrated with soils containing evolving argillic Bt horizons. The same principles apply to fragipan containing soils. Therefore, for this chapter and volume, the discussions are most applicable to those areas that have uplands mapped with fragiudepts, fragiaquepts, fragiudalfs, fragiaqualfs, hapludalfs, and haplaqualfs, namely glaciated areas of northeastern and northwestern Pennsylvania, southern New York, southern New England, portions of Ontario, and portions of the Atlantic Provinces.

ARTIFACT DISTRIBUTION IN SOIL PROFILES

Context Within Individual Horizons

Identifying and understanding the characteristics and processes associated with a particular horizon, particularly the diagnostic B horizons, are an important first step to evaluating archaeological context and the horizon-associated processes that produce artifact patterning (concentration and dispersion). When the properties and processes of individual horizons are evaluated relative to the entire arrangement of horizons that make a soil profile, then determination of archaeological context becomes a function of natural processes operative on an otherwise geomorphically stable ground surface.

O horizons are low density, organic surficial horizons, subject to oxidation, destruction by fire, and wind erosion. Because they form by accumulation of organic materials, any process that promotes the accumulation will preserve the O horizon. Upon burial, the accumulation source is cut off, and the horizon may become compressed. More important, mineral artifacts are much denser than organic materials and will ultimately not "float" in such a matrix. Any disturbance will cause artifacts to sink to a denser horizon. A horizons are biologically active, mineral surficial horizons, also high in organic matter and lower in density than underlying mineral horizons. A horizons are subject to a wide variety and intensity of biological, physical, and chemical processes. A horizons are also the most radiocarbon dated of all soil horizons.

E horizons in a profile indicate a relative lack of disturbance in the near subsurface. Soils with well-developed E horizons probably have not been plowed, or clear cut, and have little disturbance from tree tip. E horizons are the most easily destroyed soil horizon due to their light color and relative lack of pigments. They are easily blended with darker A horizons during plowing or any near surface disturbance. BE horizons, by position the top of the subsoil, are often the first in situ horizon underneath a plow zone (Ap horizon). It is often the horizon that contains the remnants of features and in which the horizontal extent of features, and the site, are defined. During investigation, at many sites, the plow zone (Ap horizon) will be mechanically stripped to expose the top of the subsoil so that features can easily be identified in the lighter-colored matrix.

B horizons are not subject to as wide a variety of inputs, particularly biological inputs, that characterize the overlying horizons. Because B horizons are subsurface horizons, artifacts in the B horizon may predate the B horizon as part of the original sediment package or, as described below, may be intruded into the B horizon at some point in its pedogenic history. BC horizons are transitional between the B and C horizons and are the deepest part of the solum. They also represent the deepest portion of most biological activity.

C horizons are outside the zones of major biological energy and solum-forming processes associated with the land surface and are only subject to minor physical and chemical modification. Interpretation of the context of artifacts in the C horizon of a soil is rather limited, most easily interpreted as originating as part of the parent-material sedimentary package, thus predating the soil. Artifacts in C horizons may also have moved down from an adjacent A horizon in a weakly developed (A-C profile), via pedoturbation, and then later isolated due to burial and developmental upbuilding with pedogenesis. This is obviously not a viable option in residual soils but is plausible, although difficult to prove, in alluvial and colluvial soils.

Artifact Distribution in Immature and Moderate Profiles

In immature profiles developed on young geomorphic surfaces, artifacts concentrate on the immediate surface (A horizon) on which they are deposited (Figure 4.5). There is insufficient time for pedoturbation vectors (Johnson and Watson-Stegner 1990) to cause significant vertical artifact movement. When vertical movement does occur, the weak or incipient cambic Bw horizon does not provide a focusing mechanism. In a multisola (containing more than one A-B horizon arrangement), buried soil situation (i.e., on a relatively young terrace), the artifacts remain closely associated with the A horizon on which they were deposited. This is the ideal situation for evaluating archaeological context and chronology (Ferring 1992). On a young upland geomorphic surface with an immature single solum soil, there is little occurrence of artifacts below the A horizon.

As geomorphic surfaces age, subsequent human occupation builds up the artifact density at the immediate surface (Figure 4.6). Pedoturbation forces resulting from the evolving ecosystem begin to move artifacts

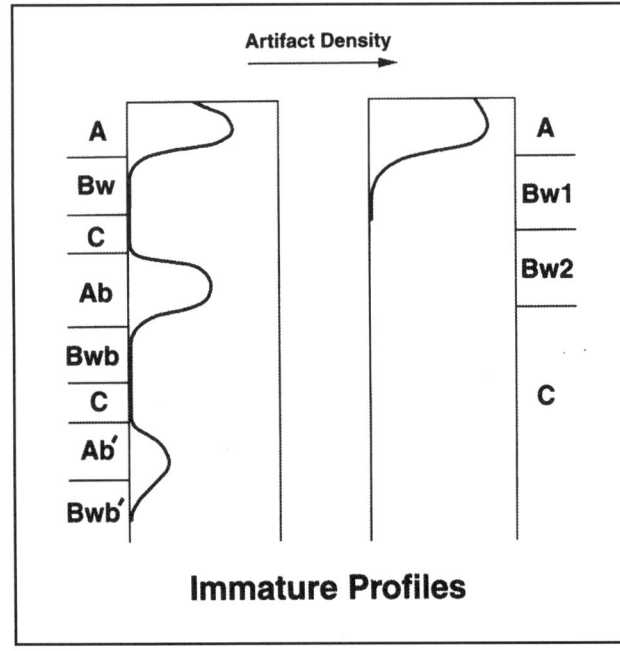

Figure 4.5. Artifact distribution in immature soil profiles.

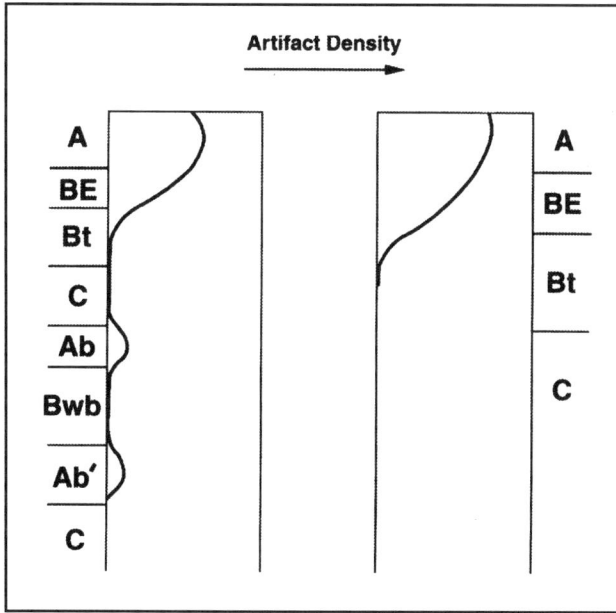

Figure 4.6. Artifact distribution in moderately formed soil profiles.

es and not, as originally thought, as evidence of a discrete occupation surface. The BE to Bt transition was characterized by a 10 to 12 percent increase in clay content, defining a well-developed argillic Bt horizon. Further migration of artifacts was minimal due to the reduced soil activity in the denser, more cohesive argillic Bt horizon.

The high level of biological activity in the A horizon coupled with its low density cause artifacts to become concentrated at the top of the increasingly dense and impenetrable B horizon (Figure 4.7). Multisola profiles, such as might be found on old terraces, can still contain cryptic evidence of former A horizons in the form of artifact concentrations even though the A horizons are no longer evident, either being lost due to oxidation or pedogenic overprinting (Cremeens et al. 1998). The mature soil profile, in undisturbed condition, contains a complete suite of upper subsoil (BE) and subsurface (E) horizons (Figure 4.8) in which artifacts concentrate as a result of high biological activity in the low-density overlying A horizon and the relatively impenetrable underlying mature B horizon. This upper subsoil and subsurface zone becomes the concentration zone being supplied by the turbation zone. Occasionally, more extreme bioturbation forces, such as a large tree tip or a large burrowing mammal, bring artifacts out of, or add artifacts to, the main body of the B horizon. In this way the B horizon becomes a semistorage zone where artifacts may indicate former surfaces (Cremeens et al. 1998) or,

deeper into the profile, away from the surface. In the multisola system, artifacts in buried A horizons still remain associated with the original A horizon. At the modern surface of multisola surfaces, and at the surface of single sola systems, artifacts begin downward movement resulting from increased pedoturbation. The subsoil starts to evolve into one of the diagnostic horizons, which in the case of the argillic Bt, fragipan Bx, and certain spodic Bhs horizons, become increasingly difficult to penetrate by the biological vectors. Thus the downward movement of artifacts slows, or in some cases even halts, in the zone of the upper B horizon.

Artifact Distribution in Mature Profiles: The Mature Soil Model

Mature soil profiles on older geomorphic surfaces have B horizons that have evolved into mature diagnostic subsoil horizons. In the case of argillic Bt, fragipan Bx, or ortstein Bhsm horizons, the subsoils become relatively difficult to nearly impossible to penetrate by biological forces (including prehistoric humans). As a result the pedoturbation mechanisms become focused above the B horizon. Paleoindian artifacts, found in the undisturbed BE horizon of a mature colluvial soil (A-E-BE-Bt-BC profile) in central Ohio, were explained by Frolking and Lepper (2001) as objects that accumulated on the stable surface and then migrated about 30 cm downward into the upper B horizon by bioturbation process-

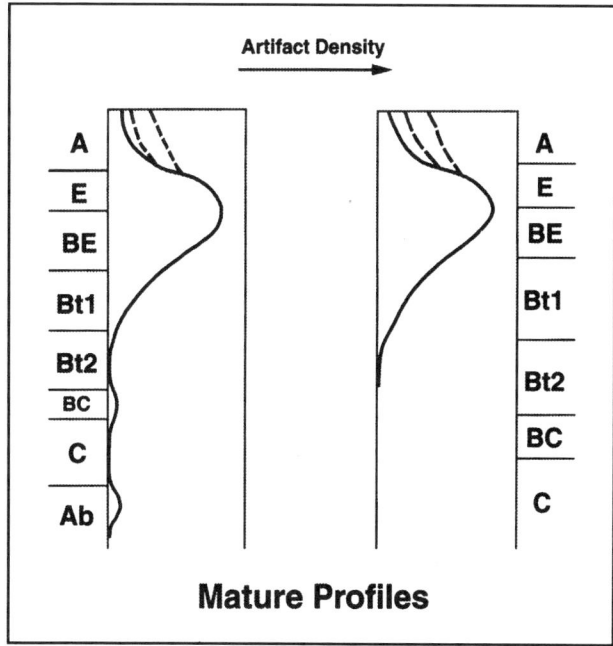

Figure 4.7. Artifact distribution in mature soil profiles.

more likely, pedoturbation forces early in the soil evolution. Large tree tips are particularly notable in that they not only leave a large pit for artifacts to fall into, but they may also provide a brief niche environment for people to exploit in some fashion. Artifacts brought upward out of the B horizon with the root plate may also fall on the contemporary surface (Waters 1992; Lothrop and Parson 1998). The C horizon, being beyond the extent of most biological vectors, functions as a storage zone where artifacts have little to no association with the overlying soil surface. Artifacts in the C horizon are most easily interpreted from a sedimentology perspective (i.e., the artifacts were deposited with the sediments either by human action or, in the case of colluvial sediments, as part of the sedimentary package).

To be sure, the seeming concentration of biological activity in the upper soil horizons may not be entirely a function of physical resistance, but may also result from greater amounts of organic matter and the associated feeding activities of many soil-dwelling organisms (Van Nest 2002).

SUMMARY AND CONCLUSIONS

At archaeological sites the evolution of soil horizons, along with the timing of prehistoric occupation episodes relative to the development stage of soils, determines the postoccupation distribution of cultural remains (Figure 4.9). Artifacts deposited on immature geomorphic surfaces and young soils are more susceptible to patterning changes resulting from the developing soil than those artifacts associated with a mature soil at the time of occupation. Artifacts buried deeply in a multisola fluvial soil (left side of Figure 4.9) may be sealed in original context if buried quickly and deep enough to be removed from pedogenic processes. Otherwise, pedoturbation vectors will begin to disperse and/or concentrate artifacts.

With increasing time and further pedogenesis on the maturing geomorphic surface, subsequent occupations will add to the archaeological record at the surface. At the same time the B horizon begins to develop, artifacts begin to move downward in response to various bioturbation vectors. The B horizon becomes increasingly resistant to penetration by the same biological forces, and as a result the artifacts begin to concentrate at the top of the B horizon in a concentration zone. This concentration mechanism continues throughout the "lifespan" of the geomorphic surface and the soil. Occasionally artifacts deeper in the profile, in the lower B horizon semistorage zone, are brought to the surface via a larger-scale mechanism, such as a large tree tip or a large animal burrowing. Artifacts can also move deeper into a profile by the same mechanisms.

Artifacts associated with a soil during its development can segregate into four zones (Figure 4.8). The deepest zone, corresponding to C horizons, is the *storage zone*, removed from pedogenic modification, where artifact distribution is most probably a function of prepedogenic sedimentary processes and retain that context with time. The B horizon or subsoil corresponds to the

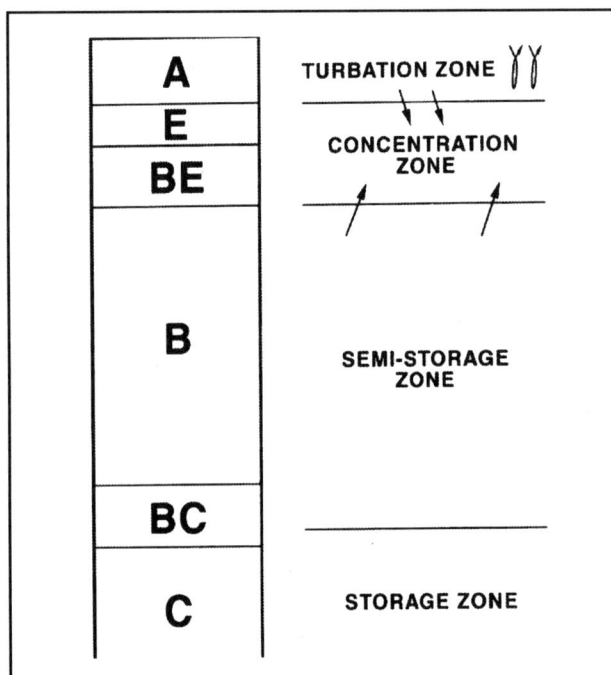

Figure 4.8. The mature soil model.

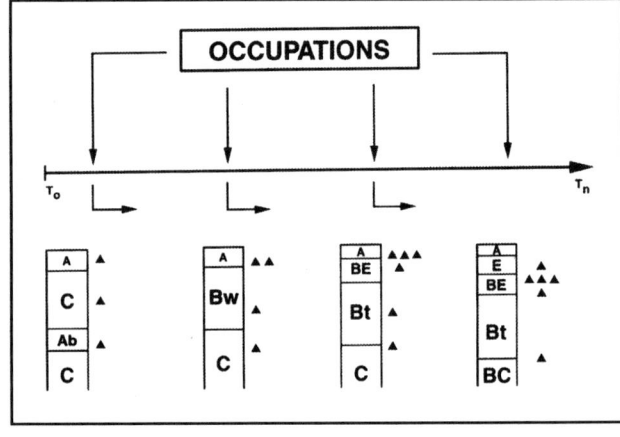

Figure 4.9. Distribution of cultural remains as a function of the timing of prehistoric occupations relative to the development stages of soils.

semistorage zone where original depositional or sedimentary context may be preserved, although large-scale pedoturbation processes (e.g., tree throw, animal burrowing) occasionally bring artifacts out of storage and/or add artifacts to storage. The top of the B horizon or immediate subsurface, represented by E, BE, or BA horizon, is the *concentration zone* where artifacts may be vertically concentrated as a result of pedogenic processes as well as human activities. Finally the *turbation zone* corresponds to A horizons of mature soils, where artifacts may not maintain their original context of deposition.

The archaeological record provides a precise stratigraphic record of Late Pleistocene and Holocene soils and stratigraphy in North America. With this record we can further subdivide events in the Late Pleistocene and Holocene and quantitatively evaluate the rates of geomorphic and pedogenic processes. To do this and to use the archaeological record as the chronological framework, the pedological context of datable artifacts must be understood. On stable geomorphic surfaces in upland areas, and on older terraces, vertical patterning in the artifact record of human occupation becomes a function of pedogenic processes. This chapter is a guide to a basic understanding of the archaeological context of a site as a result of those processes.

REFERENCES CITED

American Geological Institute. (1976). *Dictionary of geological terms* (rev. ed.). Anchor Press, Doubleday, New York.

Bettis, E.A., III. (1988). The role of geology in shaping the archaeological record. *Iowa Geology* **13**:12–15.

Bilzi, A.F., and Ciolkosz, E.J. (1977). Time as a factor in the genesis of four soils developed in alluvium in Pennsylvania. *Soil Science Society of America Journal* **41**:122–127.

Birkeland P.W. (1999). *Soils and Geomorphology.* Oxford University Press, New York.

Butler, D.R. (1995). *Zoogeomorphology, Animals as Geomorphic Agents.* Cambridge University Press, Cambridge.

Callum, K.E. (1995). Archaeology in a region of spodosols. In *Pedological Perspectives in Archaeological Research,* edited by M.E. Collins, B. J. Carter, B. G. Gladfelter, and R.J. Southard, pp. 81–94. Soil Science Society of America Special Publication No. 44, Madison, Wisc.

Ciolkosz, E.J., Day, R.L., Cronce, R.C., and Dobos, R.R. (1999). Soils (pedology). In *Geology of Pennsylvania*, edited by C.H. Shultz, pp. 692–699, Pennsylvania Geological Survey, Harrisburg, and Pittsburgh Geological Society, Pittsburgh.

Ciolkosz, E.J., Thurman, N.C., Waltman, W.J., Cremeens, D.L., and Svoboda, M.D. (1994). *Argillic Horizons in Pennsylvania Soils.* Agronomy Series No. 131, College of Agriculture, Pennsylvania State University.

Ciolkosz, E.J., Waltman, W.J., Simpson, T.W., and Dobos, R.R. (1989). Distribution and genesis of soils in the northeastern United States. *Geomorphology* **2**:285–302.

Ciolkosz, E.J., Waltman, W.J., and Thurman, N.C. (1992). *Fragipans in Pennsylvania Soils.* Agronomy Series No. 119, College of Agriculture, Pennsylvania State University.

Cremeens, D.L. (1995). Pedogenesis of Cotiga Mound, a 2,100-year-old Woodland mound in southwest West Virginia. *Soil Science Society of America Journal* **59**:1377–1388.

Cremeens, D.L., and Hart, J.P. (1995). On chronostratigraphy, pedostratigraphy, and archaeological context. In *Pedological Perspectives in Archaeological Research,* edited by M.E. Collins, B.J. Carter, B.G. Gladfelter, and R.J. Southard, pp. 15–33, Soil Science Society of America Special Publication No. 44, Madison, Wisc.

Cremeens, D.L., Hart, J.P., and Darmody, R.G. (1998). Complex pedostratigraphy of a terrace fragipan at the Memorial Park site, central Pennsylvania. *Geoarchaeology: An International Journal* **13**:339–359.

Ferring, C.R. (1992). Alluvial pedology and geoarchaeological research. In *Soils in Archaeology,* edited by V.T. Holiday, pp.1–40, Smithsonian Institution Press, Washington D.C.

Frolking, T.A., and Lepper, B.T. (2001). Geomorphic and pedogenic evidence for bioturbation of artifacts at a multicomponent site in Licking county, Ohio, U.S.A. *Geoarchaeology: An International Journal* **16**:243–262.

Holliday, V.T. (1990). Pedology in archaeology. In *Archaeological Geology in North America,* edited by N.P. Lasca and J. Donahue, pp. 525–540, Centennial Special Vol. 4, Geological Society of America, Boulder, Colo.

Johnson, D.L., and Watson-Stegner, D. (1990). The soil evolution model as a framework for evaluating pedoturbation in archaeological site formation. In *Archaeological Geology in North America,* edited by N.P. Lasca and J. Donahue, pp. 541–560, Centennial Special Vol. 4, Geological Society of America, Boulder, Colo.

Lothrop, J.C., and Parson, K. (1998). Phase II archaeological evaluation of prehistoric sites 46Cb140 and 46Cb156, Alignment 9 of the Merrick Creek connector, Cabell County, West Virginia. Site report prepared for West Virginia Department of Transportation. GAI Consultants, Inc., Monroeville, Pa.

Morrison, R.B. (1978). Quaternary soil stratigraphy: Concepts, methods, and problems. In *Third York Quaternary Symposium,* pp.77–108, Geo Abstracts, Norwich, England.

North American Commission on Stratigraphic Nomenclature. (1983). North American Stratigraphic Code. *American Association of Petroleum Geologists Bulletin* **67**:841–875.

Rapp, G.R., Jr., and Hill, C.L. (1999). *Geoarchaeology, The Earth-Science Approach to Archaeological Interpretation.* Yale University Press, New Haven, Conn.

Schiffer, M.B. (1983). Toward the identification of formation processes. *American Antiquity,* **48**:675–706.

Schiffer, M.B. (1987). *Formation Processes of the Archaeological Record.* University of New Mexico Press, Albuquerque.

Schiffer, M.B. (1988). The structure of archaeological theory. *American Antiquity* **53**:461–485.

Slater, B.K., McSweeney, K., Ventura, S.J., Irvin, B.J., and McBratney, A.B. (1994). A spatial framework for integrating soil-landscape and pedogenic models. In *Quantitative Modeling of Soil Forming Processes,* edited by R.B Bryant and R.W. Arnold, pp. 169–185, Soil Science Society of America Special Publication No. 39, Madison, Wisc.

Soil Science Society of America. (1997). *Glossary of Soil Science Terms*. Soil Science Society of America, Madison, Wisc.

Soil Survey Division Staff. (1993). *Soil Survey Manual.* Agricultural Handbook 18, U.S. Department of Agriculture, Washington, D.C.

Soil Survey Staff. (1999). *Soil Taxonomy: A Basic System of Soil Classification for Making and Interpreting Soil Surveys* (2nd ed.). Agricultural Handbook No. 436, U.S. Department of Agriculture, Washington, D.C.

Stanley, S.R., and Ciolkosz, E.J. (1981). Classification and genesis of spodosols in the central Appalachians. *Soil Science Society of America Journal* **45**:912–917.

Van Nest, J. (1993). Geoarchaeology of dissected loess uplands in western Illinois. *Geoarchaeology: An International Journal* **8**:281–311.

Van Nest, J. (2002). The good earthworm: How natural processes preserve upland Archaic archaeological sites of western Illinois, U.S.A. *Geoarchaeology: An International Journal* **17**:53–90.

Waters, M.R. (1992). *Principles of Geoarchaeology, a North American Perspective.* University of Arizona Press, Tucson.

CHAPTER 5

BLUFF TOP SAND SHEETS IN NORTHEASTERN ARCHAEOLOGY: A PHYSICAL TRANSPORT MODEL AND APPLICATION TO THE NEVILLE SITE, AMOSKEAG FALLS, NEW HAMPSHIRE

Robert M. Thorson and Christian A. Tryon

Overlook sites from upland settings comprise a significant portion of the archaeological record in the northeastern United States (e.g., Ritchie 1965; Funk 1976; McBride 1984). They range in size from isolated lithic scatters to dense accumulations from intermittent reoccupations. Stratigraphic isolation of cultural components in such settings is rare, owing to the combination of low sedimentation rates and continuous secondary disturbance by biogeochemical processes, which act together to produce complex palimpsests (e.g., Wood and Johnson 1978; Johnson 1990). Although bioturbation is most closely associated with the outward dispersion of artifacts from initially tight clusters, it can also concentrate artifacts into horizons through the effects of particle settling and tree-throw and the development of lag horizons during colluvial transport. Most of these disturbance processes occur within the A and upper B horizons of the soil, where nutrients encourage root growth and animal activity, leading to a network of macropores that control the fate of artifacts.

Burial of this biologically active zone, or even partial burial, can isolate surface archaeological components from subsequent disturbance processes, at least statistically, thereby contributing to the integrity of buried artifact horizons. Lowland contexts—beaches, floodplains, colluvial footslopes, and alluvial fans—are depositional environments where sediment accumulation can outpace bioturbation, at least during extreme events associated with floods and storms. Under normal conditions elsewhere, however, the pace of sedimentation is almost always slower than turbation processes.

West-facing bluff-top settings are an exception to this generalization, providing an additional context where burial can outpace turbation. In this setting, discrete pulses of eolian deposition ranging in thickness from a film of dust to decimeter-thick strata are often associated with geologically short intervals of bluff erosion and exposure. Unlike lowland areas, however, bluff-top deposition takes place on overlook sites that are warmer, drier, and breezier than lowland sites, making them attractive settlement locales, providing not only high visibility of the surrounding terrain, but also access to both riverine and upland resources. Burial of bluff-top archaeological components by windblown sediment from local sources is known from a number of North American sites (e.g., Thorson and Hamilton 1977; Bettis and Hajic 1995; Van Nest 2002:58). Similar settings exist in New England, although they are historically under-recognized (Colby et al. 1953) and are thus largely unreported in the local archaeological literature.

This chapter presents a qualitative model for bluff-top sedimentation that can be applied throughout the northeastern U.S. and adjacent Canada. We begin by introducing a candidate site for this depositional model—the Neville site at Amoskeag Falls, New Hampshire (Dincauze 1976), the type site for the Middle Archaic in New England (Table 5.1 and Figure 5.1), a place where massive beds of "dusty" sand separate its buried, bluff-top, archaeological components.[1] We then develop a physical model for sediment transport and archaeological burial in bluff-top settings. Next we survey the distribution and stratigraphic contexts of northeastern sites, examining their buried sediment horizons with our model in mind, giving special attention to the Neville site. Finally we speculate that the high concentration of Middle Archaic artifacts at the Neville site, the dearth of early artifacts, and the separation of artifact horizons throughout the sequence can all be explained by local conditions associated with bluff-edge retreat that intermittently enhanced the suitability of the Neville terrace for habitation.

AMOSKEAG FALLS

The Neville site is one of several archaeological sites in the vicinity of Amoskeag Falls, in Manchester, New

Geoarchaeology of Landscapes in the Glaciated Northeast edited by David L. Cremeens and John P. Hart. New York State Museum Bulletin 497. © 2003 by the University of the State of New York, The State Education Department, Albany, New York. All rights reserved.

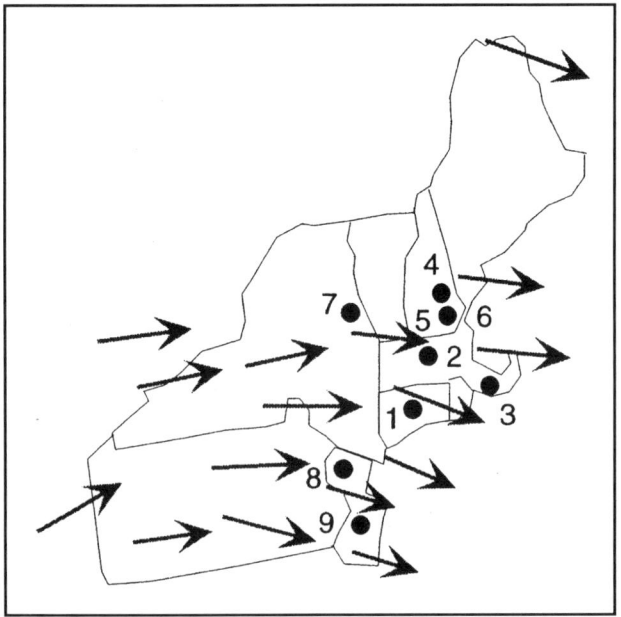

Figure 5.1. Sketch map of the northeastern states showing resultants for sand roses of winds exceeding 6 m sec^{-1}. Modified from Wells (1983). Also shown are sites from the region listed in Table 5.1.

Hampshire, which formed after drainage of glacial Lake Merrimack ca. 12,500 years ago (Ridge et al. 2001). Subsequent downcutting of the Merrimack River exposed bedrock at the falls as well as glaciolacustrine and glaciofluvial sediments, most of which are deltaic sands (Koteff personal communication, cited in Dincauze 1976; Koteff 1976). The archaeological importance of this area is demonstrated by the close proximity of three excavated sites to within 0.5 km of the falls—the Neville site (Dincauze 1971, 1976), the Smyth site (Foster et al. 1981), and the Eddy site (Bunker 1992)—and by the large numbers of artifacts recovered from these sites and others, now stored in both private and local historical society collections (Dincauze 1976:1).

The Eddy site lies on the western bank of the Merrimack River, downstream from the falls. Artifacts lie within overbank alluvial deposits and include a basal quartz-dominant lithic industry typologically assigned to the Early Archaic, with associated ^{14}C assays dating to ca. 8,000 years ago. Additional vertically discrete artifact layers span the Archaic and Later Woodland Periods. At higher elevation and across the river is the Smyth site, which occurs on top of and within Holocene soil developed on a Late Pleistocene dune overlooking the falls; most recovered artifacts span the Late Archaic through Recent Periods, indicating that eolian deposition continued, at least locally, during late Holocene time.

The Neville site lies on a glaciofluvial terrace above the eastern bank of the Merrimack River at an elevation intermediate between the Eddy and Smyth sites. The terrace tread lies about 10 m above the river under baseflow conditions and well above flood level, precluding a fluvial origin for the archaeological sediments. The tread also lies about 13 m below the upper, dune-draped bluff on which the Smyth site is located. The Neville terrace contains a 3-m-thick, plow-truncated sequence of

Table 5.1. Northeast Sites in Bluff-Top Settings

No.	Site	Age	State/Province	References
1	Red Hills	Unknown	CT	Bellantoni, personal communication (2002); personal observation
2	WMECO	Middle Archaic–Middle Woodland	MA	Thomas 1980; Dincauze 1989
3	Wapanucket	Paleoindian–Recent	MA	Robbins 1980
4	Sewall's Falls	Archaic–Woodland	NH	Starbuck 1982, 1983
5	Smyth	Paleoindian–Recent	NH	Foster et al. 1981
6	Neville	Middle Archaic–Recent	NH	Dincauze 1971, 1976
7	Upper Hudson Valley	Woodland	NY	Bender and Curtin 1990; Bender, personal communication (2002)
8	Rosenkrans Ferry	Woodland	NJ	Cross 1941:132–143
9	Abbott Farm	Paleoindian–Early Woodland	NJ	Cross 1956; Stewart 1983
10	Barnes Creek	Late Woodland	MI	Larsen 1985
11	Knechtel	Late Archaic	ON	Wright 1972

sediments, bearing artifacts spanning the Middle Archaic through Recent Periods. The Neville site was excavated in 1968 by Peter Lane, of the New Hampshire Archaeological Society, as a salvage operation prior to destruction of the area for bridge construction. Lane died prior to completion of the final report. Dena Dincauze (1971, 1976) completed subsequent laboratory analysis and reconstruction of site stratigraphy. It was the first stratified and well-dated Middle Archaic (8000–6000 B.P.) site in New England (Dincauze 1971).

Until recently salmon, shad, alewives, and other fish ran Amoskeag Falls to spawn during the late spring, providing a predictable, seasonal resource that likely provided the primary attraction for human settlement in the area; native populations were observed to congregate at the falls in spring well into the 17th century. Although faunal remains are not preserved, concentrations of iodine and mercury within sediment recovered from subsurface strata of the Neville site are consistent with the processing of anadromous fish (Dincauze 1976) throughout much of the Holocene time, which is likely responsible for the dense archaeological accumulations there. The repeated and concentrated human activity at the falls has obscured reconstruction of individual occupations (e.g., Dewar and McBride 1992) as well as social, political, and refuse-related phenomena, which often leave no discernable archaeological trace.

TRANSPORT MODEL

Owing to the importance of wind erosion and desertification, the literature on eolian sedimentation is extensive and highly quantitative. Empirical field and laboratory research, largely from engineering, focuses on the shear coupling (traction) between the airstream and the ground (fluid mechanics), the properties of sedimentary grains (density, shape, sorting, size, etc.), the mechanics of particle transport (creep, saltation, suspension, and variants thereof), and the role of roughness in causing flow separation. Simplifying assumptions associated with flat beds, uniform roughness, and spherical sediment grains are usually made to render the problems mathematically tractable. Forested bluff-top settings, however, depart from such idealized conditions largely due to the contrast in roughness associated with the forest edge (and the bluff face), the compression and acceleration of the airstream in response to the topographic constriction, the frictional loss of wind energy by chaotic processes associated with leaf and branch motion, and the mixed sediment texture of bluff-top deposits (particle-size ranges, generally 0.05–0.5 mm).

For simplicity, we restrict our bluff-top site-formation model to inland, forested terrace edges overlooking broad lowlands filled with glaciolacustrine, glaciomarine, and glaciofluvial sediments. Although mantling less than 15 percent of the Northeast by area, these waterlain deposits are widely distributed throughout the region, are associated with all important valleys (Lakes Erie and Ontario, and the Hudson, Connecticut, Merrimack, and Penobscot Rivers, etc.), and co-occur with thousands of known archaeological sites. We do not consider the case of unforested bluffs, because forest has been present during the entire human occupation interval. (An early and abrupt arrival of *Picea* and *Pinus* in the Northeast roughly coincides with the Bolling-Allerod transition (Interstade 1), dated to 12,900 B.P. [Gaudreau and Webb 1985].) We also restrict our model to settings where lake or river levels have risen and fallen through time, rather than to coastal settings where sea level has risen monotonically; in these circumstances bluff erosion and sedimentation take place during rare but dramatic geological events such as hurricanes, meander shifts, valley avulsion, and lake flooding. Finally, our model deals only with the eastern sides of major valleys, downwind from the sedimentary basins that provide both a source for sedimentary particles as well as a broad fetch along which the wind may intensify (Figure 5.1).

Boundary Conditions

We begin by considering the problem in two dimensions along a line parallel to the wind direction (Figure 5.2). Three topographic locations are considered: the *fore-bluff* of infinite width, refers to the lake, estuary, or alluvial valley below the bluff in question; small scarps may be present. The *bluff face* is a relict topographic step facing into the wind (terrace, shoreline, cliff, etc.) that was created initially by downcutting, and which is intermittently steepened by lateral erosion; the angle of the topographic step varies during steepening events, but generally remains close to the angle of repose for uncohesive sands and gravels (typically 31–37°). Steeper slopes are present only during a transient phase of lateral erosion; gentler slopes having the downhill convex-concave profile of humid, creep-dominated slopes occur during much longer intervals of soil development. The *bluff top* is the original subhorizontal terrace tread; its level is stationary but its front may erode backward during increments of bluff erosion.

Wind energy, which diminishes progressively in a downstream direction from the lowland, is lost at five different transport levels. Level 1: Topographically lowest is the distal, downwind limit of the fore-bluff, here assumed to be an unvegetated bar or beach below the bluff face; wind moves parallel to the surface. Level 2: The bluff face is both the site of sediment entrainment and a back-facing escarpment that forces convergence of

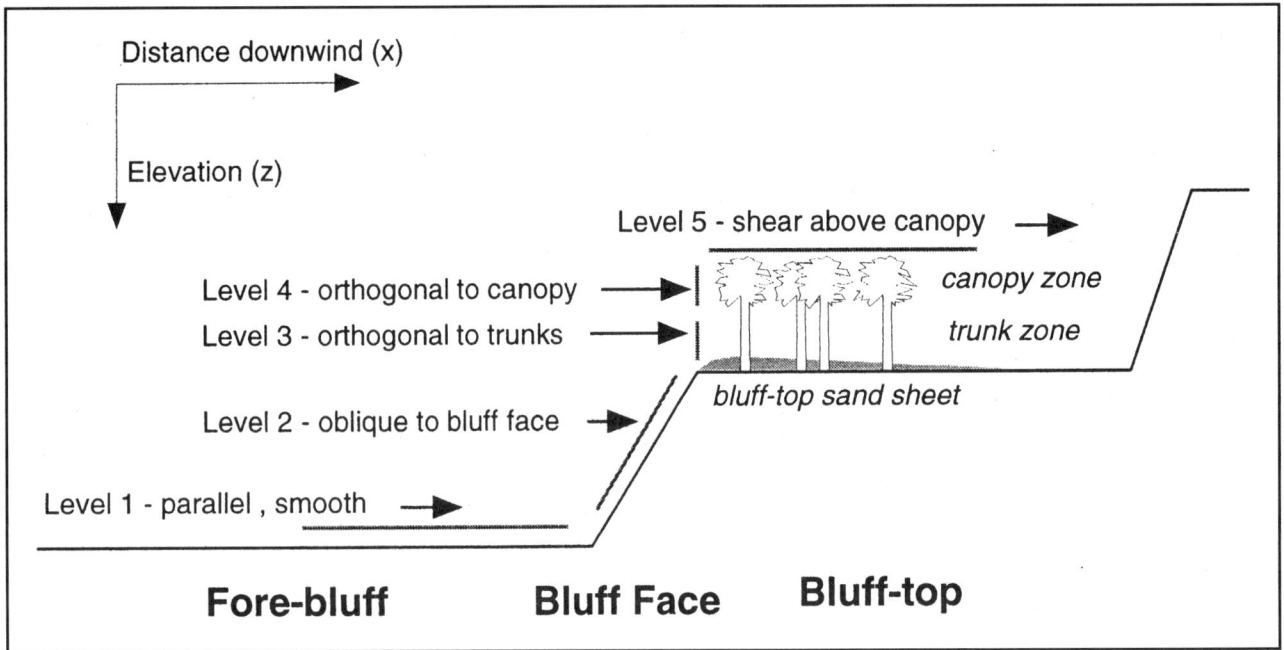

Figure 5.2. Geographic terms used in this chapter. Transport levels are defined in the text.

the airstream; this topographic step includes the height of the forest if the trees are thick enough to block the wind, but otherwise it does not include the forest. The third and fourth levels comprise the vertical front of the forest on the bluff face. Level 3, the trunk zone, is defined only when the understory is thin enough to allow wind to blow above the soil but beneath a raised canopy; in this setting, the trunks of the trees act as a comb for the wind stream moving through the canopy zone above and the surface of the bluff below. Level 4, the canopy zone, is effectively a barrier to the wind, because easier passage lies below (in the trunk zone of Level 3) and above. Level 5 is the aerodynamically rough, subhorizontal surface of the canopy; there the energy loss takes place at the base of the airstream.

Physical Processes

The processes (Figure 5.3) increase in complexity from the simplest (horizontal traction on a uniform bed) to the most complex (interactive effects from five levels).

Wind Profile on a Uniform Bed; Levels 1 and 5

On aerodynamically rough surfaces, the wind speed above the ground under steady-state conditions is given by the Prandtl-von Karman equation (Thwaites 1960)

$$U/u^* = 1/k^* \ln(z/z_o)$$

where U is the velocity at height z, u* is the velocity gradient shear velocity, k is a constant, z_o is the characteristic roughness displacement height, the point at which velocity is effectively zero. On smooth surfaces, z_o ranges from 0.5 to 10 x 10^{-4} m with d, the predicted displacement height, being negligible. For grass 0.25 to 1-m high, z_o and d rise to 0.04 to 0.2 m and < 3 m, respectively. For forest (average value for deciduous and coniferous), however, z_o and d rise to between 1.0 to 6.0 m and < 25 m, respectively. In other words, beneath a continuously forested landscape, the wind effectively blows at treetop level on a very rough surface, and the velocity is effectively zero near the forest floor.

Forest Edge with No Other Topographic Change: Contrast Between Levels 1 and 5

Consider an abrupt increase in the roughness such as that along the edge of a continuous, dense forest. Here the velocity profile is displaced upward, becoming

$$U/u^* = 1/k\,[(z-d)/z_o]$$

Under limiting conditions in which the forest is completely impermeable to wind, all of the kinetic energy is displaced upward, where it flows over a much rougher surface on top of the canopy, an action that lowers the shear velocity faster than on unforested ground, enhancing sedimentation. In this case, the forest edge effectively acts as a vertical bluff with a soft, tapered edge. Wind, forced upward, must accelerate over the forest edge at the front of the canopy. The transition from

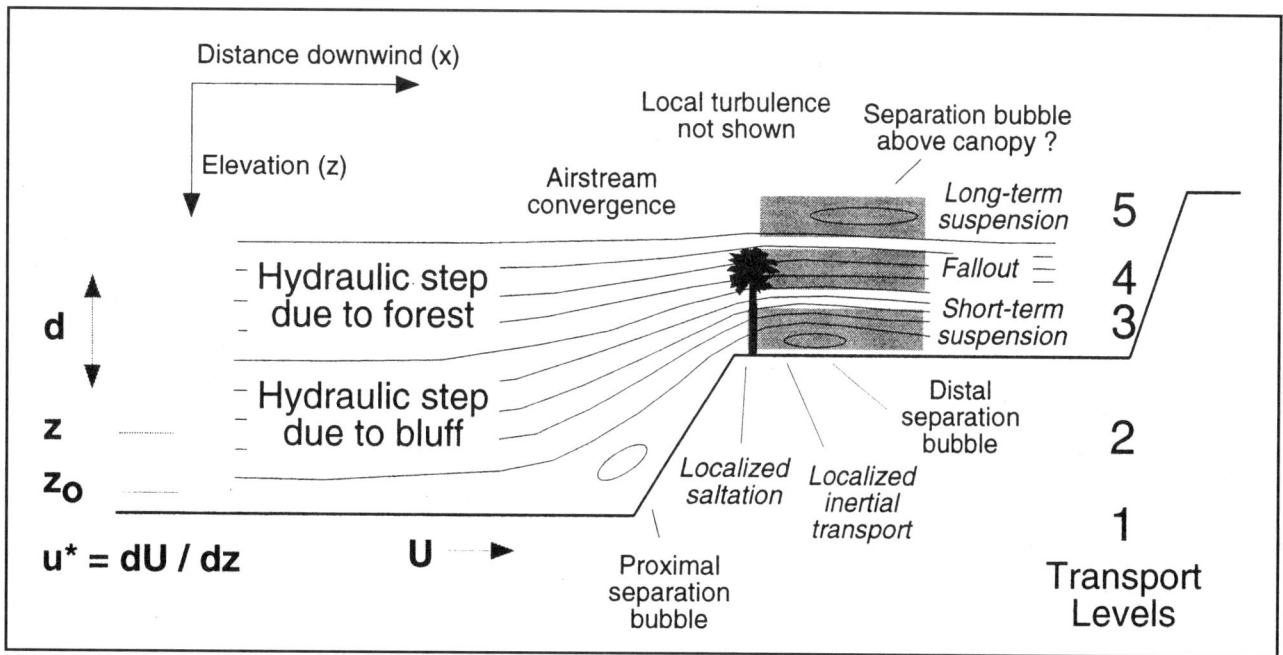

Figure 5.3. Simple hydraulic factors associated with deposition of bluff-top sand sheets. Letters indicate terms in equations 1 and 2. The hydraulic step caused by the forest is present only when the forest is thick. Separation bubbles (or low-velocity zones) form above the bluff when the forest is thin or when the canopy is either raised or very dense. Italicized terms identify sediment transport processes to the bluff-top sand sheet.

smooth bed to rough bed is not simple. The addition of a few roughness elements (trees) may actually increase local entrainment and enhance transport, because the wind is funneled around objects, causing flow separation and the formation of turbulent eddies. Above a critical concentration of trees, however, the addition of roughness elements uniformly diminishes the rate of sediment transport.

Leaking Forest Edge; Levels 3 and 4

Under normal circumstances, however, some of the wind energy is lost as it blows inward to the forest from its exposed edge. If the forest edge is thin enough (gallery forest), an outlet for the wind in a downstream direction will exist, and the wind will blow through, rather than above, the forest, especially in Level 3. If the bluff-top band of forest is thick, however, there is no outlet for wind forced horizontally into the trees, causing wind that enters to be ventilated upward through the canopy. The total kinetic energy of the wind (E_{kt}) is given by

$$E_{kt} = E_{kl} + E_{kc} - E_p - F$$

It must equal the sum leaking into Levels 3 and 4 (E_{kl}) and that being displaced upward to canopy height (E_{kc}), minus the frictional losses (F) associated with swaying and fluttering of forest components and the small gain in potential energy (E_p) associated with wind being lifted to canopy height.

Flow Separation at Topographic Breaks at Levels 2 and 3

When a viscous fluid (such as air) encounters an obstacle, it experiences flow separation, creating what are known as separation bubbles (or rollers for the two-dimensional case). The most familiar separation bubbles are stationary eddies—whirlpools—which exhibit local reversals in flow direction. In the case of bluff edges, there are two separation bubbles. The proximal bubble lies at the base of the bluff and diminishes in importance as the slope angle declines. The distal bubble originates at the bluff edge, and, if the wind is strong enough, re-attaches downwind; the shape of the resulting sediment body depends on many factors, primarily the viscosity of the fluid, the velocity, and the roughness (Hetu 1992:Figure 9A). In addition to creating a permanent distal separation bubble, the bluff also acts as a constriction for the wind, forcing the horizontal flux through a smaller cross-sectional area. As a consequence, the airstream is deflected upward and increases in velocity (which lowers the pressure through the Bernouli effect). Separation is straightforward above an unforested vertical bluff edge or under conditions when a dense forest with uniform canopy essentially raises the height of the bluff. More realistic and complex, however, is the case where the forest is "permeable" to the wind (Level 3). A highly permeable Level 3 merely complicates the flow separation process, rather than eliminates it. An impermeable Zone 3 forces flow separation to take place largely above the canopy. Under normal circumstances blufftop flow separation is present, but weakly expressed.

Sediment Source

Sediment entrainment by wind is negligible on continuously vegetated surfaces, hence a loss of vegetation cover by erosion is required. Rarely, slumping of a bluff edge can expose a surface to deflation. Normally, however, the bluff must be undercut by water; the loss of support then propagates upward, destabilizing the entire bluff face. On sand and gravel, the slopes quickly ravel to angle of repose slopes. Maintenance of the unvegetated bluff requires continuous undercutting. Sediment may be entrained either from the beach or floodplain or from the bluff itself. Erosion is limited by the threshold shear velocity (U_{*t}), which is given by

$$U_{*t} = A\ [(\rho_p - \rho_a)\ g\ D\ /\rho_a]^{0.5}$$

where ρ_p and ρ_a are the particle and air densities, respectively, and A is an empirical coefficient related to surface roughness and grain characteristics, and D is the particle-size diameter. Note that the threshold velocity is related to the square root of the particle diameter. Once motion is initiated by creep, the movement of grains enhances local turbulence and their momentum is transferred on impact, causing rapid sediment movement called saltation. Dust liberated will be carried forward by other processes.

Sediment sources on the bluff are maximized when the bluff is high (maximum contact with the deflating wind and maximum constriction of the airstream), uniformly weak in composition (smooth, continuously raveling surface without secondary roughness), and oriented favorably with respect to prevailing wind direction. Sediment sources on the lowland are maximized when the fetch is broad and the surface smooth and when waves or floods prevent the formation of a lag horizon.

Sediment Transport

Three principal modes of transport—suspension, surface creep, and saltation—are largely a function of particle size. Suspension can be divided into long-term suspension characteristic of dust storms, acting on particles generally < 20 μ (0.020 mm) and short-term suspension, a complex but widely present transport mechanism restricted to coarse silt and fine sand at shear velocities of 50–200 cm sec^{-1}. Creep involves the rolling and pushing of large particles (> 500 μ; 0.5 mm) by the bombardment of saltating particles. Saltation refers to the movement of particles of intermediate size by a series of shorts leaps or bounces. Regardless of process, particle concentration decreases hyperbolically with increasing height (z) above the ground. Bluff-top dunes, especially when eolian bedforms are present, indicate saltation as a dominant process. A thin layer of fine dust, which would eventually lead to the accumulation of loess, is dominated by long-term suspension.

Most bluff-top sand sheets are neither dunes nor loess accumulations per se but are more complex in origin. In New England they have been referred to as the "eolian mantle" (Hartshorn 1976; Thorson and Schile 1995). Similar "mixed-source" eolian deposits are common components of geoarchaeological models in early and mid-Holocene deposits throughout North America (Bettis 1995). Hetu (1992) presents an extreme case in which a sheet of windblown coarse sand and granules composed largely of shale chips (mean thickness of 11.4 mm; area of 1,200 m^2) which was deposited by a single storm.

Bluff Retreat

Implicit in the model of the exposed bluff is the idea that the bluff top retreats incrementally backward, sometime after erosion at its base (Figure 5.4). Conditions might arise, however, when erosion is restricted to the base of the bluff. Alternatively, erosion near the bluff top might lag behind that at the base, leading to slumps. Regardless, sediment entrained from the lowland or from the base of the bluff may be blown up into vegetation on, rather than above, the bluff edge, complicating the model.

Soil Development

Holocene soil formation dominated by the accumulation of organic matter, biogeochemical changes in the parent material, and bioturbation, takes place continuously, with a small addition of eolian dust. Pulses of sediment, if rapidly deposited, isolate soil horizons from the surface action.

Time Steps

Three time steps take place in sequence on a forested surface in which mature trees have raised canopies (Figures 5.3 and 5.4).

Step 1. Background conditions (10^2–10^4 years) consist of times when net sedimentation is limited to dust influx from distant source areas. Bioturbation and pedogenesis take place under continuously forested conditions.

Step 2. Bank erosion (10^0–10^2 years) associated with high lake levels, extreme floods, channel avulsion, or

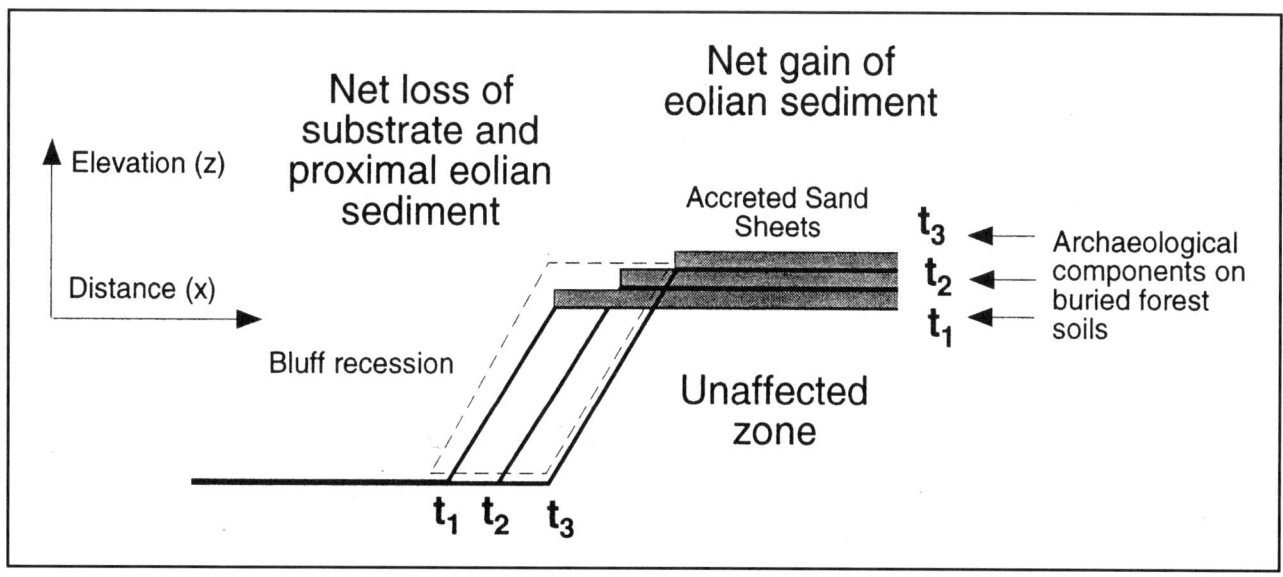

Figure 5.4. Schematic model emphasizing temporal component of bluff-edge sand sheet accretion. Pulses of erosion t_1 through t_3 lead to parallel retreat of the bluff face (dark lines), during which enhanced sediment transport and local entrainment lead to the deposition of bluff-edge sand sheets (gray). Note that erosion also removes part of the sand sheet. Artifacts are most likely to be deposited during the transition from exposed to forested conditions.

meandering shifting leads to destabilization of the slope. After the instability has propagated upward to the bluff top, an increment of erosion may take place. An exposed bluff provides a source area for deflation, or it may merely provide an aerodynamically smooth surface through which sediment entrained from the fore-bluff (Level 1) may occur. The sedimentation pulse consists of three basic facies.

Sand saltating up the bank can be accelerated into short-term suspension near the top of the bluff and be deposited immediately leeward of the bluff edge. Normally the sand will simply fall from short-term suspension into the proximal part of the separation bubble, where it is mixed with finer material blown up by other processes. In the rare cases where the sand flux is high, however, saltation can continue beyond the bluff edge, forming a true bluff-top dune the upper surface of which must conform to the local airstream; as the dune grows it usually assumes a parabolic (blowout) form with the separation bubble migrating downwind. Such dunes typically form only locally where sediment flux is maximized by bluff-parallel topography and a lack of vegetation cover.

Sand moving in turbulent gusts up the bluff face within short-term suspension carries well past the bluff edge, falling over a broad area perhaps 2–10 m wide for a thick, raised-canopy forest. Owing to inertial effects, the coarsest sand is restricted to the edge. This process is enhanced when the canopy is high, allowing fine sand to be carried in suspension through the trunk zone, forming a deposit that gradually diminishes in thickness. Sediment transport diminishes through normal fallout processes, both because the wind flux is diminished through upward loss, and because the wind shear is diminished by frictional losses from the ground and from the base of the canopy. Canopy structure will largely determine how far inland this facies penetrates.

Very fine sand and silt carried to the top of Level 4 in suspension (front edge of the canopy) will move sub-horizontally in Level 5 above a weak separation bubble. Enhanced roughness consumes wind energy forcing the deposition of material being carried low in the wind stream. Deposition may be direct to the forest floor or indirect through the rain washing of leaves. Sedimentation is supply limited, its rate dependent on bluff-face conditions.

Excluding parabolic bluff-top dunes, the shape of the sediment body resulting from these processes will likely be tabular. In theory, the body will thicken away from the bluff edge, because inertial forces force the maximum flux several meters inland from the bluff edge then thin away from the zone of maximum concentration through normal fallout processes. Bedforms and discrete beds will be absent, because traction cannot take place on a rough surface that probably remained continuously vegetated during the pulse of sediment deposition; at normal sedimentation rates, bioturbation is dominant. The bulk texture will be poorly sorted because several processes are at work. Importantly, the eolian nature of these deposits may not be apparent. If thick enough, and deposited fast enough, the bluff-edge sand sheet can isolate a former ground surface producing a buried paleosol.

Step 3. Stabilization (10^2–10^3 years). Eventually, the fore-bluff and the bluff face will become stabilized, cutting off the sediment supply, and inaugurating a phase of soil development. The bluff face, being xeric and deficient in organic matter, must experience a transient phase of ecological succession before forest can be reestablished on the bluff face. Hence a transient phase will exist when the bluff top provides ideal conditions as an overlook, because an unobstructed valley view is available from a shaded, well-drained vantage point. Therefore we should expect archaeological occurrences to lie at or near sediment interfaces or incipient soil horizons.

Our model refers to the unrealistic two-dimensional situation where there is no component of sediment motion oblique to the bluff edge. Yet any curvature of the bluff face will influence sedimentation patterns. For example, wind is often funneled up small gullies, resulting in irregular, bluff-top dunes. Sand blowing obliquely on the slope will be focused inland to the bluff top where the curvature of the scarp is toward the fore-bluff. Erosion of the bluff top by lateral processes often leaves the dense mat of roots intact; they can drape the upper edge of the bluff face, either enhancing (by macroscale smoothing) or minimizing (entrapment) sand transport. Blowdown by trees, which may be concentrated at the bluff face, will locally complicate the scenario.

Assuming the human occupants use the bluff as an overlook, the maximum concentration of artifacts will lie a short distance inland from the bluff edge. Loss of integrity to components includes three processes. The first is pervasive bioturbation, which fractionates artifacts generally downward. Second is human disturbance, which can take many forms, including trampling, site maintenance and secondary disposal, feature excavation, development of anthrosols, etc. (Schiffer 1987). Third is frontal erosion of the bluff, in which the artifacts are permanently lost to the bluff-top environment. The absence of artifacts from the lowest layers of a bluff-top site may reflect frontal erosion, rather than the absence of occupation. The artifacts may not be permanently lost from the system but may be buried in the colluvial wedge below.

APPLYING THE MODEL

Given the apparent preferential past occupation of river terraces, and the number of recent regional surveys that have focused upon major river valleys (e.g., Snow 1980; Kenyon and McDowell 1983; McBride 1984; Bender and Curtin 1990; Funk 1976), this model should be applicable to a number of sites in the Northeast. Table 5.1 presents a preliminary list of 11 such sites, in both fluvial and lacustrine contexts, from a brief review of the published literature; the model may not apply equally to all of them.

Eolian deposits and artifacts co-occur in the Connecticut Valley at the Red Hills area and the WMECO site complex in Gill, Massachusetts. The Neville, Smyth and Sewall's Falls sites in the Merrimack Valley all preserve archaeological deposits in windblown sediment on bluffs or terraces upstream of major waterfalls. A number of candidate sites exist in the Upper Hudson Valley, along the Hoosic and Battenkill Rivers where bluff-top sites are known, and eolian sedimentation suspected (S.J. Bender, personal communication 2002). Bluff-top eolian sedimentation occurred at the Rosenkrans Ferry site in upstate New Jersey, along the Delaware River and near the southern limit of the last glaciation. The Abbott Farm site complex rests on redeposited glacial outwash ("Trenton Gravels") from which fines were winnowed and blown onto the overlying bluff tops.

Climate-driven fluctuating lake levels may expose fine-grained lacustrine sediment during low lake stands, with subsequent entrainment by wind burying stable land surfaces. Examples of this from the Great Lakes region are recorded by Larsen (1985) for the Barnes Creek site and likely apply to the Knechtel 1 site as well. Smaller lakes may also provide sufficient wind velocities for eolian reworking of beach deposits. The various sites at Wapanucket occur within a large dune at the northern edge of Lake Assawompsett, Massachusetts. The dune appears to have undergone a continual process of net aggradation accompanied by localized erosion. Strong winds blow off the lake, and the adjacent bluff is undercut by wave action today (Robbins 1980). Artifacts from the Paleoindian to Recent Periods occur in the sediment, although little cultural stratigraphy has been preserved due to both the intensive (re)occupation of the site in the past and extensive looting.

We propose that the Neville site was occupied during intervals of bluff-face erosion and restabilization because of (1) the fact that grain-size coarsens upward, suggesting an increasingly proximal sediment source, such as that provided by a progressively retreating bluff edge, (2) the coincidence of high sedimentation rates with periods of intense occupation, suggesting a correlation between sediment availability and settlement suitability, and (3) the location of major occupation floors at or immediately below strata interfaces, indicating a correlation between occupation and change in sedimentary regime.

Stratigraphy

After a period of subaerial exposure, a ~6-cm-thick bed of alluvium was deposited upon the weathered surface of the original terrace tread. Subsequent deposition was of windblown sediment (Stratum 5B-2). Strata were divided both on color and textural differences as well as artifact type and density. Stratum 4 is described as a "typical greasy, black midden soil" (Dincauze 1976:19), while Strata 2 and 1 are a plow zone and 19th-century construction debris, respectively.

The 1.14-m depth marks a significant change in the site's history.[2] The Neville terrace appears to have been occupied most frequently below this depth (the Middle Archaic occupation), which also coincides with a period of increased sedimentation rates relative to the overlying strata (Dincauze 1976; 1989:3). We believe that our model provides an explanation for this co-occurrence, which may be due in part to changes in the local environment driven by a warmer, drier interval of the Holocene.

Grain-Size Analysis (Figures 5.5 and 5.6)

Sediment samples were taken at 15.2-cm intervals from a soil monolith approximately 12 m from the terrace edge (Dincauze 1976:10–12). Samples below 1.14 m (Neville 2, 3, and 4) coarsen upwards (increase in the "C" fraction of medium to very fine sand), whereas those above Stratum 4B exhibit little compositional change (Neville 5 and 9). Although the winnowing of fines by wind cannot be ruled out, this coarsening-upward sequence is consistent with an erosional retreat of the terrace edge (Time Step 2), resulting in an increasingly proximal sediment source. The bulk of the material at the Neville site is not well sorted; this, and the absence of recorded sedimentary features, is consistent with our depositional model. The bulk texture is also similar to the eolian mantle found throughout much of New England (e.g., Thorson and Schile 1995:753; Hartshorn 1967).

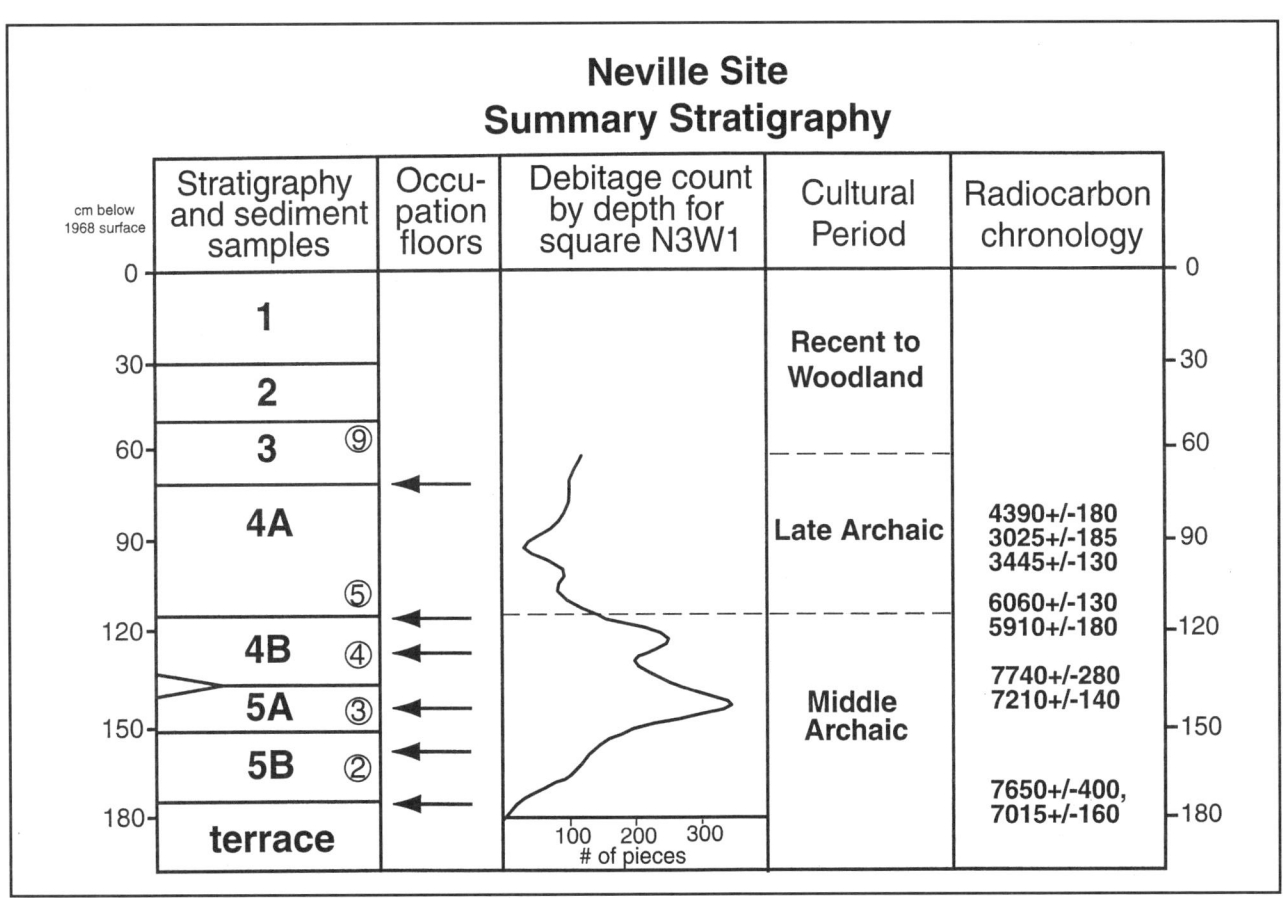

Figure 5.5. Schematic stratigraphic summary of the Neville site, adapted from Dincauze (1976). Included are locations of sediment samples (circled numbers) plotted in Figure 5.6, location of horizons of high artifact density ("occupation floors," see text for discussion), and selected radiocarbon dates plotted by depth, using a half-life value of 5570, before A.D. 1950. Debitage count modified from Dincauze (1976:Figure 7).

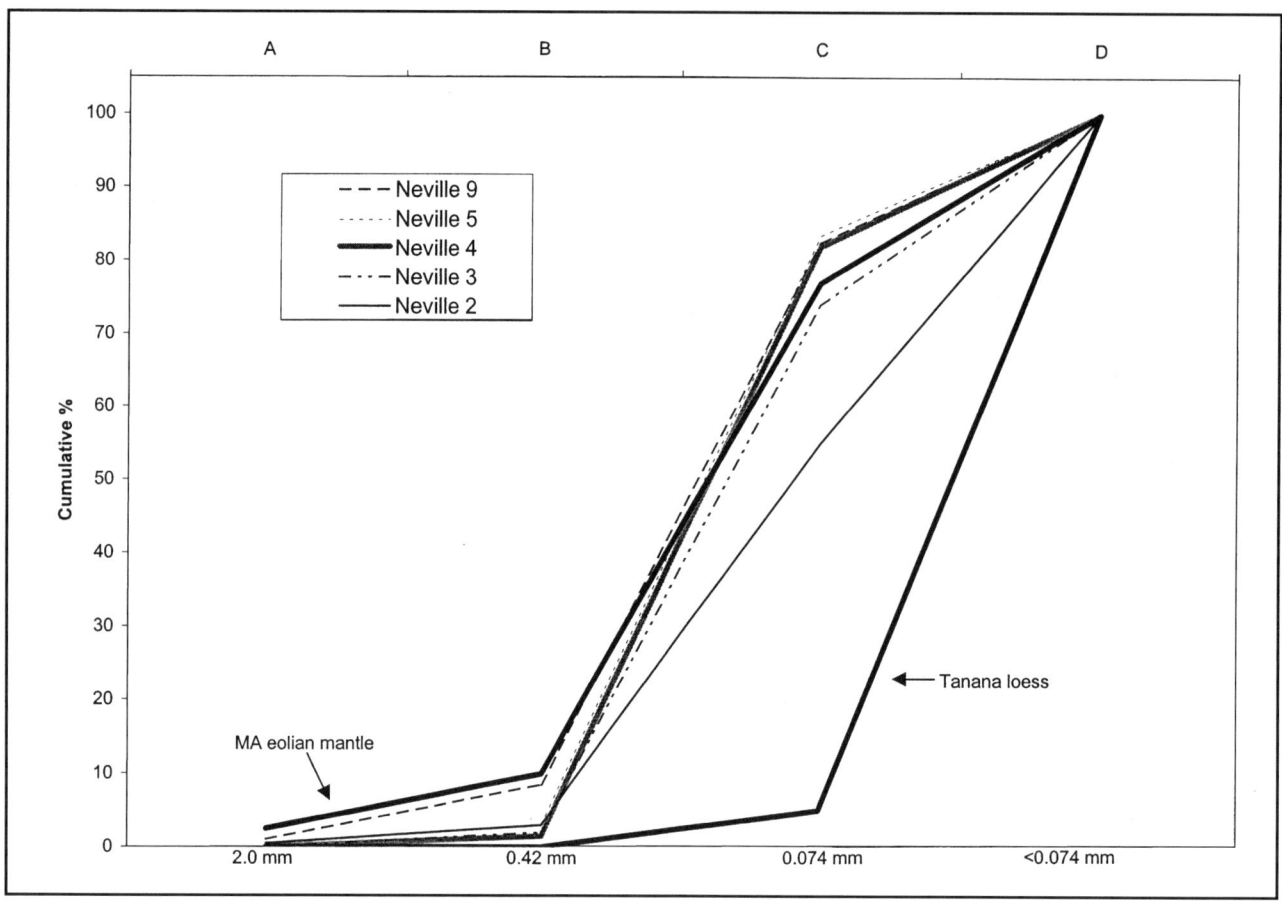

Figure 5.6. Grain-size distribution of selected samples from the Neville site, compared with a representative sample of the Massachusetts "eolian mantle" (Hartshorn 1967) and loess from the Tanana Valley, Alaska (Thorson and Hamilton 1977). Vertical axis records cumulative percentage; size fraction on the upper horizontal axis following that from the Neville site (Dincauze 1976).

Sedimentation Rates (Figure 5.5)

Comparison of sediment thickness to radiocarbon chronology suggests that Strata 5B–4B (the Middle Archaic occupation) accumulated at a much higher rate than did the overlying strata (ca. 1 cm/24 yr vs. 1 cm/52 yr) (Dincauze 1976, 1989:3). Occupation of the terrace was apparently both more frequent (as measured by the number of occupation floors, see below) and more intense (as measured by debitage density) prior to the deposition of Stratum 4A. This suggests a preferential use of the terrace when it was most subject to eolian deposition, that is, when the edge and bluff face were relatively clear of vegetation, providing good visibility, as well as a source of sediment. Strata above 1.14 m contain fewer artifacts and a single occupation floor; lower sedimentation rates and apparent terrace stability indicate the predominance of our Time Step 1 (background conditions).

Occupation Floors (Figure 5.5)

Several "occupation floors," consisting of thin, dense concentrations of artifacts, occur throughout the stratigraphic sequence of the Neville site. Such concentrations in unconsolidated sediment may result from postdepositional processes, as several artifact-refitting studies have shown (e.g., Cahen 1978; Villa 1982), although no such data is available for the Neville site. These debitage concentrations are associated with strata interfaces, which suggest formerly stable land surfaces. Successive major "occupation floors," at or immediately below these stable surfaces (Dincauze 1976), suggest a link between occupation and sedimentary regime. We interpret this as occupation during Time Step 3 (stabilization), a period following the major pulse of sedimentation (diminished amounts of windblown dust), but prior to complete reappearance of obscuring vegetation on the bluff face.

Climatic Effect?

The Middle Archaic occupation of the Neville terrace occurred during an interval of the Holocene both warmer and drier than the present (e.g., COHMAP 1988; see also papers in Bettis 1995), and climatic changes may be broadly responsible for the higher rates of sedimentation and apparent terrace instability at the Neville site during this period. Proxy records in the Northeast document lowered water tables in small lakes and bogs (Thorson and Webb 1991), an increase in pine pollen (Davis 1983), and reduced effective soil moisture (Webb et al. 1993) during ca. 8000–5000 B.P. The response of fluvial systems is variable and complex and remains unresolved (e.g., Knox 1983). Although we can only speculate on the climatic effect at the Neville site (increased frequency of floods and bank undercutting?), a generally warmer and drier climate would tend to extend the duration of Time Step 2 (erosion), prolonging sediment availability for wind transport prior to restabilization due to possible changes in the vegetative screen or a loss of cohesion of terrace sediment.

DISCUSSION

The Neville site conforms to our model of terrace-edge retreat and incremental burial of archaeological components by windblown proximal sediments, because (1) the site is located on a west-facing terrace above the flood level of a major river; (2) the bluff face fronting the site could have been destabilized by episodes of fluvial bank erosion; (3) fine-grained sediment suitable for entrainment by wind is widely available; (4) Holocene eolian deposits are widespread in the area (Koteff 1976); (5) sediment texture and structures at the site are consistent with an eolian origin; and (6) the alternative of colluvial transport is generally precluded by the low slope on the terrace tread.

Direct evidence for bluff-edge retreat is seen in the coarsening upward sequence of sediments preserved at square N0E1, located some 8 m from the bluff edge. Inertial effects dictate the settling of larger particles first; this signal is interpreted as increasingly proximal sediment source provided by the retreat of the terrace edge.

Most compelling, however, is the presence of five discrete archaeological horizons within 75 cm of sediment that apparently formed in ca. 1,500 years; each archaeological horizon occurs at or immediately beneath the interfaces below Strata 5B–4B. Although pedogenetic and bioturbative processes can result in artifact burial, stratification, and secondary concentration (e.g., Johnson 2002; Van Nest 2002), we interpret the stratigraphic separations as due principally to increments of eolian sedimentation followed by occupation during Time Step 3 of our model (stabilization), a period following the major pulse of sedimentation (diminished amounts of windblown dust), but prior to complete reappearance of the obscuring vegetation on the bluff face.

Reliable sedimentation rates cannot be calculated for the Neville site deposits. This is based on theoretical grounds (Ager 1981; Anders et al. 1987; Van Andel 1981) as well as the impact of soil formation and cultural disturbances at the site. However, comparison of the available radiocarbon chronology to sediment thickness at the Neville site suggests increased sediment availability and deposition during the Middle Archaic occupation of the site. Thus the ~75 cm of Strata 5B–4B accumulated in ca. 1,500 radiocarbon years, whereas age estimates for the 1.14 m of Stratum 4A-1 indicate some 6,000 years of accumulation. We argue that this is significant because the Middle Archaic dates to a warmer and drier interval of the Holocene, which may have increased the incidence of bluff-edge destabilization and eolian transport.

Our bluff-edge retreat model may also account for the scarcity of Paleoindian and Early Archaic artifacts. Dincauze (1976:118) interprets the few artifacts from these periods as curios brought to the site by later inhabitants, suggesting that, with the Amoskeag Falls not yet formed, there was little in the way of a persistent environmental attraction (anadromous fish) to drive continued, frequent occupation of the area (Dincauze 1976:9; 1993:17; personal communication 2002). Although we cannot rule out such cultural and historical explanations, the simplest explanation for the absence at the Neville site is that the earliest occupations were concentrated along a bluff edge that has since been eroded away.

Our model suggests that terrace edges will be most frequently selected for occupation following rare erosive events and subsequent bluff-face restabilization, the frequency of which may be controlled by broader regional environmental conditions. Occupational hiatuses at a regional scale may be due, in part, to an incomplete understanding of the geological processes that lead to the formation, burial, and ultimate recovery of archaeological remains (e.g., Petersen and Putnam 1992; Van Nest 1993; papers in Bettis 1995).

Future testing of our model will require excavation of a multicomponent archaeological site in an upwind-facing, bluff-top setting. Excavations will be required along a transect perpendicular to the bluff face, which will provide an opportunity to examine and sample the gradients in bulk texture, sedimentary structure, and pedogenic modification from proximal-to-distal local wind regimes (e.g., Thorson and Hamilton 1977)

END NOTES

1. Artifact concentrations ("occupation floors") at or immediately beneath strata interfaces suggest a formerly stable surface, although soil development may have been quite weak.

2. This corresponds to the 45-in depth noted by the excavators; all measurements converted to metric.

REFERENCES CITED

Ager, D.V. (1981). *The Nature of the Stratigraphical Record, 2nd ed)*. Wiley, New York.

Anders, M.A., Krueger, S.W., and Sadler, P.M. (1987). A new look at sedimentation rates and the completeness of the stratigraphic record. *The Journal of Geology*: **95**:1–14.

Begin, C., Filion, L., and Michaud, Y. (1995). Dynamics of a Holocene cliff-top dune along Mountain River, Northwest Territories, Canada. *Quaternary Research*: **44**:392–404.

Bender, S.J,. and Curtin, E.V. (1990) *A Prehistoric Context for the Upper Hudson Valley: Report of the Survey and Planning Project*. Manuscript on file at the New York State Office of Parks, Recreation, and Historic Preservation. Albany.

Bettis, E.A. III. (editor). (1995). *Archaeological Geology of the Archaic Period in North America*. Special Paper No. 297. Geological Society of America, Boulder, Colo.

Bettis, E.A. III, and Hajic, E.R. (1995). Landscape development and the location of evidence of Archaic cultures in the Upper Midwest. In *Archaeological Geology of the Archaic Period in North America*, edited by E.A. Bettis III, pp. 87–113. Special paper—Geological Society of America 297, Boulder, Colo.

Bunker, V. (1992) Stratified components of the Gulf of Maine Archaic tradition at the Eddy site, Amoskeag Falls. In *Early Holocene Occupation in Northern New England*. edited by B.S. Robinson, J.B. Peterson, and A.K. Robinson, pp. 135–147. Occasional Publications in Maine Archaeology, No. 9. Maine Historic Preservation Commission, Augusta.

Cahen, D. (1978). New excavations at Gombe (ex-Kalina) Point, Kinshasa, Zaire. *Antiquity* **52**:51–56.

COHMAP Members. (1988) Climatic changes of the last 18,000 years: Observations and model simulations. *Science* **241**:1043–1052.

Colby, W.G., Light, M.A., and Bertinuson, T.A. (1953). The influence of wind-blown material on the soils of Massachusetts. *Soil Science Society Proceedings* **17**:395–399.

Cross, D. (1941). *The Archaeology of New Jersey, Volume 1*. The Archaeological Society of New Jersey and the New Jersey State Museum, Trenton.

Cross, D. (1956). *The Archaeology of New Jersey, Vol. 2: The Abbott Farm*. The Archaeological Society of New Jersey and the New Jersey State Museum, Trenton.

Davis, M.B. (1983). Holocene vegetation history of the eastern United States. In *Late-Quaternary Environments of the United States, Vol. 2: The Holocene*, edited by H.E. Wright, Jr., pp. 166–181. University of Minnesota Press, Minneapolis.

Dewar, R.E., and McBride, K.A. (1992). Remnant settlement patterns. In *Space, Time, and Archaeological Landscapes*, edited by J. Rossignol and L. Wandsnider, pp. 227–255. Plenum Press, New York.

Dincauze, D.F. (1971). An Archaic sequence for southern New England. *American Antiquity* **36**:194–197.

Dincauze, D.F. (1976). *The Neville Site: 8,000 years at Amoskeag*. Peabody Museum Monographs No. 4. Harvard University Press, Cambridge, Mass.

Dincauze, D.F. (1989). Geoarchaeology in New England: An early Holocene heat spell? *The Review of Archaeology* **10**:1–4.

Dincauze, D.F. (1993). Antecedents and ancestors, at last. *The Review of Archaeology* **14**:12–22.

Follmer, L.R. (1982). The geomorphology of the Sangamon surface: Its spatial and temporal attributes. In *Space and Time in Geomorphology*, edited by C.E. Thorn, pp. 117–146. George Allen & Unwin, London.

Foster, D.W., Kenyon, V.B., and Nicholas, G.P. II. (1981). Ancient lifeways at the Smyth Site, NH-38-4. *The New Hampshire Archaeologist* **22**:1–91.

Funk, R.E. (1976). *Recent Contributions to Hudson Valley Prehistory*. New York State Museum Memoir No. 22. The University of the State of New York, Albany.

Gaudreau, D.C., and Webb, T. III. (1985). Late-Quaternary pollen stratigraphy and isochrone maps for the northeastern United States. In *Pollen Records of Late-Quaternary North American Sediments*, edited by V.M. Bryant, Jr., pp. 247–280. American Association of Stratigraphic Palynology Foundation, Austin, Tex.

Hartshorn, J.H. (1967). *Geology of the Taunton Quadrangle, Bristol and Plymouth Counties Massachusetts*. Geological Survey Bulletin No. 1163-D. U.S. Geological Survey, Washington, D.C.

Hartshorn, J.H. (1976). Quaternary problems in southern New England. In *Quaternary Stratigraphy of North America*, edited by W.C. Mahaney, pp. 91–92. Dowden, Hutchinson & Ross, Stroudsburg, Pa.

Hetu, B. (1992). Coarse cliff-top aeolian sedimentarion in northern Caspesie, Quebec (Canada). *Earth Surface Processes and Landforms* **17**:95–107.

Johnson, D.L. (1990) Biomantle evolution and the redistribution of earth materials and artifacts. *Soil Science* **149**:84–102.

Johnson, D.L. (2002). Darwin would be proud: Bioturbation, dynamic denudation, and the power of theory in science. *Geoarchaeology* **17**:7–40.

Kenyon, V.B., and McDowell, P.F. (1983). Environmental setting of Merrimack Valley prehistoric sites. *Man in the Northeast* **25**:7–23.

Knox, J.C. (1983) Responses of river systems to Holocene climates. In *Late-Quaternary Environments of the United States, Vol. 2: The Holocene*, edited by H.E. Wright, Jr., pp. 26–41. University of Minnesota Press, Minneapolis.

Koteff, C. (1976). Surfical geologic map of the Nashua North Quadrangle, Hillsborough and Rockingham Counties, New Hampshire. Geologic Quadrangle Map—U.S. Geological Survey, Report: GQ-1290. U.S. Geological Survey, Reston, Va.

Larsen, C.E. (1985). Geoarchaeological interpretation of Great Lakes coastal environments. In *Archaeological Sediments in Context*, edited by J.E. Stein and W.R. Farrand, pp. 91–110. University of Maine, Orono.

McBride, K.A. (1984) Prehistory of the Lower Connecticut River valley. Unpublished doctoral dissertation, University of Connecticut, Storrs.

Petersen, J.B, and Putnam, D.E. (1992). Early Holocene occupation in the central Gulf of Maine region. In *Early Holocene Occupation in Northern New England*, edited by B.S. Robinson, J.B. Peterson, and A.K. Robinson, pp. 13–62. Occasional Publications in Maine Archaeology, No. 9, Maine Historic Preservation Commission, Augusta.

Ridge, J.C., Canwell, B.A., Kelly, M.A., and Kelley, S.Z. (2001). Atmospheric ^{14}C chronology for late Wisconsinan deglaciation and sea level change in eastern New England using varve and paleomagnetic records. In *Deglacial History and Relative Sea-Level Changes, Northern New England and Adjacent Canada*, edited by T.K. Weddle and M.J. Retelle, pp. 243–270. Geological Society of America, Boulder, Colo.

Ritchie, W.A. (1965). *The Archaeology of New York State*. The Museum of Natural History Press, Garden City, N.Y.

Robbins, M. (1980). *Wapanucket*. Trustees of the Massachusetts Archaeological Society, Inc., Attleboro, Mass.

Schiffer, M.B. (1987). *Formation Processes of the Archaeological Record*. University of Utah Press, Salt Lake City.

Snow, D.R. (1980). *The Archaeology of New England*. Academic Press, New York.

Starbuck, D. (1982). Excavations at Sewall's Falls (NH 31-30) in Concord, New Hampshire. *The New Hampshire Archaeologist* **23**:1–20.

Starbuck, D. (1983). Survey and excavation along the upper Merrimack River in New Hampshire. *Man in the Northeast* **25**:25–41.

Stewart, R.M. (1983). Soils and the prehistoric archaeology of the Abbott Farm. *North American Archaeologist* **4**:27–49.

Thomas, P. (1980). The Riverside District, the WMECO Site, and suggestions for archaeological modeling. In *Early and Middle Archaic Cultures in the Northeast*, edited by D.R. Starbuck and C.E. Bolian, pp. 73–96. Occasional Publications in Northeast Anthropology, No. 7. George's Mill, N.H.

Thorson, R.M., and Hamilton, T.D. (1977). Geology of the Dry Creek site: A stratified Early Man site in interior Alaska. *Quaternary Research* **7**:149–176.

Thorson, R.M., and McBride, K.A. (1988). The Bolton Spring site, Connecticut: Early Holocene human occupation and environmental changes in southern New England. *Geoarchaeology* **3**:221–234.

Thorson, R.M., and Schile, C.A. (1995). Deglacial eolian regimes in New England. *Geological Society of America Bulletin* **107**:751–761.

Thorson, R.M., and Webb, R.S. (1991). Postglacial history of a cedar swamp in southeastern Connecticut. *Paleolimnology* **6**:17–35.

Thwaites, B. (1960). *Incompressible Aerodynamics*. Oxford University Press, Oxford.

Van Andel, T.H. (1981). Consider the incompleteness of the geological record. *Nature* **294**:397–398.

Van Nest, J. (1993). Geoarchaeology of dissected loess uplands in western Illinois. *Geoarchaeology* **8**:281–311.

Van Nest, J. (2002). The good earthworm: How natural processes preserve upland Archaic archaeological sites of western Illinois, U.S.A. *Geoarchaeology* **17**:53–90.

Villa, P. (1982) Conjoinable pieces and site formation processes. *American Antiquity* **47**:278–290.

Webb, R.S., Anderson, K.H., and Webb, T. III. (1993). Pollen response-surface estimates of late-Quaternary changes in the moisture balance of the Northeastern United States. *Quaternary Research* **40**:213–227.

Wells, G.L. (1983). Late-glacial circulation over North America revealed by aeolian patterns. In *Variations in the Global Water Budget*, edited by A. Street-Perrott, M. Bernan, and R. Ratcliff, pp. 317–330. Reidel Publishers, Boston.

Wood, W.R., and Johnson, D.L. (1978). A survey of disturbance processes in archaeological site formation. In *Advances in Archaeological Method and Theory, Vol. 1*, edited by M.B. Schiffer, pp. 315–381. Academic Press, New York.

Wright, J.V. (1972). The Knechtel I Site, Bruce County, Ontario. National Museum of Man, National Museums of Canada, Ottawa.

CHAPTER 6

LIFE IN A POSTGLACIAL LANDSCAPE: SETTLEMENT-SUBSISTENCE CHANGE DURING THE PLEISTOCENE-HOLOCENE TRANSITION IN SOUTHERN NEW ENGLAND

Brian D. Jones and Daniel T. Forrest

Two recently discovered sites on the Mashantucket Pequot Reservation in Connecticut document the effects of the Pleistocene-Holocene transition on the region's early hunter-gatherer population. The adjacent sites lie along the margins of a glacial lake basin that today holds the Great Cedar Swamp (Figure 6.1). Both sites share approximate elevations, distance to fresh running water, and proximity to the wetland. Both rest within deep sandy soils deposited during the retreat of glacial ice ca. 15,000 years ago. The sites differ markedly, however, in size, duration of occupation, raw material use, and degree of reoccupation. We propose that these differences reflect divergent foraging strategies—each an effective response to the unique ecologic conditions prevailing at the time of occupation.

The Hidden Creek site is a small, short-term Late

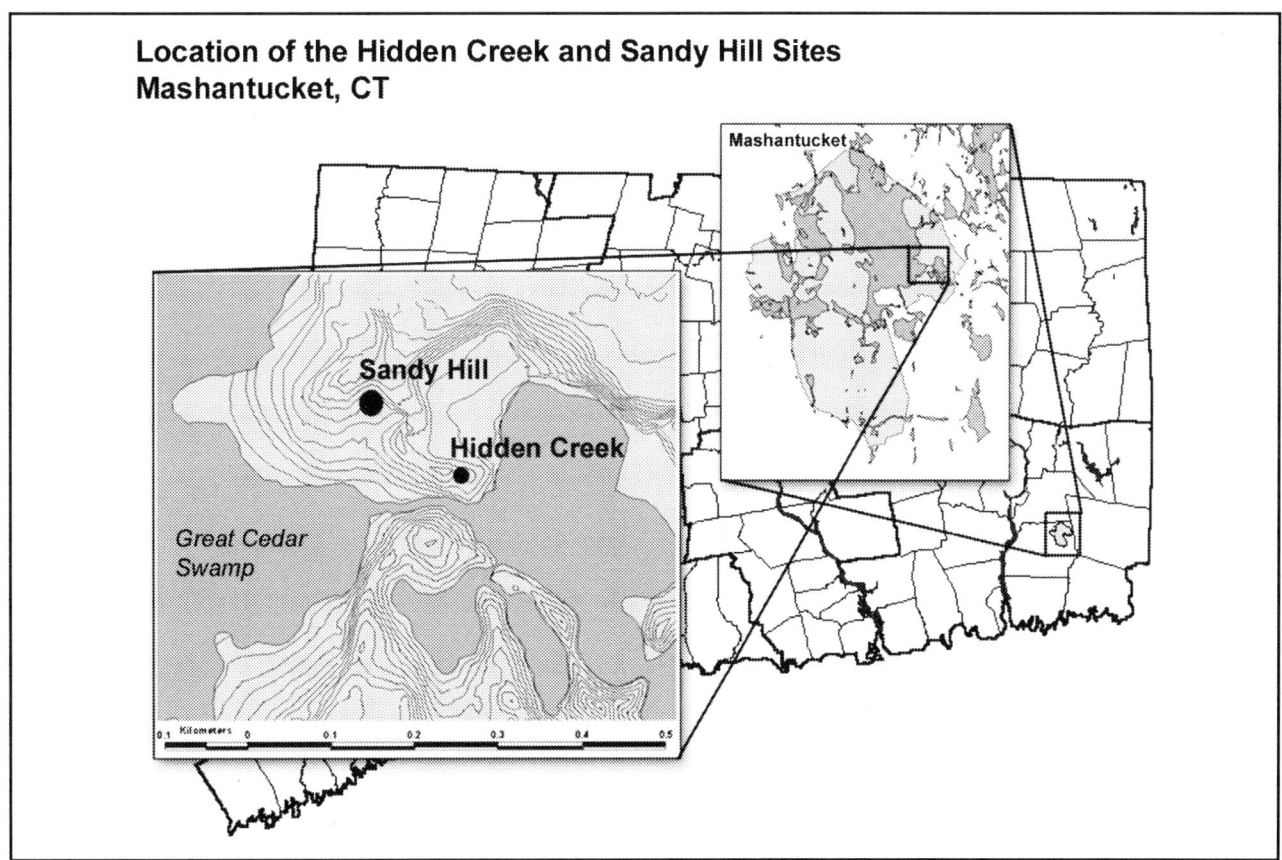

Figure 6.1. Location of the Hidden Creek and Sandy Hill sites on the Mashantucket Pequot Reservation, Ledyard, Connecticut.

Geoarchaeology of Landscapes in the Glaciated Northeast edited by David L. Cremeens and John P. Hart. New York State Museum Bulletin 497. © 2003 by the University of the State of New York, The State Education Department, Albany, New York. All rights reserved.

Paleoindian foraging camp used during the Pleistocene-Holocene transition, between ca. 10,250 and 9,150 radiocarbon years ago. Sandy Hill is a large Early Archaic seasonal base camp repeatedly occupied between 9,000 and 8,500 radiocarbon years ago. The period during which these two adjacent sites were occupied was marked by dramatic changes in both climate (e.g., Jacobson et al. 1987) and the local environment (McWeeney 1994a, 1994b; Thorson and Webb 1991). During this time the Great Cedar Swamp was shifting from a shallow, predominantly open water body to a diverse wetland that included forested swamp, marsh, and open water microhabitats (McWeeney 1994b). Evidence from the sites indicates that these changes coincided with a dramatic reorganization of settlement and subsistence patterns within the local watershed and adjacent areas. In particular, adaptation to the developing postglacial landscape appears to have resulted in changes in patterns of site reuse, duration of occupation, group size, residential mobility, and the lithic economy.

This chapter examines differences between Hidden Creek and Sandy Hill in terms of the prevailing environmental conditions during their periods of occupation. We suggest that the Cedar Swamp wetland likely offered a variety of plant and animal food resources attractive to hunter-gatherers during both periods. The predictability of these resources may have differed markedly during the two periods of site use, however. Thus resource predictability, rather than abundance, may have been a key limiting factor affecting prehistoric land use during the terminal Pleistocene when Hidden Creek was occupied. The data from Sandy Hill, however, suggest that resources were both predictable and abundant enough to support both long-term and redundant site use after 9,000 radiocarbon years ago. Anthropogenic disturbance to the local environment during the intensive Early Archaic occupation of Sandy Hill certainly resulted in an altered landscape. Several plant species identified within the Sandy Hill botanical assemblage may have benefited from these anthropogenic changes, both indirectly through nutrient influx and edaphic disturbances and directly through selective harvesting. Recent research suggests that the initiation of horticulture and plant domestication is complex—the result of long-term process rather than instantaneous invention. Sites such as Sandy Hill prompt us to look for the roots of this process in evermore remote parts of the past.

ENVIRONMENTAL SETTING

Cedar Swamp is located in southeastern Connecticut on the Mashantucket Pequot Reservation roughly between the cities of Norwich and Groton. The swamp covers an area of 1.9 km^2 within a glacially eroded bedrock basin. The modern vegetation surrounding the swamp comprises a mixed hardwood forest dominated by oak, hemlock, and pines. The wooded swamp itself consists primarily of red maple, white pine, hemlock, and white cedar with abundant shrub growth including dense stands of rhododendron (Thorson and Webb 1991:19). Environmental and geological studies have focused on the Cedar Swamp wetland basin since the mid-1980s (Thorson 1993; Thorson and McWeeney n.d.; Thorson and Webb 1991; McWeeney 1994a, 1999).

Detailed palynological, plant macrofossil, and sediment analyses have documented environmental and climate change at Cedar Swamp since deglaciation. Pollen stratigraphy and macrofossils indicate the establishment of a tundra environment that included sedge, dryas, dwarf birch, herbaceous willow, and bilberry between ca. 15,000 and 13,000 B.P. (McWeeney 1994a; Thorson and Webb 1991). During this phase, a deep proglacial lake impounded by a large recessional moraine to the south filled the Cedar Swamp basin. The retreating glacial front left a series of ice contact deltas between 50 and 60 m of elevation near the basin margins (Thorson and McWeeney 1994). As the ice continued to retreat to the north, a western outlet was exposed ca. 12,700 B.P., causing an abrupt drop of ca. 15 m in lake level. Pollen and macrofossils postdating the drainage of the lake indicate a boreal environment dominated by a developing open parkland of spruce, fir, larch, white pine, and birch. Wetland species within and bordering the pond included sedges, grass, pond weed (*Potamogetan*), naiads *(Najas)*, and sphagnum (McWeeney 1994a).

Between 12,000 and 10,000 years ago, the shallow pond began to develop into a vegetated swamp. The transition occurred first at the southern end of the basin and progressed slowly northward, in response to postglacial isostatic uplift (Thorson and McWeeney n.d.). The result was a complex mixture of microenvironments ranging from open water and marsh to thickly vegetated shrubby bog. Cattail (*Typha*), sweetgail (*Myrica*), and water lily (*Nuphar*) appear at this time within the basin. Regional pollen studies suggest that after 12,000 years ago the terrestrial parkland included more temperate deciduous elements such as ironwood, oak, *Juglans* (walnut family), ash, beech, and even chestnut, albeit in very low numbers (McWeeney 1994a:86–88, 1994b). About 11,000 years ago a brief increase in alder, birch, and spruce likely indicates a return to cooler climatic conditions during the Younger Dryas (see pollen profiles in Thorson and Webb 1991). This transition is marked in the Cedar Swamp cores by an abrupt return to inorganic sedimentation most likely attributable to a higher water table within the basin (Thorson and McWeeney nd:12). This is in accord with other regional studies that indicate fluctuating water levels during this very unstable climatic phase (Mayewski et al. 1993). After 9,000 B.P., warmer, drier conditions prevailed, and cores at Cedar Swamp show the development of a thick zone of woody peat.

Between ca. 10,000 and 8000 B.P., the Cedar Swamp basin contained a complex mosaic environment. The terrestrial flora consisted of pine and oak, with lesser quantities of larch, birch, hemlock, and heath. The southern swamp basin was likely thickly overgrown with shrubs and water-tolerant trees that provided good cover and browse for both large and small game animals. The center of the swamp was marshy and offered a broad range of wetland resources, including cattail and other emergent species. The northern edge of the swamp continued to hold open water as indicated by water lily seed macrofossils dated to ca. 10,000 and 9,000 years ago (McWeeney 1994a). There is some evidence that regional water tables rose between 10,000 and 9,000 years ago but may have begun to drop or fluctuate thereafter (e.g., McWeeney 1999:8).

Sometime shortly after 8,000 B.P., conditions in the Cedar Swamp basin changed markedly. Between 7,500 and ca. 5,000 years ago, all cores show a level of decomposed peat and charcoal associated with the onset of warm-dry hypsithermal conditions and a lowered water table (McWeeney 1994a; Thorson and Webb 1991). Similar indicators of mid-Holocene lowered regional water tables are in evidence in the Northeast as a whole (e.g., Webb et al. 1993). A lowered water table would have transformed the prior complex mosaic wetland into a brushy meadow. This simplification of the basin's dominant landform is presumed to have resulted in a lowered diversity and abundance of resources useful to humans at this time (Jones 1999). Only after about 5,000 years ago did local water tables rise and the basin regain its wetland character. Since that time a forested bog fed primarily by rainwater (Thorson and Webb 1991) has characterized the Great Cedar Swamp at the heart of Mashantucket.

HIDDEN CREEK

The Late Paleoindian Hidden Creek site (72-163) is located on a small sandy terrace in the southeast corner of Cedar Swamp (Jones 1997, 1998). The site likely dates to the close of the Pleistocene, ca. 10,000 radiocarbon years ago. Precise dating of the site has proven difficult, however. An accelerator mass spectrometry (AMS)-dated carbonized hazelnut shell fragment has provided an age estimate of 10,260 ± 70 B.P. (Beta-126817), whereas a carbonized cattail root fragment returned an assay of 9150 ± 40 B.P. (Beta-149920). Both botanical fragments lay in close association with each other and Paleoindian artifacts from the site. Technological aspects of the tool assemblage suggest a date close to 10,000 radiocarbon years ago based on comparisons to related sites.

At the time of occupation, the Cedar Swamp basin contained a wetland system in a state of rapid transition. The southern end of the basin likely held a wooded swamp while water lily and cattail were available from the swamp's center north (McWeeney 1994b:Figures 3-6). Terrestrial vegetation consisted primarily of pine, spruce, birch, and alder, with some larch, oak, fir, and hazel as well. The climatic instability of the terminal Pleistocene is well documented (Mayewski et al. 1993; Peteet et al. 1990). The period marking the transition from the close of the Younger Dryas to the onset of the Holocene was characterized by especially pronounced vegetational changes (Jacobson et al. 1987). Shifts in vegetation and water table elevation at Cedar Swamp could likely be measured on time scales of human generations or even years. These fluctuations likely reduced resource predictability for mobile human foragers.

Hidden Creek is a small site; most of the assemblage was recovered from a 3- by 4-m area (Figure 6.2). The site was nevertheless rich in artifacts, containing approximately 60 Paleoindian tool fragments and more than 4,000 waste flakes. Artifacts were concentrated roughly 50 cm below the surface, buried by sediments composed primarily of extremely fine-grained redeposited aeolian sands. Bioturbation and other pedogenic processes resulted in as much as 50 cm of vertical displacement of some small artifacts. More detailed site stratigraphy and geomorphology are discussed in Jones (1998).

Thorson (1996) applied data from the Hidden Creek site to his model of slope decline processes. He hypothesized that the distribution of artifacts from Hidden Creek might be explained through natural, rather than cultural, processes. Thorson's model was unable to explain the nonuniform clustering of artifacts observed at the site. In his conclusion Thorson (1996:31) states, "The diffusion model predicts a nonclustered distribution similar to the matrix cells for concave slope. Some other factor must be involved." In fact, the nonuniform horizontal clustering of tool, debitage, and material types suggest primarily human rather than natural site formation processes (Jones 1997, 1998:200-202). Although natural agents have acted to alter the original distribution of artifacts to a notable degree, Jones' interpretation is that this disturbance is limited to less than a meter in vertical and horizontal distance. This disturbance has blurred, but not destroyed, original cultural patterning.

Tools recovered from the site include spear points

Figure 6.2. View of the northwest corner of the 1994 excavation block at the Hidden Creek site, 0–80 cm below ground surface.

broken in use and manufacture, end scrapers, side scrapers, utilized flakes, and expedient choppers. Lithic raw materials are dominated by gray-green cherts of probable Hudson Valley origin (Normanskill Shale Formation). Smaller quantities of locally available quartz and quartzite were used for rough chopping and scraping tasks. The debitage recovered from the site could have been produced in a matter of hours by one or two experienced knappers. The dozen discarded and heavily used end scrapers indicate a longer span of occupation, however. The number of artifacts and their distribution together suggest that a family-sized group of people used the site for a short period of time, perhaps less than a week.

The site's occupants probably settled where large game, such as moose, elk, woodland caribou, or deer, was known to be present, but were likely dispersed in poorly predictable patch locations. Environmental reconstruction suggests that the site offered access to a variety of small game and plant resources as well. Recent microscopic plant identifications by David Perry indicate the presence of cattail (*Typha*), water plantain (*Alisma*), possible water lily (*Nymphaeaceae*), hickory (*Carya*), ground nut (*Apios*), beet family (chenopodiaceae), Indian cucumber (*Medeola*), Common Yellow Cress (*Rorippa*, likely *palustris*), as well as fern (*Dryopteris*) and clubmoss (*Lycopodium*) (Table 6.1; Perry 2000; for a discussion of Perry's methods, see Perry 1997, 1999 and Hather 1991, 1993). Such broad-based forager (sensu Binford 1980) subsistence economies are best described as opportunistic. Resource exploitation was probably focused on a diverse assortment of locations, each selected through a combination of extensive resource monitoring and anticipated cycles of resource abundance. It should be noted that the subsistence economy and settlement organization interpreted from the remains at the Hidden Creek site reflect a temporary seasonal adaptation to local resource conditions. The settlement-subsistence system at other times of the year might have been quite different.

Table 6.1. SEM Carbonized Plant Identifications from Hidden Creek and Sandy Hill

Taxa	Common Name	Number of Specimens		Possible Uses
		Hidden Creek	Sandy Hill	
Sparaganium spp.	Bur-reed		4	Food, Medicine
Typha spp.	Cattail	3	25	Food, Matting
Alisma plantago-aquatica	Water Plantain	1	3	Food
Saggitaria sp.	Arrowhead		2	Food, Medicine
Chenopodiaceae	Beet-spinach-chenopod family	3		Food
Cyperus esculentus	Yellow Nutsedge (also chufa or nut grass)		14	Food, Medicine
Apios	L. Groundnut	1		Food
Scirpus L.	Bulrush		10	Food, Matting, Basketry
Rorippa	Common Yellow Cress	1		Food
Calla L.	Wild Calla		2	Medicine, Poison
Medeola L.	Indian Cucumber	2	2	Food, Medicine
Polygonatum Mill.	Solomon's Seal		2	Food, Medicine
Iris L.	Blue Flag	1	2	Medicine, Cordage
Nymphaea L.	White Water-Lily	2?	4	Food, Medicine

Note. All identification made by David Perry (Perry 1998, 2000).

The distant source of the cherts left at this camp and the dependence on bifacial tools (Andrefsky 1998:150–153) suggest a high degree of residential mobility. The site occupants had probably last quarried stone materials along the Hudson River valley, and they may have acquired other stone types from even greater distances afield. They refurbished damaged hunting gear and processed recently killed game and other foods before moving on. Jones suggests that the occupants of the site used this part of southern New England sporadically (rather than regularly), and that the site lay within the southern extension of an annual foraging territory focused primarily on more northern latitudes (Jones 1997, 1998). The lack of comparable Paleoindian residential sites around the Cedar Swamp basin suggests that resources in the basin were either not rich or predictable enough to support redundant occupations at this time. The paleoenvironmental reconstruction of Cedar Swamp indicates that although resources were potentially abundant and varied, they may have shifted rapidly and unpredictably throughout the Pleistocene-Holocene transition. Thus resource *predictability* may have been the key limiting factor to Paleoindian use of the Cedar Swamp basin during this period of dramatic climatic and environmental change.

SANDY HILL

Lying just 150 m north of Hidden Creek, the Sandy Hill site (72-97) is one of the most intriguing sites recently discovered on the reservation (Forrest 1999). The site is located along the southwestern margin of a broad delta that prograded during the high-lake-level phase between 15,000 and 12,700 B.P. A rich Early Archaic component of the site is associated with numerous quartz cores, scrapers, and abundant quartz knapping debris. The notable lack of bifaces is typical of the Gulf of Maine Archaic tradition (Robinson and Petersen 1993). Pieces of ground hematite and graphite, ground stone tool fragments, and numerous tabular "choppers" were also recovered. The latter might have been used in part for plant food processing tasks and perhaps digging. The quartz used by the makers of these tools was available locally from stream cobbles, as well as from nearby outcrops at Lantern Hill, once the location of a large silica mine. Typologically the site compares well to others of the Gulf of Maine Archaic tradition (Petersen and Putnam 1992; Robinson 1992; Robinson and Petersen 1993; Maymon and Bolian 1992). At the time of occupation the Cedar Swamp basin contained a complex mosaic wetland system composed of wooded swamp, marsh, and open-water microhabitats. Water lily and cattail were particularly abundant in the northern open-water area of the wetland (McWeeney 1994a).

A pine-oak-hemlock forest with abundant heath growth dominated the terrestrial landscape.

In the fall of 1996, archaeological investigations at the site were undertaken by the Public Archaeology Survey Team, Inc. (Forrest 1999). During machine-assisted stripping of topsoils along the base of a steep sandy hillside in the eastern portion of the site, a broad area of black sandy sediments was exposed directly overlying undisturbed interbedded glaciodeltaic sands (Figure 6.3). Archaeological excavation of the black sands yielded an abundance of quartz unifacial tools, quartz debitage, and charred botanical fragments, including large numbers of hazelnut (*Corylus* sp.) shell fragments, suggesting the sediments were of cultural origin. Approximately 10 g of charred hazelnut shell fragments were collected near the base of the black sands, just south of the machine excavation and submitted for a conventional radiometric date (8920 ± 100 B.P., Beta-102564). Examination of the machine-cut edge along the base of the hillside suggested the black sands extended into the slope beneath an uneven layer of colluvium capping the stratigraphic sequence. A series of 1-m-wide trenches were excavated into the hillside to provide more detailed stratigraphic observations and to increase the archaeological sample.

The hillside trenches exposed a series of large intersecting features, ranging in color and texture from black loamy fine sand (10YR2/1) to light olive brown sandy gravels (2.5Y5/4) between the underlying glaciodeltaic sands and the colluvium (Figure 6.4). Formation of the features through natural processes has been ruled out. Fluvial erosion of the slope edge is highly unlikely as there are no streams or other surface waters in the immediate vicinity. Extensive excavations in the area did not expose any buried channel deposits or other Holocene-age fluvial sediments in the area. Localized rotational slumping of the slope edge was also ruled out as no slip planes were present along the feature boundaries or within the underlying deltaic sediments. Finally, wind erosion of the slope face was also discounted, as

Figure 6.3. Black sand features at site 72-97 exposed by a machine cut. Excavated trenches extend up to 7–m upslope into the hillside.

the paleoenvironmental record for the basin suggests that effective vegetational cover would have precluded such extensive sediment transport during the time the features were formed (McWeeney 1994a; Thorson and Webb 1991).

Examination of the stratigraphic sections by the authors and Dr. Robert Thorson of the University of Connecticut Department of Geology and Geophysics led to the conclusion that the features were the direct result of human activity. This conclusion is based on several observations. The sedimentary matrices of the features are characterized by very high artifact densities, often exceeding 1,000 artifacts/m^3. Large quantities of wood charcoal and charred plant fragments were also recovered from the features. A number of plant specimens recovered from the features were later identified as aquatic or emergent species (Perry 1998) that could not have grown on the excessively drained sands in the area after the initial proglacial lake drained ca. 12,700 B.P. AMS dates on identified wetland plant fragments all fall between 8750 and 8490 B.P. (Forrest 1999), minimally 4,000 years after the delta margin was exposed. The bedded deltaic sands at the base of the exposed section are clearly truncated by the deepest feature strata, indicating the original soils on the hillside had been removed prior to the accumulation of feature sediments. The darkest of the feature strata are generally level and extend over several meters in north-south section. The positions and orientation of these distinctive stratigraphic units suggest the features were formed by repeated, episodic excavations into the slope face, starting at the base and working upslope through time.

Further excavations undertaken between 1998 and 2001 of the area immediately south and west of the hillside trenches exposed similar features below the slope edge. Horizontal block excavations of one feature revealed a large stain, measuring roughly 4 by 5 m, with the long axis oriented north-south (Figure 6.5). A possible hearth and several post molds were identified within the feature. These features are currently interpreted as the remains of semisubterranean pit houses dug into the slope to take advantage of the favorable solar orientation of the hillside and insulating properties of the earth (Forrest 1999). This type of habitation may have been favored during the early Holocene as a specific adaptation to significantly colder winter temperatures characterizing this period (Kutzbach 1987:426, McWeeney 1999:8).

Five concentrations of overlapping pit houses have been tentatively identified based on the presence of quartz concentrations and distinctive black sandy deposits. Areas examined to date appear to contain multiple house floors. Depressions in the floors are associated with concentrations of charcoal and may represent small hearths, posts, or other locations of cultural activity that accumulated anthropogenic sediments. A series of dates on other floor deposits suggests repeated occupation spanning the first half of the ninth millennium B.P. (Table 6.2). The oldest date from the site, 9340 ± 60 B.P. (Beta-122013), was obtained from a small fragment of unidentified wood charcoal and may not accurately reflect the age of the Early Archaic occupations at Sandy Hill (Forrest 1999:90). Several Late Paleoindian tools were recovered from within the pit house feature matrix in close proximity to the wood sample, including a small chert scraper with a graving spur and a chert Holcombe-like point base (Forrest 1999:Figure 3). These are most likely associated with an earlier site component that was disturbed by the construction of pit houses during the Early Archaic.

Soil profiles indicate that the deepest pit houses were buried by similar features higher in the sequence, sometimes separated by a clean layer of sandy fill, indicating reoccupation of the same location. The five expected structure localities may in fact represent well over double the number of individual habitation structures. The size and depth of the pit houses suggest a substantial investment of human labor in their manufacture and maintenance. The amount of cultural debris recovered (more than 100,000 pieces of quartz debitage are estimated from existing excavations) also suggests relatively long-term occupation, as might occur over a season or more (Forrest 1999). Several factors suggest the pit houses were occupied during the winter. These include the recovery of hazelnut and cattail (foodstuffs well suited to storage), the concentration of debitage within the pit houses as opposed to the area surrounding the structures, and their location on a south-facing hillside. There are also indications that site use may have extended into the warmer months. Identified plant remains include several species available during the summer and fall. Table 6.1 summarizes plant identifications recently made by D. Perry (1998; Forrest 1999). These clearly indicate a robust foraging economy that used a variety of local wetland and terrestrial plant species. The relative contribution of animal resources to the diet of the people living at Sandy Hill is currently difficult to assess. Fragments of calcined bone were recovered from nearly all excavated areas, but their numbers are relatively small. Unfortunately the excessively drained character of the glacio-deltaic sands underlying the site is not conducive to bone preservation. Thus it is quite possible that the slight faunal assemblage reflects taphonomy, not economy. Further evaluation of animal exploitation awaits completion of the faunal analysis, now underway.

The extensive Early Archaic component at Sandy Hill suggests repeated, multiseasonal use of this location

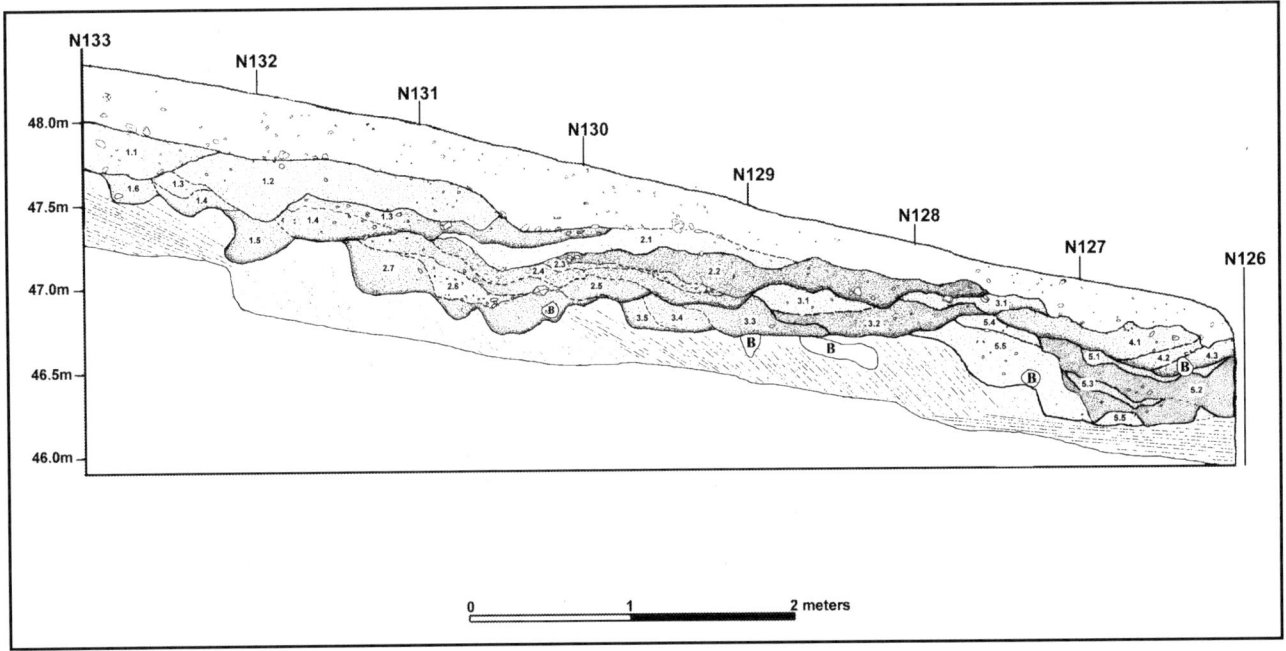

Figure 6.4. North-south section along the West 2-m line at the Sandy Hill site showing complex overlapping black sand feature strata.

Soil Descriptions:

Holocene
Colluvium: Dark yellow brown to yellow brown medium to fine sand with gravel (clasts up to 5 cm in diameter), small cobbles up to 10 cm in diameter more common near interface with feature strata. Base of unit irregular in morphology suggesting uppermost feature strata have been eroded—10YR6/6 to 6/8.

1.1: dark yellow brown slightly loamy fine sand with gravel (clasts up to 6 cm in diameter)—10YR4/6.

1.2: dark yellow brown slightly loamy fine to medium sand with small gravel (clasts < 3 cm in diameter)—10YR4/6.

1.3: dark yellow brown fine foamy sands with gravel and diffuse wood charcoal fragments (clasts < 4 cm in diameter)—10YR3/6.

1.4: very dark brown fine sandy loam with some gravel and large quantity of microdivided wood charcoal (clasts < 3 cm in diameter)—10YR2/2.

1.5: dark yellow brown fine sandy loam—10YR4/6.

1.6: dark yellow brown fine slightly loamy sand with fine gravels (clasts < 2 cm in diameter)—10YR4/6.

2.1: dark yellow brown fine to medium slightly loamy sands with coarse gravels (clasts up to 5 cm in diameter)—10YR4/6.

2.2: very dark yellow brown fine loamy sand with scattered gravel and wood charcoal (clasts up to 6 cm in diameter)—10YR3/3.

2.3: dark yellow brown fine slightly loamy sand with fine gravels and wood charcoal (clasts < 1 cm in diameter)—10YR4/4.

2.4: dark yellow brown fine slightly loamy sand with small gravel and scattered wood charcoal (clasts up to 2 cm in diameter)—10YR4/4 to 4/6.

2.5: dark yellow brown fine slightly loamy sand with gravel and scattered wood charcoal (clasts up to 3 cm in diameter)—10YR4/4.

2.6: very dark gray brown medium to coarse gravelly sand with scattered wood charcoal and moderate quantity of microdivided charcoal (clasts up to 6 cm in diameter)—10YR3/2.

continued

Figure 6.4. *continued*

2.7: very dark brown slightly loamy fine to medium sand with scattered wood charcoal and abundant microdivided charcoal—10YR2/2.

3.1: yellow brown fine to medium sand with fine gravel (clasts up to 2 cm in diameter)—10YR5/8.

3.2: dark yellow brown fine slightly loamy sand with scattered fine gravel (clasts < 2 cm in diameter) and wood charcoal, diffuse microdivided charcoal—10YR4/6.

3.3: dark yellow brown fine sand – very few pebbles or granules, diffuse microdivided charcoal—10YR4/6.

3.4: dark yellow brown fine to medium sand with fine gravel (< 3 cm in diameter) and wood charcoal concentrated at base—10YR4/4 to 4/6.

3.5: light olive brown fine sand with scattered wood charcoal, no gravel—2.5Y5/4.

4.1: yellow brown fine slightly loamy sand with gravel concentrated to south (clasts up to 3 cm in diameter)—10YR5/6.

4.2: dark yellow brown fine loamy sand with some fine gravel and wood charcoal (clasts < 2 cm in diameter)—10YR4/4.

4.3: dark yellow brown fine slightly loam sand and fine gravel (clasts < 2 cm in diameter), scattered wood charcoal near base—10YR4/6.

5.1: very dark gray brown fine slightly loam sand with scattered very fine gravel (clasts < 0.5 cm in diameter)—diffuse microdivided charcoal—10YR3/2.

5.2: very dark brown fine sandy loam with some gravel concentrated near center of unit, diffuse microdivided charcoal and wood charcoal fragments—10YR2/2.

5.3: dark yellow brown fine sandy loam with common, distinct mottles of very dark yellow brown—no gravel—10YR4/4 to 2/2.

5.4: dark yellow brown slightly loamy sand with fine gravels and scattered wood charcoal fragments (clasts < 2 cm in diameter)—10YR4/6.

5.5: yellow brown fine sand with scattered granules (clasts < 2 cm in diameter)—10YR5/6.

Glaciodeltaic
Sands: finely bedded to interbedded fine to medium sands—no gravel. Grain size appears to increase to the north. Beds inclined downward to south between N133 and N128—possible deltaic foresets. Subhorizontal beds between N128 and N126 may be remnant bottom sets. Very light gray to light olive gray (2.5Y7/1 to 2.5Y7/3) alternating with olive gray and light red brown (2.5Y5/6, 5YR5/8).

9,000 to 8,500 years ago. The data indicate a higher level of social and economic complexity and settlement sedentism than was formerly anticipated for this period. The Gulf of Maine Archaic in northern New England is associated with a rich mortuary tradition (Robinson 1996), suggesting stronger bonds between local populations and specific geographic areas. Interestingly, a possible rise in socioeconomic complexity does not appear to be related to *regional* population pressure as is most often the case (Kelly 1995:302–321). Although there are significant recognition issues hindering the identification of many Early Archaic sites in the Northeast (e.g., Robinson and Petersen 1993), sites such as Sandy Hill are far too uncommon in the Northeast to suggest substantial population increase at this time. Rather, the reorganization of the early Holocene environment may have encouraged the concentration of populations in ecologically favorable settings (Nicholas 1988). It is this increase in *local* population density that may have fostered subsistence intensification, increased sedentism and possibly a diversification in social roles and behaviors. Although it is unclear how much time lapsed between reoccupations, the fact that the site *was* reoccupied on multiple occasions strongly suggests that resources in the Cedar Swamp basin were both abundant and predictable at this time.

This contradicts previous models of early Holocene settlement that went as far as to suggest regional emigration at this time (e.g., Fitting 1968; Ritchie 1965:16–19, 1979). These models were grounded in the belief that the

Figure 6.5. Excavation of a pit-house feature at the Sandy Hill site during the 2000 field season. Black feature strata were exposed at 20 cm below the machine-stripped surface and extended to 60 cm. The 1995 test pit that first identified this feature is visible left of the balk.

resources of the early Holocene were too poor to support significant human populations. The recovery of carbonized nutshell, cattail, and a variety of other plant food resources helps in part to explain the discrepancy with previous models of Early Archaic settlement in the Northeast. Nut fruits and tubers provide a very rich (high-caloric), predictable, and abundant food source that can be stored after processing. The fact that many of the plant species recovered thrive in disturbed environments suggests that a feedback loop may have been established after initial settlement which promoted reoccupation. The focused harvest of cattail, bulrush, hazelnut, and nutsedge in particular may have encouraged aggressive vegetative regrowth of plant stands in the following season. The exploitation of trees and shrubs for firewood may also have played an important role in changing the composition of the local forest. The large quantities of hazelnut recovered from the Early Archaic features at Sandy Hill indicate that these trees represented a valued resource. Simply sparing hazelnut from the firewood harvest would have reduced the populations of competing species and encouraged the spread of this economically important shrub.

DISCUSSION

The above data indicate striking differences in the use of the Cedar Swamp basin by hunter-gatherers between 10,000 and 9,000 years ago. The differences are primarily expressed in the scale of foraging and residential activities, and they appear to relate at least in part to the character of the local resource base. At the close of the Pleistocene, ecological communities were in a stochastic transitional state. The most profound effect of the dynamic climate and environment is assumed to have been one of lowered resource predictability, despite intermittent periods of potentially high yields. There is adequate evidence in the geological, palynological, and macrobotanical records, as in the archaeological remains of

Table 6.2. Radiocarbon Dates from the Hidden Creek and Sandy Hill Sites

Site Name	Site No.	Lab Number	Measured ^{14}C Age	$^{13}C/^{12}C$ Ratio	^{13}C Adjusted Age	Dated Carbonized Material
Sandy Hill	72-97	Beta-113499	8490 ± 60	-24.10	8510 ± 60	*Typha* sp.
Sandy Hill	72-97	Beta-162837	8570 ± 60	-25.10	8570 ± 60	*Corylus* sp.
Sandy Hill	72-97	Beta-162872	8630 ± 50	-23.40	8660 ± 50	*Corylus* sp.
Sandy Hill	72-97	Beta-126812	8660 ± 60	-26.60	8640 ± 60	Unid. plant tissue
Sandy Hill	72-97	Beta-126816	8680 ± 60	-26.10	8660 ± 60	Unid. plant tissue
Sandy Hill	72-97	Beta-113498	8710 ± 60	-27.70	8670 ± 60	*Typha* sp.
Sandy Hill	72-97	Beta-102564	8920 ± 100	Est. -25.00	8920 ± 100	*Corylus* sp.
Sandy Hill	72-97	Beta-162920	8960 ± 40	-24.70	8960 ± 40	Unid. plant tissue
Sandy Hill	72-97	Beta-122014	9020 ± 60	-25.80	9000 ± 60	*Corylus* sp.
Sandy Hill	72-97	Beta-122013	9340 ± 60	-25.20	9340 ± 60	Unid. wood
Hidden Creek	72-163	Beta-121846	9160 ± 50	-25.60	9150 ± 50	*Pinus* sp.
Hidden Creek	72-163	Beta-149920	9190 ± 40	-27.20	9150 ± 40	*Typha* sp.
Hidden Creek	72-163	Beta-126817	10250 ± 70	-24.60	10260 ± 70	*Corylus* sp.

Note. All represent AMS dates with the exception of Beta-102564.

charred plant food remnants from this period to suggest the presence of a diverse and abundant food base. Nevertheless, Hidden Creek is the largest Paleoindian camp known at Mashantucket and in eastern Connecticut as a whole. Much larger Paleoindian sites are well documented in most other New England States, suggesting that the resource base elsewhere (and at other times during this period) was able to support larger residential group sizes, longer encampments and probably episodic site reoccupation.

Jochim (1991, 1996) argues that resource predictability is critical to understanding hunter-gatherer settlement organization over periods longer than seasons. Where resource predictability is low, site reuse should be uncommon, and intersite assemblage variation high as foragers follow an opportunistic subsistence pattern. Jochim emphasizes the importance of diversity in human behavioral response to rapid climatic and environmental changes (Jochim 1991:308), a point that has also been made regarding northeastern Paleoindians by Curran (1987). From the perspective of evolutionary ecology, variability in the environment is strongly correlated with behavioral flexibility (Baldwin and Baldwin 1981:65; Jochim 1991:311). Pianka states:

> ... in a temporally varying environment, selective pressures vary from time to time and the phenotype of highest fitness is always changing. There is inevitably some lag in response to selection, and organisms adapted to tolerate a wide range of conditions are frequently at an advantage ... Indeed, in unpredictably changing environments, reproductive success may usually be maximized by the production of offspring with a broad spectrum of phenotypes ... (Pianka 1995:130)

Socially, small human groups with fewer decision makers can express more behavioral flexibility than larger groups (Jochim 1991; Kelly 1995:210–213), suggesting that during periods of rapid environmental reconfiguration, smaller, highly mobile foraging groups were at an advantage. Hidden Creek appears to represent an excellent example of the material traces left by a small group of terminal Pleistocene opportunistic foragers with the competence to survive in a particularly dynamic environment. That the occupants of the site were capable of such behavioral flexibility suggests that the settlement and subsistence organization expressed at Hidden Creek were already in the repertoire of prior Paleoindian adaptive strategies. Had they not been, regional emigration to more familiar or stable environmental regions would have prevailed. Thus the environment did not determine the pattern of behavior expressed at Hidden Creek; rather it limited the range of its potential expression.

By 9,000 years ago something very different was happening at Mashantucket. The rich archaeological remains at Sandy Hill indicate intensive, redundant site use over a period of at least 500 years. A reliance on local raw materials and the informal lithic toolkit are themselves suggestive of an increased degree of sedentism (Andrefsky 1998:213–214). Profound changes in social, residential, and subsistence practices occurred between the use of Hidden Creek and Sandy Hill. Often such notable social-settlement change is attributed to population pressure (e.g., Kelly 1995:302–305). The rarity of Early Archaic finds in the Northeast suggests that this was not likely a major contributing factor. How, then, is the data from Sandy Hill best explained?

A decade before the discovery of Sandy Hill, George Nicholas contributed a number of important articles concerning the potential early Holocene use of glacial wetland basins (Nicholas 1987, 1988, 1991). These papers represent a watershed in the way northeastern archaeologists came to view the environment of the early Holocene. Before Nicholas published his work, and despite increasing evidence to the contrary, the Ritchie-Fitting hypothesis remained the dominant explanation of the poor visibility of the Early Archaic in the Northeast (Fitting 1968; Ritchie 1965, 1979). Grounded in more recent palynological and ecological studies, Nicholas suggests that some early Holocene glacial lake wetland systems supported productive, predictable, and diverse resources that would have attracted the region's human foragers (Nicholas 1988). Nicholas also stresses the importance of land-use versus single-site studies, that is, studies of archaeological landscapes as a whole that might better express the variability of past social, economic, and subsistence systems (Nicholas 1988:262). He notes that land-use studies resolve questions concerning changes in social, economic, and subsistence patterns over time better than single-site studies (Nicholas 1988:264; see also Jochim 1991).

Nicholas emphasizes the dynamism of past northeastern environments and suggests that earlier models of hunter-gatherer relations with the environment were overly simplistic. This reflects a recent shift in ecological studies away from strict, community-based successional models toward a more dynamic understanding of the responses of individual taxa to the ever-changing environment (Nicholas 1988:265–266; Botkin 1991). In particular, Nicholas sees "glacial lake basin mosaic wetlands" such as Cedar Swamp as important resource areas during the early Holocene. He suggests that these habitats were more productive (in terms of net primary productivity and available biomass) and more stable than most nonwetland regions during the early Holocene. He sees glacial lake mosaic wetlands as capable of supporting a more diverse suite of resources than less heterogeneous upland or valley bottom locations (Nicholas 1988:268–270). Nicholas also provides an impressive list of potential wetland resources that includes plants and animals for food, utilitarian, and medicinal use (Nicholas 1991).

Based on the above reconstructions, Nicholas suggests that glacial lake mosaic wetlands were focal places in Early Archaic settlement, social, economic, and subsistence activities. He anticipated such core areas to have supported both redundant base camp locations, of possibly long-term and intensive habitation, as well as a variety of smaller peripheral support locations (Nicholas 1988:281–284). Nicholas summarized his modeled early Holocene glacial lake mosaic wetland land-use patterns and their resultant archaeological manifestations as follows (after Nicholas 1988:282).

Expected land-use patterns:
1. A general foraging subsistence pattern develops that is not limited to a small number of seasonally specific resources.
2. The wetland landscape system stabilizes and becomes physically distinct and culturally separable from other areas.
3. Wetland basins become places on the landscape as a result of their increased use as "core" areas of settlement, subsistence, economic, and social activities.
4. There is a shift toward more formal, structured, and redundant patterns of land use; core and periphery areas become more clearly defined.

Archaeological visibility:
1. Increased diversity in artifact assemblages or a more generalized tool kit as a response to the breadth of the resource base.
2. Presence of both base camps and special-function sites within a watershed due to a significant portion of the annual social and economic system being focused within or adjacent to the basin.
3. Presence of burials related to a long-term pattern of land use or similar commitment (e.g., territoriality) to this location on the landscape.
4. Presence of storage pits and caches, and other reused features reflecting an expected return to the sites within a relatively short period.
5. Changes in degree of site "openness" and other indicators of site structure reflecting shifts in site function, types of social activities, group size, and intra- or intergroup relations.
6. Diversity in seasonal representation and length of individual site occupation representing a longer and perhaps nonseasonally specific use of the basin.
7. Evidence of site reuse and of increased site density as use of the basin intensifies over time.
8. Increase in midden deposits representing longer and more intense use of sites.

Although these expectations were surprising to many archaeologists at the time of publication, it is remarkable how well they describe observations made at the Sandy Hill site. The missing elements (i.e., cemeteries and substantial storage features) may have more to do with the limits of excavation than their true absence, as neither would be a surprise if encountered. Cemeteries are, in fact, well documented elsewhere in New England as an established component of the Gulf of Maine Archaic tradition (Robinson 1992, 1996). Nicholas' expectation that large wetland basins became targeted resource areas, or "places on the landscape" during the early Holocene appears justified, as is his suggestion that this was primarily a result of their predictable and abundant resources in an otherwise low-productivity environment.

We would like to take Nicholas' model one step further and suggest that during this episode of relatively intensive seasonal site use, humans soon learned that the disturbance they caused to the environment had certain beneficial side effects. These included the promotion of some edible plant species such as cattail, nutsedge, and hazelnuts (Forrest 1999). Nicholas himself has noted that cattail thrives in disturbed environments (Nicholas 1999:34). Nutsedge, too, would benefit from opening the forest canopy in areas heavily used by humans (Forrest 1999). Forest fires, whether intentional or not, would have helped to maintain such open patches. Finally, hazel shrubs grow much more densely when regularly pruned, resulting in richer nut harvests. Early Holocene Mesolithic foragers of Europe are now known to have purposefully coppiced hazel shrubs at regular intervals to increase nut harvests and to gather the useful straight hazel wands that develop after cutting back the bush (T.D. Price and D. Perry, personal communication 2000).

The Early Archaic foragers who established Sandy Hill as their home for much of the year altered their environment in a way that potentially increased the harvest of both wetland and terrestrial food resources. Although potentially deliberate, this alteration of the local environment is assumed to have been predominantly inadvertent. Nevertheless, there can be no doubt that after more than 500 years of residence, the landscape around the Sandy Hill site, and perhaps much of the Cedar Swamp basin, was largely anthropogenic. The area surrounding the site was probably transformed by at least 500 years of repeated long-term occupation. The harvesting of local trees for firewood and wild fires, whether intentionally set or of natural cause, likely shifted the species composition and canopy density within the site vicinity. This would have opened patches within the surrounding woodlands creating sunny edge environments; walking paths must have been strung along the adjacent hills and waterways to connect gathering locations; deadwood was likely a rare commodity for miles around, because most had been consumed as fuel; the vegetation within and around Cedar Swamp itself had likely been changed by consistent human activities such as wading and selective plant harvesting; finally, cattail in particular would have benefited from the run-off of human waste and other phosphorous-rich camp byproducts. Unlike the light footprint of the visitors to the Hidden Creek site, the occupants of Sandy Hill had altered the landscape of Cedar Swamp for centuries to come.

CONCLUSIONS

This chapter has examined two adjacent archaeological sites created by some of the region's earliest hunter-gatherers. Despite superficial similarities in location, the Hidden Creek and Sandy Hill sites are remarkably different. The period separating their use was marked by dramatic changes in climate and the environment—arguably the most profound to occur since glacial retreat ca. 18,000 years ago. The occupation of the Hidden Creek site occurred during a phase of heightened climatic instability at the close of the Pleistocene. The small size of the site, the heavy reliance on distantly quarried lithic raw materials, and the broad-spectrum harvest of plant foods all indicate the presence of a small group of individuals following an opportunistic, highly mobile foraging strategy. We have argued that this subsistence and settlement organization represents a flexible adaptive strategy able to meet the demands of a rapidly shifting, poorly predictable resource base.

The Sandy Hill site was repeatedly occupied between 9,000 and 8,500 radiocarbon years ago. By this time, climatic conditions in the Northeast had become milder and less chaotic. Despite marked changes in floral composition, the Cedar Swamp wetland and adjacent uplands provided an array of plant food resources similar to that which had existed during the earlier occupation of Hidden Creek, as testified to by many parallels in the botanical remains from both sites. However, the scale and redundancy of site use at Sandy Hill indicate that the resource base was now abundant and predictable enough to support much more intensive human occupation. Much of what has been observed at Sandy Hill was anticipated a decade earlier by models of early Holocene wetland use by George Nicholas.

We have argued, following initial suggestions by Forrest (1999), that the focused use of the Sandy Hill site must have resulted in notable alterations to the natural environment. In particular, certain food resources such as cattail, hazelnuts, and nutsedge might have actually been promoted by human harvesting and other "disturbances" to the landscape. We hope this discussion will promote further thinking about the potentially complex relationships between human foragers and their environments. In particular, it is important to view such ecological relationships as an early aspect of the ties between humans and their plant food resources that would eventually lead to horticulture. We also hope that this paper raises awareness of the fact that the archaeology of the Northeast's earliest occupants can provide valuable information concerning such globally significant issues.

REFERENCES CITED

Andrefsky, W., Jr. (1998). *Lithics: Macroscopic Approaches to Analysis*. Cambridge University Press, Cambridge.

Baldwin, J.D., and Baldwin, J.I. (1981). *Beyond Sociobiology*. Elsevier, New York.

Binford, L.R. (1980). Willow smoke and dog's tails: Hunter-gatherer settlement systems and archaeological site formation. *American Antiquity* **45**:4–20.

Botkin, D.B. (1991). Rethinking the environment. *The Wilson Quarterly* **15**:60–72.

Curran, M.L. (1987). *The Spatial Organization of Paleoindian Population in the Late Pleistocene of the Northeast*. Doctoral dissertation, University of Massachusetts. University Microfilms, Ann Arbor, Mich.

Curran, M.L. (1999). Exploration, colonization, and settling in: The Bull Brook phase, antecedents, and descendants. In *The Archaeological Northeast*, edited by M.A. Levine, K.E. Sassaman, and M.S. Nassaney, pp. 3–24. Bergin & Garvey, Westport, Conn.

Cwynar, L.C., and Levesque, A.J. (1995). Chironomid evidence for late-glacial climatic reversals in Maine. *Quaternary Research* **43**:405–413.

Fitting, J.E. (1968). Environmental potential and the postglacial readaptation in eastern North America. *American Antiquity* **33**:441–445.

Forrest, D.T. (1999). Beyond presence and absence: Establishing diversity in Connecticut's early Holocene archaeological record. *Bulletin of the Archaeological Society of Connecticut* **62**:79–98.

Hather, J.G. (1991). The identification of charred archaeological remains of vegetative Parenchymous tissue. *Journal of Archaeological Science* **18**:661–675.

Hather, J.G. (1993). *An Archaeobotanical Guide to Root and Tuber Identification, Vol. I: Europe and Southwest Asia*. Oxbow Monograph 28, Oxford, England.

Jacobson, G.L., Webb, T., and Grimm, E.C. (1987). Patterns and rates of vegetation change during the deglaciation of eastern North America. In *The Geology of North America*, Vol. K-3. The Geological Society of America, Boulder, Colo.

Jochim, M.A. (1991). Archaeology as long-term anthropology. *American Anthropologist* **93**:308–321.

Jochim, M.A. (1996). Surprises, recurring themes, and new questions in the study of the late glacial and early postglacial. In *Humans at the End of the Ice Age: The Archaeology of the Pleistocene-Holocene Transition*, edited by L.G. Straus, B.V. Eriksen, J.M. Erlandson, and D.R. Yesner, pp. 357–363. Plenum Press, New York.

Jones, B.D. (1997). The Late Paleoindian Hidden Creek site in southeastern Connecticut. *Archaeology of Eastern North America* **25**:45–80.

Jones, B.D. (1998). *Human Adaptation to the Changing Northeastern Environment at the End of the Pleistocene: Implications for the Archaeological Record*. Doctoral dissertation, University of Connecticut. University Microfilms, Ann Arbor, Mich.

Jones, B.D. (1999). The Middle Archaic Period in Connecticut: The view from Mashantucket. *Bulletin of the Archaeological Society of Connecticut* **62**:101–123.

Kelly, R.L. (1995). *The Foraging Spectrum: Diversity in Hunter-Gatherer Lifeways*. Smithsonian Institution Press, Washington, D.C.

Kutzbach, J.E. (1987). Model simulations of the climatic patterns during the deglatiation of North America. In *North America and Adjacent Oceans During the Last Deglaciation*, edited by W.F. Ruddiman and H.E. Wright, Jr., Vol. K-3, pp. 425–446. Geological Society of America, Boulder, Colo.

Mayewski, P.A., Meeker, L.D., Whitlow, S., Twickler, M.S., Morrison, M.C., Alley, R.B., Bloomfield, P., and Taylor, K. (1993). The atmosphere during the Younger Dryas. *Science* **261**:195–197.

Mayle, F.E., and Cwynar, L.C. (1995). Impact of the Younger Dryas cooling event upon lowland vegetation of maritime Canada. *Ecological Monographs* **65**:129–154.

Mayle, F.E., Levesque, A.J., and Cwynar, L.C. (1993). Accelerator-mass-spectrometer ages for the Younger Dryas event in Atlantic Canada. *Quaternary Research* **39**:355–360.

Maymon, J., and Bolian, C. (1992). The Wadleigh Falls site: An Early and Middle Archaic Period site in southeastern New Hampshire. In *Early Holocene Occupation in Northern New England*, edited by B.S. Robinson, J.B. Petersen, and A.K. Robinson, Occasional Publications in Maine Archaeology No. 9. Maine Archaeological Society and the Maine Historic Preservation Commission, Augusta.

McWeeney, L. (1994a). *Archaeological Settlement Patterns and Vegetation Dynamics in Southern New England in the Late Quaternary*. Doctoral dissertation, Yale University, University Microfilms, Ann Arbor, Mich.

McWeeney, L. (1994b). Environmental reconstruction using plant macrofossil analyses: The Mashantucket Pequot's Cedar Swamp, Ledyard, Connecticut. Unpublished manuscript submitted to the Mashantucket Pequot Museum and Research Center.

McWeeney, L. (1999). A review of late Pleistocene and Holocene climate changes in southern New England. *Bulletin of the Archaeological Society of Connecticut* **62**:3–18.

Nicholas, G.P. (1987). Rethinking the Early Archaic. *Archaeology of Eastern North America* **15**:99–124.

Nicholas, G.P. (1988). Ecological leveling: The archaeology and environmental dynamics of early postglacial land use. In *Holocene Human Ecology in Northeastern North America*, edited by G.P. Nicholas, pp. 257–288. Plenum Press, New York.

Nicholas, G.P. (1991). Putting wetlands into perspective. *Man in the Northeast* **42**:29–38.

Nicholas, G.P. (1998). Wetlands and hunter-gatherers: A global perspective. *Current Anthropology* **39**:720–731.

Nicholas, G.P. (1999). A light but lasting footprint. In *The Archaeological Northeast*, edited by M.A. Levine, K.E. Sassaman, and M.S. Nassaney, pp. 25–38. Bergin & Garvey, Westport, Conn.

Perry, D.W. (1997). *The Archaeology of Hunter-Gatherers: Plant Use in the Dutch Mesolithic*. Doctoral dissertation, New York University. University Microfilms, Ann Arbor, Mich.

Perry, D.W. (1998). Interim report on the analysis of vegetative plant remains from sites 72-97, 72-91, and 72-66. Unpublished report on file at the Mashantucket Pequot Museum and Research Center.

Perry, D.W. (1999). Vegetative tissues from Mesolithic sites in the Netherlands. *Current Anthropology* **40**:231–238.

Perry, D.W. (2000). Vegetative plant remains from site 72-163. Unpublished report on file at the Mashantucket Pequot Museum and Research Center.

Peteet, D.M., Vogel, J.S., Nelson, D.E., Southon, J.R., Nickmann, R.J., and Heusser, L.E. (1990). Younger Dryas climatic reversal in northeastern USA? AMS ages for an old problem. *Quaternary Research* **33**:219–230.

Petersen, J.B., and Putnam, D.E. (1992). Early Holocene occupation in the central Gulf of Maine region. In *Early Holocene Occupation in Northern New England*, edited by B.S. Robinson, J.B. Petersen, and A.K. Robinson. Occasional Publications in Maine Archaeology No. 9. Maine Archaeological Society and the Maine Historic Preservation Commission, Augusta.

Pianka, E.R. (1994). *Evolutionary Ecology* (5th ed.). HarperCollins College Publishers, New York.

Pianka, E.R. (1995). Evolution of body size: Varanid lizards as a model system. *American Naturalist* **146**:398–414.

Ritchie, W.A. (1965). T*he Archaeology of New York State*. The Natural History Press, Garden City, N.Y.

Ritchie, W.A. (1979). Some regional ecological factors in the prehistory of man in the Northeast. In *The Bulletin of the New York State Archaeological Association* **75**:14–23.

Robinson, B.S. (1992). Early and Middle Archaic Period occupation in the Gulf of Maine region: Mortuary and technological patterning. In *Early Holocene Occupation in Northern New England*, edited by B.S. Robinson, J.B. Petersen, and A.K. Robinson. Occasional Publications in Maine Archaeology No. 9. Maine Archaeological Society and the Maine Historic Preservation Commission, Augusta.

Robinson, B.S. (1996). Archaic period burial patterning in the Northeast. *The Review of Archaeology, Special Issue* **17**:33–44.

Robinson, B.S., and Petersen, J.B. (1993). Perceptions of marginality: The case of the early Holocene in northern New England. *Northeast Anthropology* **46**:61–75.

Thorson, R.M. (1993). Pieces of the Pequot past. Unpublished manuscript submitted to the Mashantucket Pequot Museum and Research Center.

Thorson, R.M. (1996). The five-square strategy for excavating colluvial slopes. *The Review of Archaeology* **17**:25–32.

Thorson, R.M. and McWeeney, L. (1994). Non-climatic controls on lake-level changes near the Laurentide limit. Unpublished manuscript in possession of the authors.

Thorson, R.M., and McWeeney, L. (n.d.). Lake-level changes near the southeastern Laurentide limit. Unpublished manuscript in possession of the author.

Thorson, R.M., and Webb, R.S. (1991). Postglacial history of a cedar swamp in southeastern Connecticut. *Journal of Paleolimnology* **6**:17–35.

Webb, T. III, Bartlein, P.J., Harrison, S.P., and Anderson, K.H. (1993). Vegetation, lake levels, and climate in eastern North America for the past 18,000 years. In *Global Climates Since the Last Glacial Maximum*, edited by H.E. Wright, Jr., J.E. Kutzbach, T. Webb, W.F. Ruddiman, F.A. Street-Perrott, and P.J. Bartlein, pp. 415–467. University of Minnesota Press, Minneapolis.

CHAPTER 7

GLACIAL GEOLOGY AND PREHISTORIC SENSITIVITY MODELING FORT DRUM, NEW YORK

Laurie W. Rush, Randy Amici, James Rapant, Carol Cady, and Steve Ahr

The Fort Drum Cultural Resources Program is using an increasingly sophisticated understanding of the geomorphology of the glacial Lake Iroquois basin to develop a working sensitivity model for prehistoric occupation patterns in the Fort Drum section of Jefferson and Lewis Counties. Critical variables include elevation, glacial landform, and proximity to ravines or fossil waterways. The model was tested using a protocol of 20-m interval shovel test grids placed in both high- and low-sensitivity areas. The model proved to be a powerful predictor of prehistoric site density and has dramatically increased the success of Fort Drum's prehistoric archaeological survey in finding prehistoric archaeological sites. This chapter will describe the sensitivity model and then discuss two sites that illustrate differences in preservation integrity within dune landforms.

GEOLOGIC AND PHYSIOGRAPHIC SETTING

Fort Drum is located in northwestern New York, about 13 km east of Lake Ontario and 48 km south of the St. Lawrence River (Figure 7.1). The military reservation

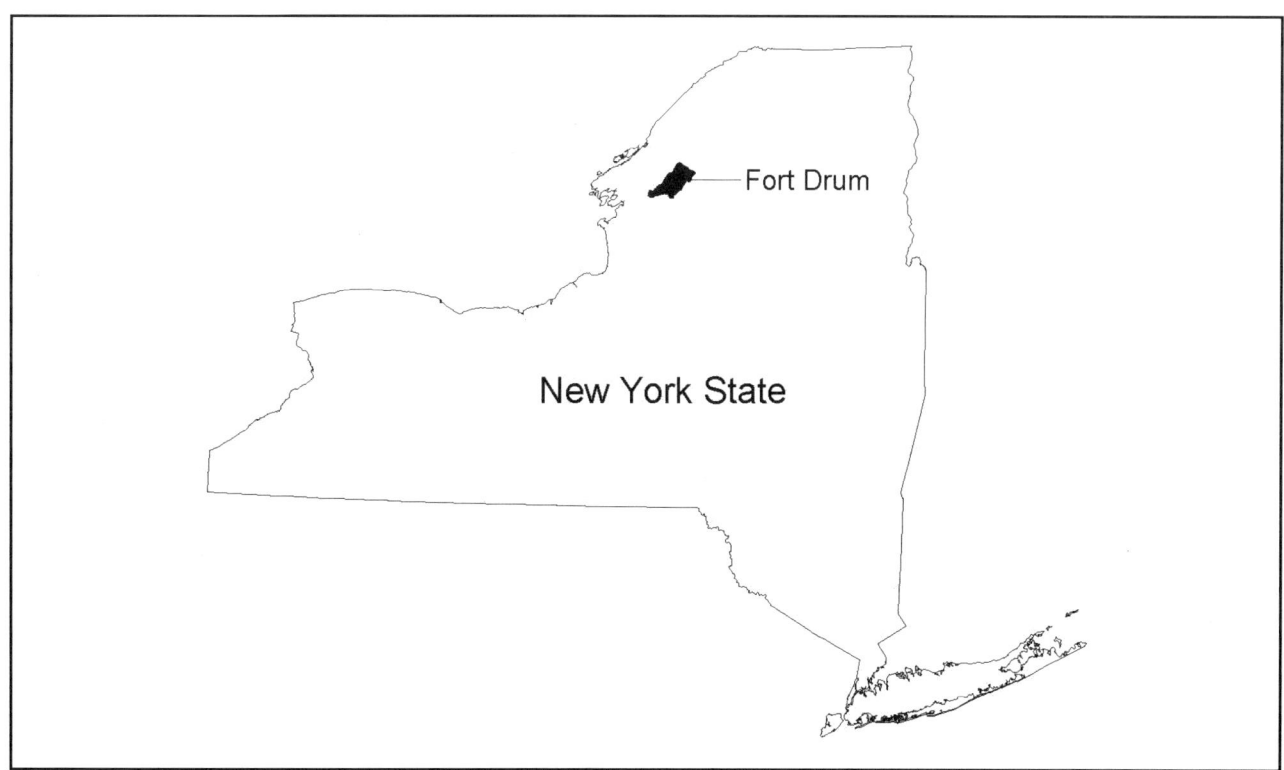

Figure 7.1. Location of Fort Drum in New York State.

Geoarchaeology of Landscapes in the Glaciated Northeast edited by David L. Cremeens and John P. Hart. New York State Museum Bulletin 497. © 2003 by the University of the State of New York, The State Education Department, Albany, New York. All rights reserved.

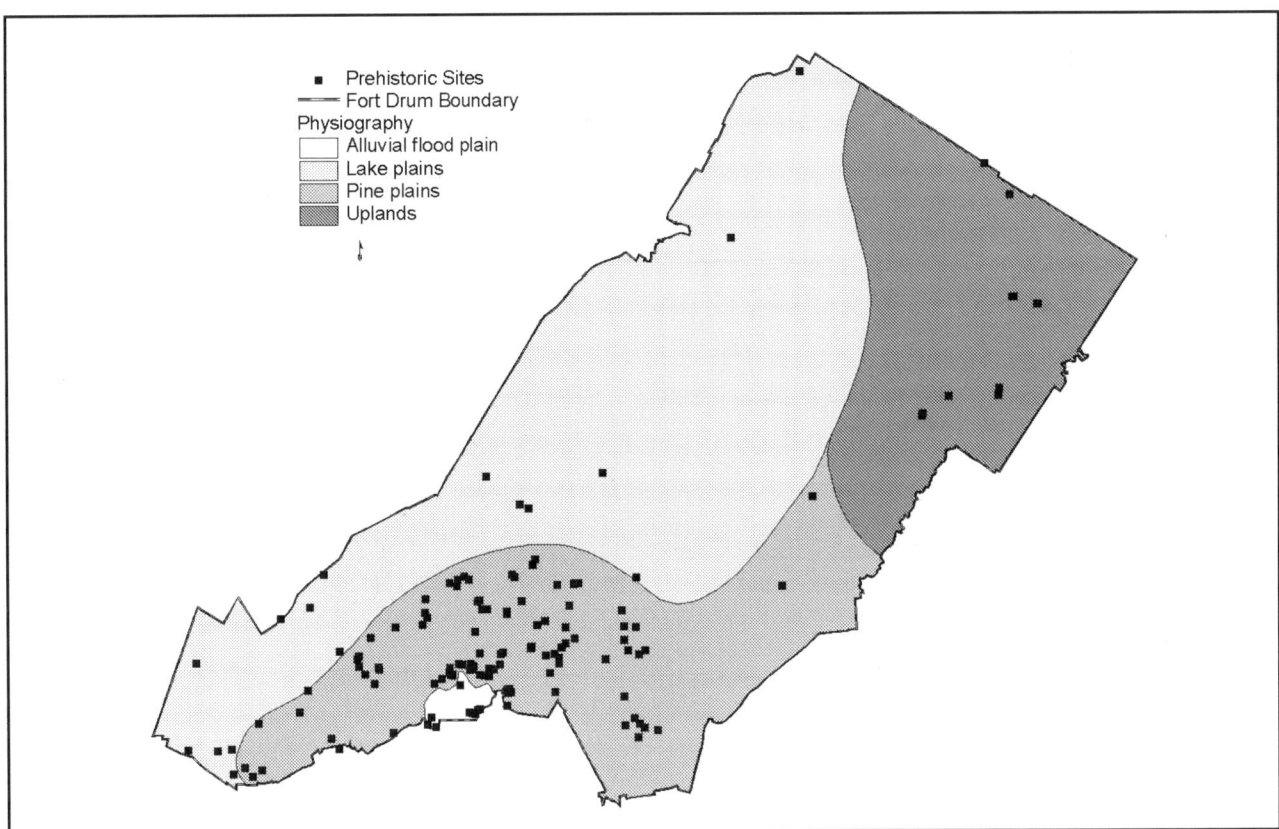

Figure 7.2. Fort Drum physiographic zones.

encompasses portions of Jefferson and Lewis Counties and contains important prehistoric and historic archaeological sites. Fort Drum lies within the Canadian Biotic Province and encompasses four main physiographic regions (Figure 7.2): the Grenville and Adirondack Foothills (uplands), Ontario-St. Lawrence Lowlands (lake plain), the Pine Plains (delta), and Black River Valley (alluvium). The two regions selected for development and testing of the prehistoric sensitivity model were the lake plain and delta landforms.

Ontario-St. Lawrence Lowlands (Lake Plain)

Flat topography, poorly drained clay soils, and extensive wetlands are typical of the lake plain. Younger Cambrian Potsdam Sandstone and Theresa Formations dominate the geology of the lake plain, although some older Precambrian granite gneiss units are present along the western boundary. Nearly all bedrock units are covered in remnant fine-grained glaciofluvial lacustrine clays and silts from former glacial Lake Iroquois. Soils within the lake plain region are composed of poorly drained lacustrine silts and clays, with peat and muck present over much of the area. Till piles up to 15 m high composed of glacial gravels occur intermittently in this landform as well.

Pine Plains Sands (Delta)

The topography of the Pine Plains in the southern part of Fort Drum is flat, with little variation until one reaches the northernmost end. Oak and white pine forest and grassland mosaic characterize the landscape. Glacial till and approximately 30 m thick, sandy Quaternary-aged surficial deposits, the Pine Plains Sands, overlie Precambrian granite and Ordovician sedimentary bedrock formations. Sandy, well-drained horizons have developed within these soils. The typical undisturbed profile is characterized by a dark, well-developed A horizon 10–20 cm in thickness. The B horizon is high in iron content and is yellowish/orange. Underlying this horizon is unaltered parent material consisting of lenses and laminae of glaciofluvial and glaciolacustrine sands and silts.

The Pine Plains Sand Formation exhibits lobate shapes and contains numerous radiating branches. In some places the sands may achieve thickness of up to 40 m (Reynolds 1986). Currently the hydrology of this area consists of intermittent streams, which flow southeast to northwest across the delta. Springs occur at the heads of the ravines along the northern edge of the formation. One of the most important characteristics of the Pine Plains physiographic region is the abundance of water. The springs are only one manifestation of one of the most productive aquifers in the region.

The Pine Plains is currently classified geomorphologically as a fan-delta (cf. Ritter et al. 1995). The early hypothesis concerning formation of these sands is that the Pine Plains delta may have formed at the terminal Pleistocene Epoch with the Black River flowing into the standing water of glacial Lake Iroquois. As the flow velocity of the Black River decreased, it would have emptied its coarser suspended sediment load at the mouth of the river, creating the sandy delta. Finer-grained clasts such as silts and clays remained in suspension longer and were deposited farther from the river mouth, near the prodelta and lake bottom of glacial Lake Iroquois (cf. Buddington 1934; Ritter et al. 1995).

An alternative hypothesis is that these sands are the result of a glacial outwash apron with the 500–740-ft contour line area representing a frontal moraine formed at the ice-lobe limits. The presence of large boulders in outlying areas of the sands, transported by break-away icebergs, support this hypothesis (MacClintock and Stewart 1965). Regardless of the original deposition, it is generally agreed that these sands were beaches and shorelines for glacial Lake Iroquois, glacial Lake Frontenac, and maybe the subsequent Gilbert Gulf Marine Event during the final phases of the Pleistocene Epoch.

PATTERNS OF PREHISTORIC ARCHAEOLOGICAL SITE LOCATION ON FORT DRUM

Previous Model

In the 1980s a prehistoric site predictive model was established for Fort Drum (U.S. Army 1985; Hasenstab and Resnick 1990). This model established low-, moderate-, and high-sensitivity areas based on similar models for the Mohawk Valley, using known site distribution patterns, distance to water, soil drainage, and slope as variables. This model predicted that prehistoric archaeological sites would occur within 250 m of water, on well-drained soils, on slopes of less than 5 percent (Hasenstab and Resnick 1990).

There were problems with all of the variables in the previous model. *Water* was not defined, resulting in overemphasis of navigable and flowing waterways as the operative form of the variable. There was a temporal problem with the definition of *water* as well. "Water" in the previous model referred only to water present now. Therefore the model failed to take into consideration vast hydrologic variation in the region over time. The drainage variable was rendered meaningless within the physigraphically defined landforms, because all of the soils in the Pine Plains are well-drained sands, and most of the soils in the lake plain are poorly drained silts and clays. In addition, use of the slope variable was made problematic by behavior of the crew. Crew members developed a tendency to stop transects not only when the ground began to slope at all, but also avoided saddles between slopes and bottoms of ravines. This pattern resulted in systematic exclusion of ravine edges and dips in the landscape from subsurface investigation.

As a result of the problems with the variables in the previous model, few sites were discovered, and a high percentage of the discovered sites were in the predicted low-sensitivity areas, especially in the Pine Plains Sands (U.S. Army 1998:98). One reason for discovery of these sites was that exposed prehistoric materials were easily identified during low-sensitivity pedestrian survey across eroded sands. In summary, the previous model was a poor predictor of site location and resulted in tens of thousands of negative survey shovel tests.

Development of Current Landform-Based Model

In 1998 the Fort Drum Cultural Resources Program decided to reevaluate the sensitivity model and survey strategy. The process began with review of the prehistoric site database and analysis of location patterns of known sites. The Geographical Information System (GIS) at Fort Drum offered 2-ft resolution aerial photos of the entire installation and coverage of more than 200 environmental variables and geographic attributes. The first step was to visually compare the 120 known prehistoric site locations on Fort Drum with patterns in the landforms. This analysis revealed a proximity to ravines. To determine the presence or absence of a meaningful relationship between site location and proximity to a ravine, GIS provided a hillshade model (Figure 7.3).

The model was a 10-m resolution raster data set with pixel values that represented degrees of shade ranging from zero (shadow) to 255 (full sunlight). The relief model was generated with Arc/Grid's HILLSHADE command, using as input a 10-m resolution digital elevation model (DEM) of the installation. The digital elevation model was created by interpolating from 5-ft interval digitized contours for the entire Fort Drum installation acreage.

The visual relationship was striking. Nearly 85 percent of the previously identified sites on Fort Drum that

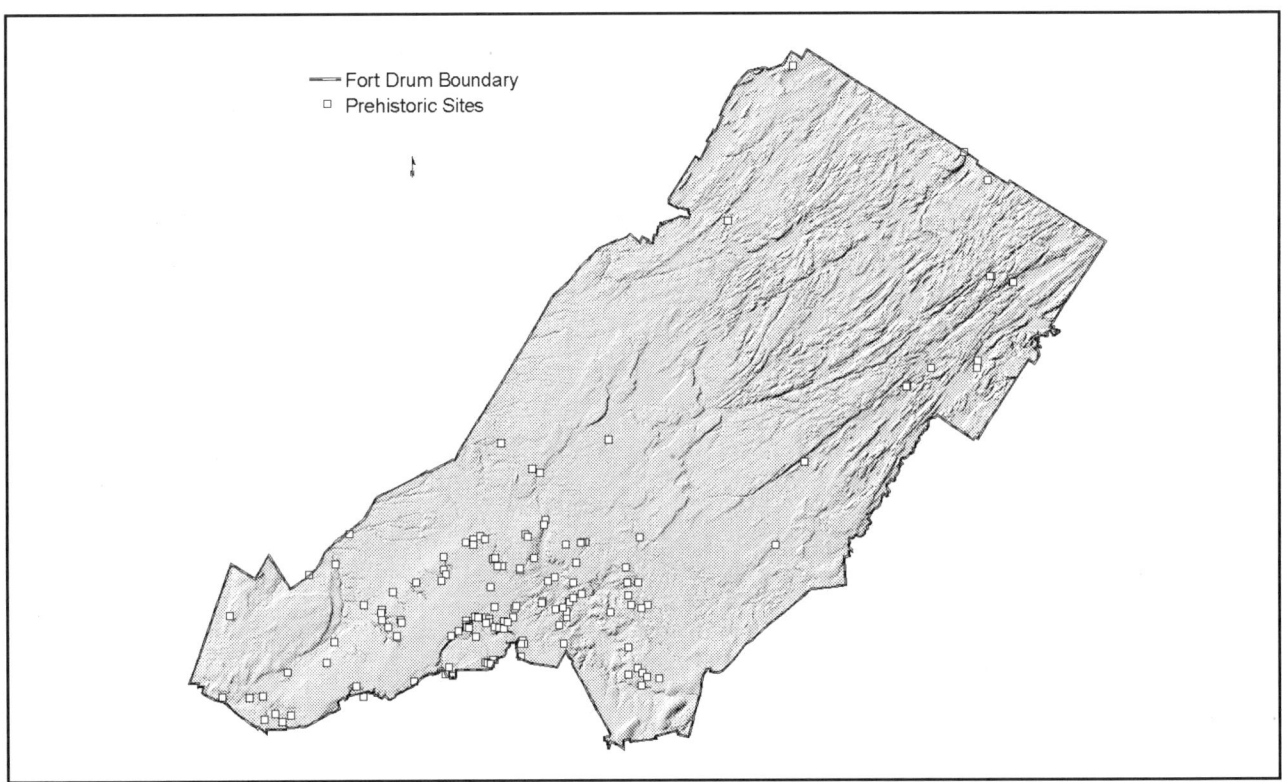

Figure 7.3. Fort Drum hillshade model. Note the clustering of prehistoric archaeological sites on the edges of ravines.

are not associated with a navigable waterway were located on the banks of a ravine or a fossil waterway. As a result it was decided to test the ravine edges as a highly sensitive landform beginning in the 1999 field season. This strategy resulted in a very successful survey in terms of new prehistoric sites identified. One way to quantify this success is to compare 229 positive tests for prehistoric material out of 4,590 total tests versus 20 positive out of 10,100 tests in 1998 using the previous model. The percentage of positive tests increased from 0.002 to 4.99.

Not only were new sites identified, but also a pattern of positive tests emerged from the successful survey. The highest frequencies of positive tests clustered around the 600-ft elevation line within the survey areas. In response, the elevation data were used to further analyze the observed pattern of site location. When the 600-ft line was compared with the maps and literature, it became clear that this elevation provided a precise delineation of glacial Lake Frontenac. This stage dates to approximately 11,200 ± 190 B.P. (Pair and Rodriques 1993). This delineation fit precisely with discovery of a Paleoindian Gainey-like fluted point (J. Holland, personal communication 2000) along the relict shoreline in 1995 (Figure 7.4). The preliminary Fort Drum results fit Rapp and Hill's (1998:72) observations of archaeological sites often being associated with fossil beaches along the margins of glacial lakes (Figures 7.5, 7.6, and 7.7).

Fort Drum has a total of three Paleoindian sites associated with glacial Lake Frontenac. In addition to the

Figure 7.4. Clovis point. This point was discovered along the glacial Lake Iroquois shoreline at the Frontenac level. The elevation was 600 ft above sea level. The estimated date for the landform as an active shoreline would be 11,300 B.P.

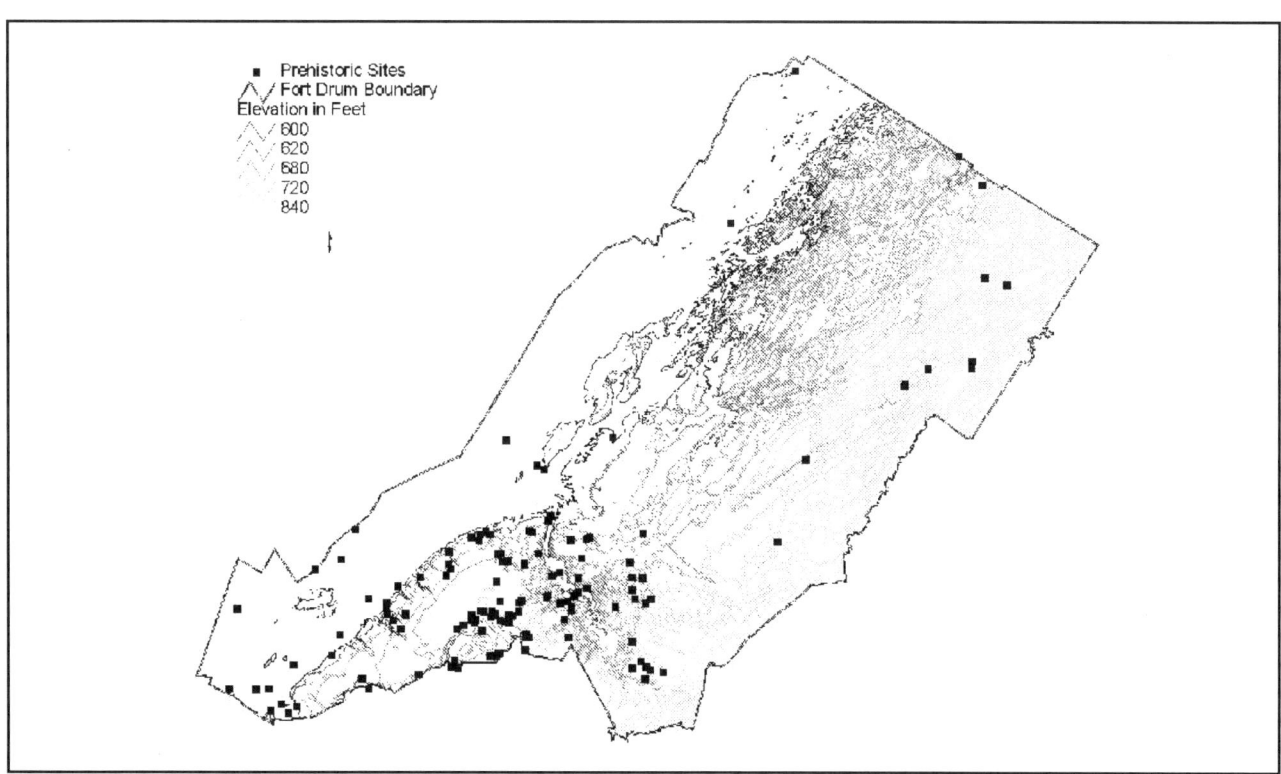

Figure 7.5. Fort Drum elevations and prehistoric archaeological site distribution.

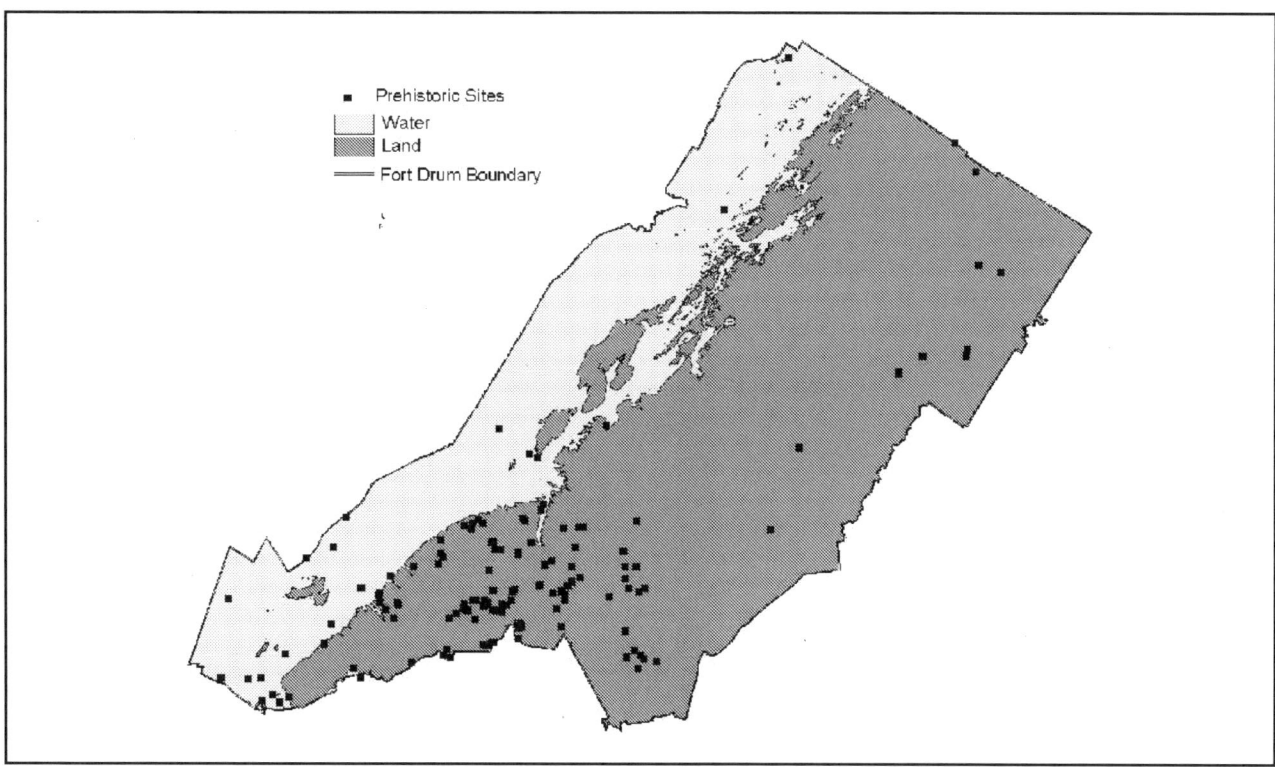

Figure 7.6. Glacial Lake Iroquois, Frontenac-level shoreline defined by 600-ft elevation line. Note that some of the prehistoric archaeological sites are occurring on fossil island landforms.

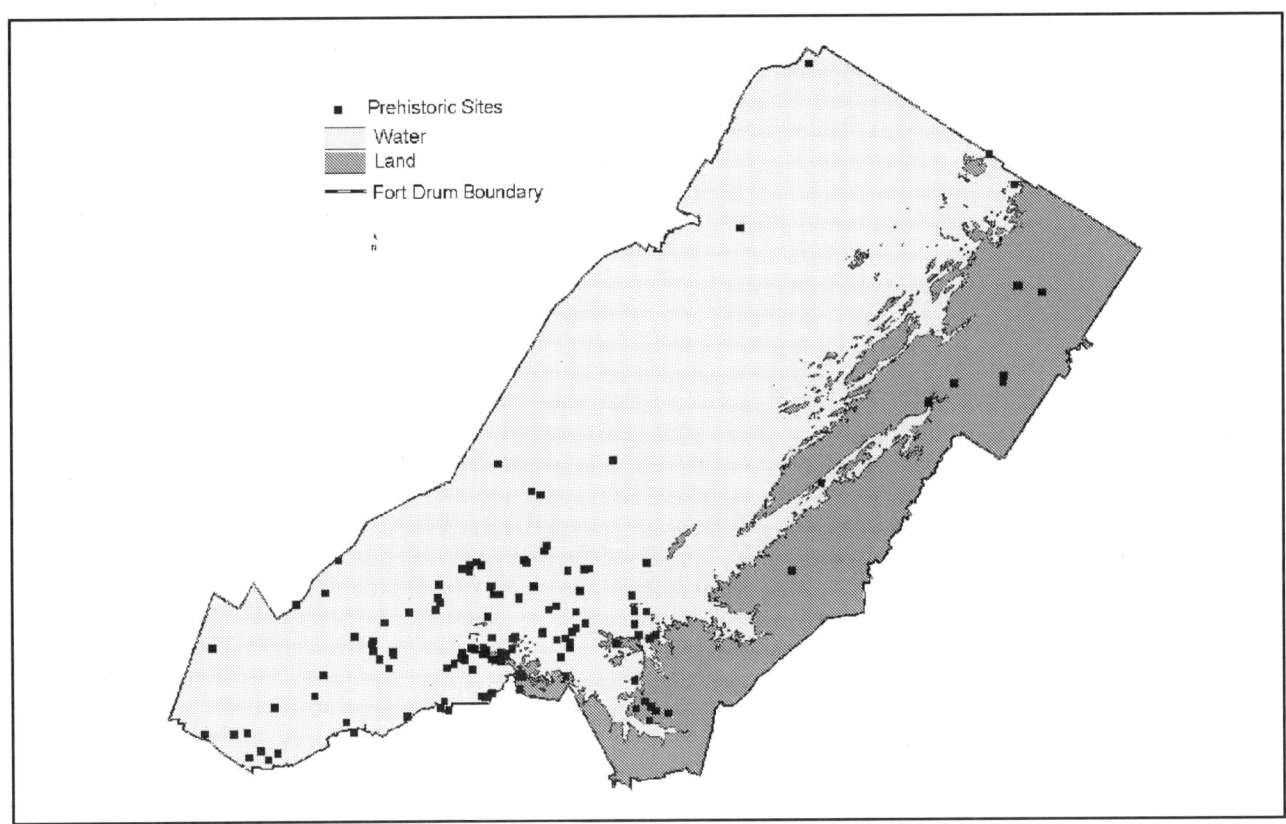

Figure 7.7. Fort Drum glacial Lake Iroquois shoreline defined by 700-ft elevation line.

Middle-Paleoindian Gainey fluted point, two other sites along the shoreline each yielded one Late-Paleoindian unfluted lanceolate. It may be that this late unfluted lanceolate tradition may represent a Paleo extension into the Early Archaic in our region. (For a complete discussion, see Dincauze 2000). In addition, during the 2002 season, the Fort Drum survey discovered superimposed hearths dating to 8090 ± 60 B.P. (Beta-168146) (cal. 2σ 7302 [7063] 6831 B.C.)[1] and 8010 ± 60 B.P. (Beta-168145) (cal. 2σ 7079 [7044] 6689 B.C.) along a fossil shoreline at 700 ft. The Fort Drum discoveries fit a chronological progression established by discoveries of two early Debert/Vail-type Paleo points (Ritchie 1957) found in other areas of Central New York associated with earlier glacial lakes, Albany and Glenfield (Isachsen et al. 1991).

Positive shovel tests and test excavation units along the 600-ft contour showed a distinct layer of pebbly sand that lies just above varve deposits from the glacial lake. The most frequent artifact category occurring in this layer was blocky debitage of Onondaga chert. There were also fragments of mineralized faunal remains. The possibility of Paleoindian occupation of the beach line was further reinforced by the discovery of cultural material in the form of quartz flakes at the 600-ft contour line on what can be classified as fossil island landforms rising out of the glacial lake plain in the northwestern corner of the Fort Drum installation. The Frontenac stage may be the most productive shoreline of the glacial lake stages in terms of prehistoric archaeological sites. There may be two reasons for the observed site density at this elevation. The first is that due to hydrologic shifts and isostatic uplift, this shore appeared stable for the longest period during the changes in the glacial lakes. Second, due to this stabilization, this shoreline has left a distinct margin across the landscape.

After a relationship was identified between patterns of site occurrence and the 600-ft contour line, it was decided to examine higher beach lines from the glacial lake. Glacial Lake Iroquois drained slowly from approximately 780 ft above sea level in its main phase 12,500 B.P. to the 600-ft-level glacial Lake Frontenac and possible subsequent Gilbert Gulf shoreline. This slow drainage left behind a series of shorelines. These shorelines are reflected in the contours along the edges of the Pine Plains delta landform between elevations of 600 and 780 ft. Using the contour lines and ARCVIEW software, water was graphically added to the image of glacial Lake Iroquois, 20 ft at a time (Figure 7.7). The visual effects made it easy to discover additional patterns of prehistoric sites occurring along fossil shorelines at 20-ft intervals varying between 600 and 780 ft above sea level. In some cases archaeological sites occurred directly across glacial bays from each other at identical elevations.

Using this visual system, it is possible to create a series of maps where water is graphically added at either 5- or 20-ft intervals. Each map then reveals an iteration of the glacial lake, complete with fossil shorelines and islands. The fossil islands and beaches generated by these maps offer landforms that can be specifically targeted for intensive survey.

As a result, another logical way to test the model was to place shovel test grids systematically on the higher elevation shorelines. This strategy in pilot test surveys yielded exotic lithics on fossil islands at the 700–740-ft elevations. It also further increased the positive shovel test rate in high sensitivity acreage from 4.99 percent to 7.1 percent or 883/12,409 during the 2000 season.

The predictive power of the model appeared to vary with slope and soil type. The model was tested again between the 600 and 620-ft contour lines on the northern edge of the delta in an area where the slope is gradual. These areas were negative for cultural material. The high organic content of the soils in this region may represent fossil wetlands from the glacial time period. The wetlands would not have been as attractive for occupation as the beaches. The model has potential for further refinement, using slope and soil types as additional variables.

Considering soil types also provides an opportunity to use some of the thousands of negative shovel tests to help further refine the predictive capabilities at Fort Drum. All shovel test data including soil layers are entered into GIS using ESRI ARCVIEW with custom grid and labeling software tied to an ACCESS database platform. As a result there is the potential to compare soil types as they relate to prehistoric site distribution patterns and preservation potential. All customized codes used at Fort Drum are public domain and can be accessed at the Colorado State University Center for the Environmental Management of Military Lands Web site: www.cemml.colostate.edu.

In addition to the discovery of possible Paleoindian occupation areas, the new prehistoric predictive model proved to be a powerful predictor of all other phases of prehistoric occupation. It appears that Native American occupation of the ravine edges and beach landforms has been nearly continuous since Paleoindian times. Radiocarbon dates from sites discovered using the model range from 8000 to 390 B.P. Freshwater springs continued to feed the proglacial lake basins providing a habitat for a vast variety of flora and fauna. Once the glacial lakes were gone, these areas even with low water levels have continued to be adjacent to wetlands. In some cases, beaver impoundments may have maintained water at levels higher than water levels today, meaning that fossil shorelines could still have been waterfront intermittently through time.

This predictive model has resulted in or aided the discovery of at least four highly significant prehistoric occupation areas including a 10-acre, stratified-occupation, Middle Woodland site with evidence of seed processing and clay-lined earth ovens or storage pits and a Middle Archaic Lamoka occupation dating to 3540 ± 60 B.P. (Beta-145752) (cal. 2σ 2031 [1882, 1836, 1834] 1692 B.C.).

Lake Plain Landform

The lake plain landform offers a striking contrast to the delta not only in potential for prehistoric cultural remains but also in depositional history and soil type. The lake plain, in essence, offered the control acreage for testing low sensitivity as defined in the model. In more than 15,000 acres surveyed in the lake plain, there have been no prehistoric archaeological sites discovered on the lake bottom landform that were not directly associated with a flowing or potentially navigable waterway. In the sites that were discovered along the waterways, the archaeological material occurred in the top 15 cm of soil, indicating relatively recent deposition within the depositional history.

To further test the model, 12 points were selected at random across the lake plain landform. A series of 100-m grids with shovel tests at 20-m intervals were placed using each of the random points as a southwest datum. All shovel tests in these grids were negative for prehistoric archaeological material. When combined with the fact that all previous shovel tests away from the water in systematic surveys in the lake plain area have also been negative, these results are not surprising.

However, the discovery of cultural deposits at the summits of till piles in the lake plain section of the Fort Drum installation offered a significant variation. These landforms were the only areas in the region that were above the 600-ft elevation line and would have been islands in glacial Lake Iroquois and Frontenac. As a result these discoveries further support the proposed sensitivity model and suggested a maritime component to the cultures represented.

APPLICATIONS OF GEOARCHAEOLOGY FOR DETERMINING DATA RECOVERY POTENTIAL

In addition to developing geologically based predictive models, we are finding that understanding and evaluating dune environments offered by a geophysical

Figure 7.8. Overview of Fort Drum Prehistoric Site 1093. Note that the site appears to be eroded and deflated. Early Woodland pottery lies on the surface, side by side with 20th century material, at this site.

perspective is extremely helpful in evaluating surface finds for stratification and data-recovery potential. The contrast in deposits identified at Fort Drum Prehistoric Sites 1093 and 1154 pointed out the challenges created by varying levels of deflation in dune environments. Visual inspection of eroded areas is deceiving. A first glance visual inspection of the two sites might create the expectation that 1093 is the more disturbed with more potential for deflation (Figures 7.8 and 7.12). However, in reality, visual inspection was at best insufficient and at worst deceiving. Site 1093 was the more intact of the two by far with test unit results indicating the potential for at least two intact and stratified cultural contexts.

Formation Processes

The physical environment in the Pine Plains is in a constant state of flux. Sites within this area have been subjected to numerous physical and biological disturbance processes. At some point in the prehistory or history of the Pine Plains delta, large areas were subjected to clearing with disruption of the sod layer, possibly even prior to military occupation. The hypothesis that Native Americans may have devegetated occupation areas should be tested as well. An aerial photograph taken in 1927 shows devegetation and sod disruption in some areas that are still barren on Fort Drum today. The loss of vegetation created a large source of erodable material, and the shearing and lifting forces of the wind allowed the entrainment and transport of sands. Some of these processes have acted to destroy some sites, while potentially preserving others through burial.

Understanding dune vegetation can also be used to the archaeologist's advantage. In devegetated areas, vegetation returns differentially by growing better in organic cultural deposits. The new vegetation will trap blowing sand, may eventually be smothered, and will provide additional organic material for more vegetation (Pendergast 1999). Several sites at Fort Drum show patterns of vegetated "mounds" that may have been formed by this process. Closer examination in the field resulted in the discovery of fire-cracked rock eroding out of the bases of some of these landscape features. These vegetation patterns are now used as markers for potential features by the Fort Drum archaeological survey team.

Site FDP 1093 in Training Area 4A is almost completely denuded of vegetation. Portions of the overlying A horizon have been scoured and truncated by wind erosion exacerbated by low-flying helicopters. Prior to discovery and protection, military vehicles and foot soldiers crossed the site routinely. Although 20th-century material lay side by side with Rocker Dentate ceramics dating to 1930 ± 40 B.P. (Beta-137376) (cal. 2σ 36 B.C. [A.D. 75] 205), under further investigation, the site was found to have retained much of its prehistoric integrity. A burned charcoal layer dating to 1090 ± 100 B.P. (Beta-137377) (cal. 2σ A.D. 780 [979] 1030) covered much of the site, with the most productive artifact layers beneath dating to as early as 5420 ± 40 B.P. (GX-28498-AMS) (cal. 2σ 4344 [4323, 4289, 4254] 4054 B.C.). A hearth dating to 1510 ± 60 B.P. (GX-28497) (cal. 2σ A.D. 419 [544, 549, 558] 657) was discovered nearly intact in a different area of the site as well. Evidence of a burned tree root at the top layer of this hearth 150 ± 60 B.P. (Beta-159582) reinforced the importance of vegetation as a marker for organic features. Discovery of fragments of a nearly complete metate (Figure 7.9) within 1 m of each other as well as the discovery and recovery of multiple friable clay features (Figure 7.10) illustrated the intact nature of the deposits.

Figure 7.10. Friable clay features found at FDP 1093. The discovery of these features also demonstrated the potential for data recovery at a site that appeared to be completely deflated in the initial walkover.

Figure 7.9. Metate discovered at FDP 1093. In spite of the fact that the site appeared to be severely eroded and disturbed, the fragments of this metate were found within 1 m of each other.

This site also yielded post molds and hearths. At least two of the discovered hearths had been protected by vegetated mounds of sand. Profiles of test units included evidence of the potential of two intact layers of culture-bearing material. Relatively sudden burial of FDP 1093 under a thick layer of sterile sand with the erosion occurring recently would explain the good preservation confirmed by the recent investigations and discoveries.

Site 1093 illustrates that the same erosional, eolian processes that have deflated portions of sites at Fort Drum also have the potential to bury and preserve archaeological material. In denuded areas of the Pine Plains, sand erosion has resulted in the formation of active dune fields. These dunes have potentially capped archaeological deposits soon after abandonment. The rate of dune formation can be quite substantial. For example, trench profiles in the dune field revealed 1-m-thick windblown sands overlying early 20th-century refuse (Figure 7.11). The layer of historical trash was resting upon a relatively stable surface of the Pine Plains and had been subsequently covered under a newly formed dune. This phenomenon creates a need for consideration of deep testing during prehistoric site survey in active dune fields.

In contrast, we considered another of the sites discovered using the predictive model, adjacent to a ravine at the 660-ft elevation, FDP 1154. This site yielded a range of exotic lithics and a high density of scattered fire-cracked rock. Artifacts include Meadowood bifaces mixed with crudely worked LeRay chert bifaces. In initial visual inspection, the site looked relatively undisturbed (Figure 7.12). However, test unit excavations

Figure 7.11. Rapid aeolian deposition. Mid-20th-century materials were found along the buried A horizon to the right in this profile.

failed to discover or document any intact features. This site may be deflated. Deflation, in this case, refers to features and fine sand deposits eroding away, leaving site artifacts on a "desert pavement" where erosional equilibrium was reached due to particle size.

However, a test unit in vegetated soils adjacent to the deflated area revealed that a layer bearing cultural material extended from the exposed surface of the eroded and deflated area as a buried layer into less eroded adjoining deposits. The relative elevations confirmed the possibility that the currently deflated and exposed area may have been an occupation surface. One way to test this hypothesis would be to flat shovel across the site to intensify the search for features. A second test would be to open a series of test excavations at the interface of the eroded surface and the less disturbed edges of the site. In any case, it is extremely important to thoroughly investigate sites in dune landforms before judging their stratigraphic integrity and site potential. First impressions can be very misleading.

Figure 7.12. Overview FDP 1154. This site initially appears to be less disturbed than FDP 1093. In actuality, it may be completely deflated with the only potential for intact deposits occurring on the edges.

SUMMARY

Several formation processes are actively transforming the archaeological record at Fort Drum. Because landscapes are dynamic, geoarchaeology and sensitivity modeling have provided a much needed means to interpret sites, site patterns, and, ultimately, human behavior. Proper management of large acreages bearing important cultural resources hinges on being able to inventory efficiently and evaluate discoveries accurately. It is therefore important that archaeologists employ detailed analyses of soils and landforms in which cultural materials were deposited to develop a more sophisticated understanding of the relationship between the people and the land and water through time. As a response to the success of the geoarchaeological approach with development of this model, Fort Drum is seeking funding for more formal geoarchaeological consultation. Program goals include expansion of model building into the other landforms on the installation acreage. Such modeling, analysis, and information management becomes even more critical when attempting to efficiently integrate cultural resources requirements and agency goals—the situation found on military reservations (see Zeidler 1998).

Acknowledgments

The authors wish to acknowledge: Fort Drum Garrison Command, The Tenth Mountain Division, and U. S. Army Forces Command.

END NOTE

1. All calibrations done with CALIB 4.3 (Stuiver et al. 1998).

REFERENCES CITED

Buddington, A.F. (1934). *Geology and Mineral Resources of the Hammond, Antwerp, and Lowville Quadrangles*. New York State Museum Bulletin No. 296, The University of the State of New York, Albany.

Dincauze, D.F. (2000). Special Report: The Earliest Americans: The Northeast. *Common Ground: Archeology and Ethnography in the Public Interest*.

Hasenstab, R.J., and Resnick, B. (1990). GIS in Historical Predictive Modeling: The Fort Drum Project. In *Interpreting Space: GIS and Archeology*. K.M.S. Allen, S.W. Green, and E.B. Zubrow, pp. 284–306. Taylor & Francis, New York.

Isachsen, Y.W., Landing, E., Lauber, J. M., Rickard, L.V., and Rogers, W.B. (1991). *Geology of New York: A Simplified Account*. New York State Museum Education Leaflet No. 28. The University of the State of New York, Albany.

MacClintock, P., and Stewart, D.P. (1965). *Pleistocene Geology of the St. Lawrence Lowland*. New York State Museum and Science Service Bulletin No. 124. The University of the State of New York, Albany.

Pair, D.L., and Rodrigues, C.G. (1993). Late Quaternary deglaciation of the southwestern St. Lawrence Lowland, New York and Ontario. *Geological Society of America Bulletin* **105**:1151–1164.

Rapp, G., Jr., and Hill, C.L. (1998). *Geoarchaeology: The Earth Science Approach to Archaeological Interpretation*. Yale University Press, New Haven, Conn.

Reynolds, R.J. (1986). *Hydrogeology of the Fort Drum Area, Jefferson, Lewis, and St. Lawrence Counties, New York*. Water Resources Investigations Reports 85–4119. U.S. Department of the Interior Geological Survey, Albany.

Ritchie, W.A. (1957). *Traces of Early Man in the Northeast*. New York State Museum and Science Service Bulletin No. 358. The University of the State of New York, Albany.

Ritter, D.F., Kochel, R.C., and Miller, J.R. (1995). *Process Geomorphology*. Wm. C. Brown Publisher, Boston.

Stuiver, M., Reimer, P.J., Bard, E., Beck, J.W., Burr, G.S., Hughen, K.A., Kromer, B., McCormac, F.G., van der Plicht, J., and Spark, M. (1998). INTCAL98 radiocarbon age calibration 24,000–0 cal BP. *Radiocarbon* **40**:1041–1083.

U.S. Army. (1985). *An Archeological Overview and Management Plan for Fort Drum*. Prepared by Envirosphere under contract with the National Park Service, Mid-Atlantic Region, Philadelphia, Pa., for the U.S. Army, Fort Drum. Final report on file at the Environmental Division, Fort Drum, N.Y.

U.S. Army. (1998). *Cultural Resource Investigations at Fort Drum, New York: FY 1997*. Prepared by S.W. Ahr, CEMML Colorado State University for the U.S. Army, Fort Drum. Final Report on file at Environmental Division, Fort Drum, N.Y.

Zeidler, J.A. (1998). *A Rationale for Technical Guidelines on Predictive Locational Modeling of Archaeological Resources on U.S. Army Installations*. U.S. Army Corp of Engineers, CERL Technical Note 98/88.

CHAPTER 8

BEHAVIORAL CONTINUITY ON A CHANGING LANDSCAPE

Douglas Frink and Allen Hathaway

In 1980 Stephen Loring, then at the University of Massachusetts at Amherst, introduced a hypothesis to explain the settlement and procurement patterns expressed by Paleoindian-Period archaeological sites in Vermont. Loring (1980:15) observed that

> All of the known Paleo Indian components in the Champlain Valley and many of the fluted point find-spots are associated with Champlain Sea landforms. The circumstantial evidence of association allows for the possibility that Paleo Indians might have adapted to a maritime-based economy for at least a part of their seasonal round.

Figure 8.1, displaying Loring's data and the reconstructed limits of the Champlain Sea marine maximum, ca. 12,000 B.P. (cal. 14,000 B.P.[1]), reveals that 42 percent of the sites (11 of 26) in the sample lack specific provenience. The recorded site proveniences varied, with site locations referenced to an individual landowner, or a town, or a county. Of the 15 sites with specific geographic provenience, only one-third (5 of 15) are physically associated (within 200 m), but not necessarily temporally associated, with the reconstructed Champlain Sea Marine Limit (Figure 8.1).

Although never formally tested, Loring's (1980) hypothesis and its premises have been adopted by the state of Vermont as the dominate model for early Native American archaeological site sensitivity. The underlying assumptions of this model are as follows: (1) that postglacial weather patterns underwent a slowly moderating evolution between 14,000 and 9000 B.P. (cal. 16,000–11,000 B.P.); (2) that the margins of the Champlain Sea provided open terrain for hunting large ungulates and proximity to estuarine resources; and (3) that the Champlain Sea constituted the dominant geomorphological feature attractive to earliest human colonization.

Over the past two decades, Paleoindian-Period sites have been located in the Champlain Valley as a result of Cultural Resource Management (CRM) archaeological surveys. However, many of these newly identified sites are located outside areas predicted by the Loring model. These results have led some researchers to conclude that additional variables need to be considered in site location models for this time period. This local situation is mirrored in a hemispheric context where environmental, geomorphological, and archaeological studies are challenging the validity of many former premises and are leading to a reformulation of archaeological site locational models for this time period.

Recent research of Late Pleistocene and Early Holocene climates and its implications for modeling early Native American settlement of the Champlain Valley in Vermont are presented below. Through the use of computer modeling, we demonstrate that a wide variety of landforms and resources existed in the past that are not obvious today. The relationships among these landforms, the environmental niches they might provide, and Native American site locations are presented, and a new predictive model is proposed that demonstrates a continuity of settlement and procurement strategies between early and later Native American people in Vermont.

POSTGLACIAL, LATE PLEISTOCENE ENVIRONMENT IN VERMONT: PERCEPTIONS PAST AND PRESENT

The Draft Prehistoric Theme for the Vermont State Historic Preservation Plan (Thomas 1990:8) introduces Native American prehistory with the following environmental reconstruction:

> Based on some radiocarbon dated sites from elsewhere in northern New England, it seems that our earliest people—called Paleoindians—began to move into Vermont by about 9000 BC [radiocarbon years], at the end of the last Ice Age. The environment was similar to what we see in today's Arctic regions: a barren tundra which gradually gave way to a park tundra of spruce, fir and birch that sustained mastodons, woolly mammoths and large herds of caribou. (p. 8).

Geoarchaeology of Landscapes in the Glaciated Northeast edited by David L. Cremeens and John P. Hart. New York State Museum Bulletin 497. © 2003 by the University of the State of New York, The State Education Department, Albany, New York. All rights reserved.

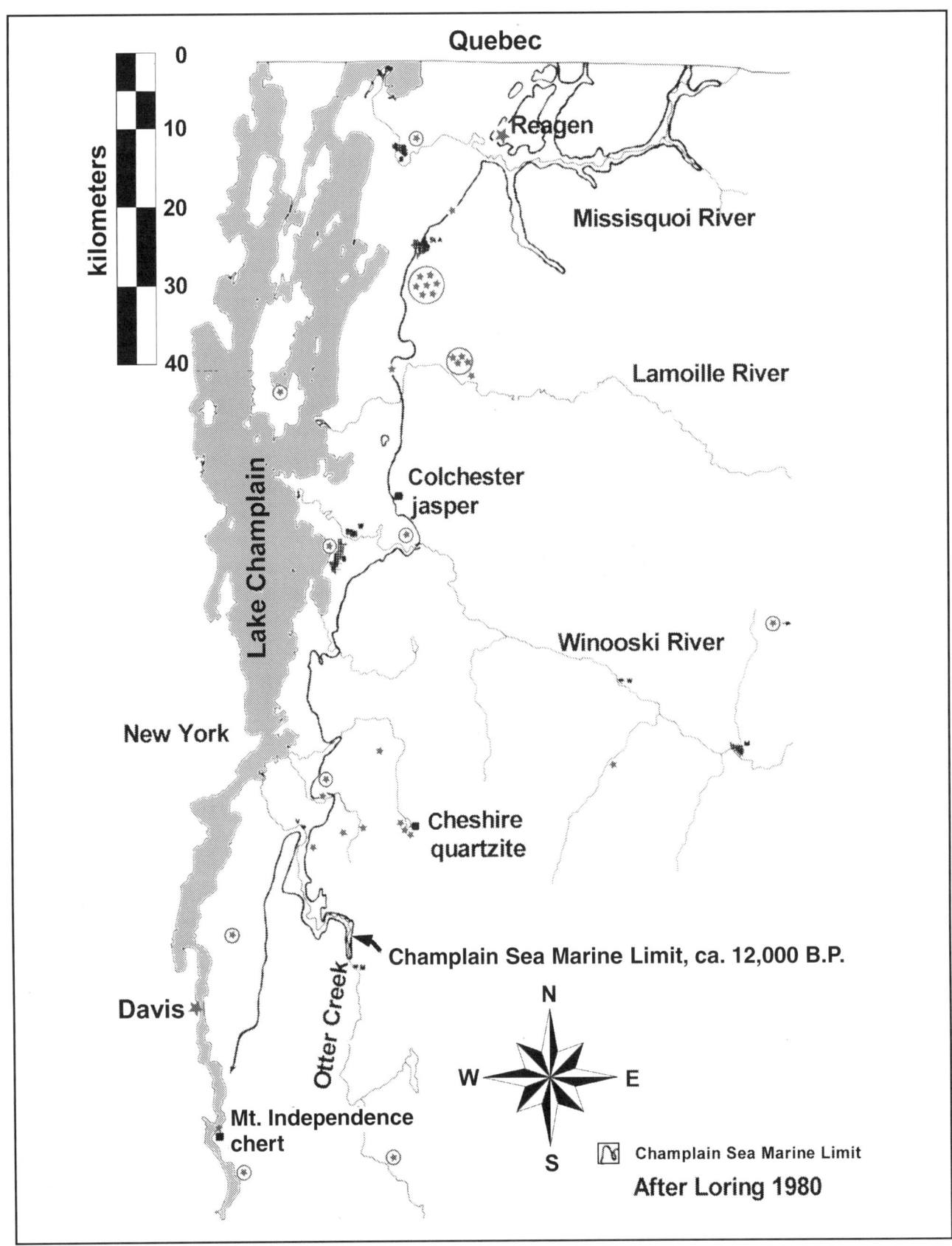

Figure 8.1. Reproduction of Loring's 1980 map of Paleoindian sites in the Champlain Valley. Circles around the object means the location is approximate.

This vision of the Late Pleistocene environment is supported by fossil pollen assemblages identified in cores recovered from ponds and bogs throughout the region.

Pollen core profiles, however, have been shown to bias critical environmental reconstructions. First, differing plant populations do not produce pollen in equal quantities, nor do they distribute pollen in the same manner. Evergreens generally produce greater amounts of pollen than deciduous trees, and trees, in general, produce more pollen than do grasses and sedges. Although many species distribute pollen by wind, some do not (e.g., chestnut, *Castaneadentata* spp., and maple, *Acer* spp.). Second, preservation of pollen varies among species within littoral environments. Pine (*Pinus* spp.) pollen is more likely to survive in a bog environment than is the pollen of many deciduous species. Third, plant species growing in and near the littoral environment are likely overrepresented in a pollen profile obtained from a pond or subsequent bog in comparison to those plant species better adapted to drier environments (Nicholas 1987). Finally, pollen specimens, which represent a minority component in the profile, are commonly considered outliers or contaminants, and researchers frequently discount their presence in reported pollen profiles (McWeeney 1995).

As early as 1980, Roger Moeller demonstrated this incongruency between environmental reconstructions based on pollen core profiles and carbonized wood specimens recovered from the archaeological context. Pollen profiles suggest that the forest environment in southern New England around 10,000 radiocarbon years ago, the age of the Templeton site (6LF21), would be characterized as a forest tundra gradually evolving to spruce woodlands. The species of plants, red oak (*Quercus* spp., Erthrobalanus sub genus), and possibly eastern red cedar (*Juniperous virginiana*), described and radiocarbon dated from carbonized wood found within the site suggests a considerably more temperate environment (Moeller 1980). These tree species are commonly found in the Northern Hardwoods forest community that dominates the region today. Referencing a work by Eisenberg (1978), Moeller (1980:37) states that

> Pollen profiles dating to the time of Paleo-Indian have occasionally shown the presence of oak, but it was dismissed as a mere contaminant from the downward movement of pollen from subsequent layers, or from a statistical aberration reflecting the decline in other types of pollen and not a true increase in the amount of oak.

McWeeney (1995) uses macrofossil evidence of plant species (leaves, seeds, and wood or charcoal fragments) present in area bogs and alluvial fans for reconstructing paleo environments of southern New England at the time of the earliest Native American settlements. Her findings demonstrate a similar incongruence between the pollen-profile-generated environmental reconstructions and the reconstructions based on macrofossils. These results suggest that the Early Holocene can be characterized as having temperate climate similar to that experienced today. Nicholas (1988) proposes a glacial lake mosaic model to interpret early site locations in New England. This model uses detailed, large-scale paleoecological reconstructions that consider the physiographic features of northeastern North America (Gaudreau 1988). The glacial lake mosaic model suggests that climatic extremes are moderated near relict postglacial lakes, and vegetational patterns reflect these milder climates. These conditions would have attracted early settlers to the region.

These regional studies have been substantiated on a more global scale by recent ice-core data obtained from ice sheets in Greenland (Alley et al. 1993; Mayewski et al. 1993), Patagonia (Rabassa et al. 1996; McCulloch and Bentley 1997), and East Antarctica (Chappellaz et al. 1990), and from studies of littoral sediments (Allen et al. 1999; Taylor et al. 1993). Comparison of the data from these diverse sources suggests that the climate worldwide changed abruptly, in as little as 50 years (Alley et al. 1993), at the end of the Younger Dryas (ca. cal. 11,600 B.P.), and not gradually, as previously assumed. Recent studies of ice cores from Upper Fremont Glacier, Wyoming, reveal a similarly abrupt (10 yr) warming following the "Little Ice Age" (ca. cal. 1400–1800 or cal. 550–150 B.P.) (Schuster et al. 2000). In New England this climate change would have initiated a consequent change in vegetation communities from the spruce parkland community envisioned in earlier models to a Northern Hardwoods–Mixed Pine forest similar to that present today. Based on rates of plant community changes documented for southern Europe at this time (Allen et al. 1999), this change in forest vegetation would have taken a few hundred years at most. Thus by cal. 11,000 B.P., a Northern Hardwoods–White Pine forest community would likely have been established over most of New England, including the Champlain Valley.

PALEOINDIAN-PERIOD SITES IN THE CHAMPLAIN VALLEY

The Vermont Archeological Inventory (VAI) contains reports of numerous Paleoindian-Period sites within the Champlain Valley region. However, many of these lack meaningful locational and, on occasion, diagnostic data sufficient for analysis. Specific locational information is available for 29 sites that constitute the archaeological

Table 8.1. Site and Date Data Used in Figure 8.2

State	Site	Feature	Sample No.	14C Date	Range	Cal. B.C.	Cal. B.P.	Prob	Reference
PA	Sawnee-Minisink		W2994a	10,590	± 300	10,993	12,943 12,097	0.966	McNett 1977
PA	Sawnee-Minisink		W2994b	11,263	± 600	11,263	13,213 11,697	0.939	McNett 1977
CT	Templeton		W3931	10,190	± 300	10,385	12,335 11,338	0.966	Moeller 1980:31
NH	Lake Winnipesaukee		NA-1	9615	± 210	9247	11,197 10,686	0.957	referenced in Haviland and Power 1981:27
NH	Whipple (Weighted mean of two samples)		NA-2	8730	± 400	8315	10,265 9397	0.918	Curran 1984
ME	Vail	F-1	Beta 1833	11,120	± 180	11,256	13,206 12,921	0.815	Gramly 1982:60
ME	Vail	F-1	SI-4617	10,300	± 90	10,392	12,342 11,922	0.844	Gramly 1982:60
ME	Michaud	F-7	Beta 13833	9010	± 210	8455	10,405 9865	0.934	Spiess and Wilson 1987:83–84
ME	Michaud	F7a	Beta 15660	10,200	± 620	10,907	12,857 11,159	1.000	Spiess and Wilson 1987:86–84
NS	Debert (average of 11 dates)		NA-3	10,604	± 45	10,906	12,856 12,720	0.524	MacDonald 1985:53

database used in this study. These sites have been defined by identified artifact assemblages and recovered isolated projectile points. Reported, but unconfirmed, collectors' sites, and collections lacking site-specific proveniences, have not been included in this study. These early Native American sites have been assigned to the Paleoindian Period based on artifact styles (fluted projectile points, spurred scrapers) except for VT-CH-679, which yielded an Oxidizable Carbon Ratio (OCR) date of 10,182 ± 305 B.P. (Frink et al. 1996 [ACT No. 1710]). Fluted projectile points basal fragments and debitage were recovered during excavations of this site prior to commercial development.

Although the limited data does not define when people first settled in Vermont's Champlain Valley, archaeologists have commonly assumed linear migration from south to north, and hypothesized age ranges based on data from sites to the south within the Northeast (Haviland and Power 1981; Loring 1980; Thomas 1990). Using data from the Shawnee-Minisink site in Pennsylvania (McNett et al. 1977), the Templeton site in Connecticut (Moeller 1980), the Lake Winnipesaukee site in New Hampshire (Haviland and Power 1981), the Whipple site in New Hampshire (Curran 1984), the Vail site in Maine (Gamly 1982), the Michaud site in Maine (Spiess and Wilson 1987), and the Debert site in Nova Scotia, Canada (MacDonald 1985), we have compared site ages to latitude to determine the feasibility of this hypothesis (Table 8.1, Figure 8.2). Although the sample size may be arguably too small for statistical purposes, no relationship between the latitude and age of a site is apparent.

Anderson and Gillam (2000) propose a model for the colonization of North America based on least-cost solution pathways. They suggest that flat terrain, found predominantly along the coastal margins in the Northeast, would have provided primary paths for initial settlement. These paths may have been more important to early settlers than riverine corridors, particularly those in the high-relief interiors of upstate New York, New Hampshire, and Vermont. Thus the interior of the Northeast may have been settled significantly later than the flat coastal areas. Furthermore, this model suggests that the Champlain Valley was likely colonized from the north by people following the flat terrain along the Atlantic coast and the St. Lawrence Valley, rather than migrating north through the interior as traditionally proposed. The geographic trends suggested by the data in Figure 8.2 reflect Anderson and Gillam's hypothesized settlement pattern. Sites along the coastal regions (above the trendline for dates by latitude) are older than those of the interior regions (below the trendline for dates by latitude) of New Hampshire and the Champlain Valley of Vermont. Although there is not enough data to be mathematically rigorous, it suggests that set-

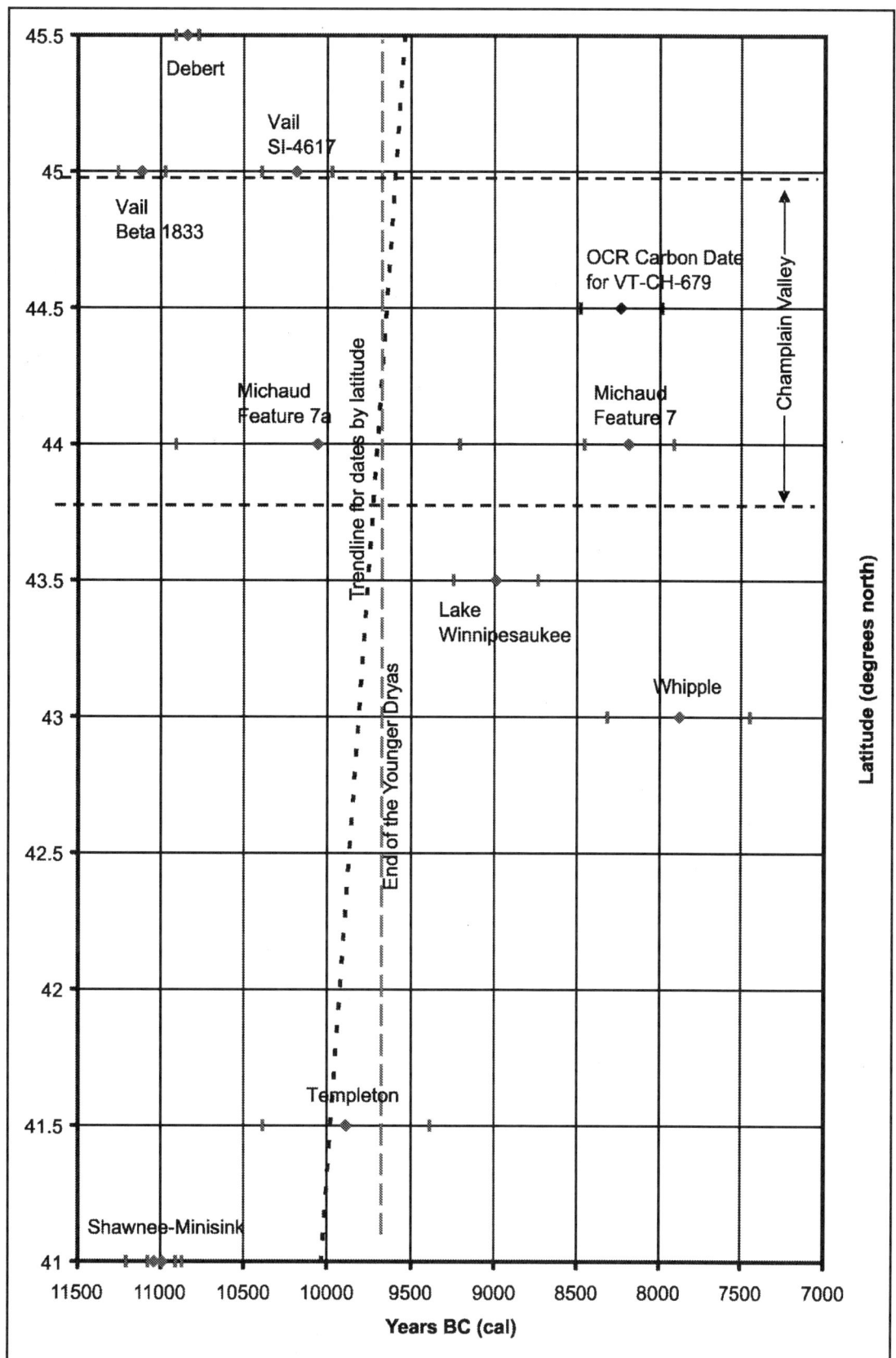

Figure 8.2. Plot of Paleoindian-Period site dates versus latitude.

tlement of the Champlain Valley by Native Americans postdates settlement of coastal areas in the Northeast.

Viewed from this perspective, the temporal data from throughout the Northeast strongly suggests that early Native American settlement of interior regions, including the Champlain Valley, did not occur until several hundred years after the climatic change at the end of the Younger Dryas (ca. cal. 11,600 B.P.). When early Native Americans settled in the Champlain Valley, the weather patterns, seasonality, forest environments, and associated floral and faunal resources would have approximated those that existed throughout the Holocene Period. Thus climatic conditions and the consequent forest environment during the early settlement of the Champlain Valley would not have differed significantly from the conditions encountered by later Native American cultures.

LATE PLEISTOCENE AND EARLY HOLOCENE LANDFORMS

During the late Pleistocene, much of North America was covered by continental glaciers. Although climatic conditions during early Native American settlement of the Champlain Valley may have approximated those that existed throughout the Holocene, the landscape was vastly different. Postglacial lakes, the incursion and eventual subsidence of the Champlain Sea, and the early stages of the present freshwater Lake Champlain created unique landscapes that existed only during the late Pleistocene and early Holocene Periods. Geomorphic and pedomorphic analyses may be applied to reconstruct the landscapes that were encountered by early Native American settlers. Despite landscape transformations, continuities are demonstrated in Native American settlement and procurement strategies by tracing Native American site locations on the reconstructed late Pleistocene and early Holocene landscapes.

The effects of the final glacial meltback on the developing Champlain Valley landscape may be examined using existing maps and databases. The hillshading base map and the U.S. Geological Survey digital elevation models (DEM) used in this study were obtained from the Vermont Geographic Information System (1998a, 1998b). Data on isostatic rebound tilting of the landform during the glacial meltback was derived from shoreline feature data given in Chapman (1937), Cronin (1977), Stewart and MacClintock (1969), and Wagner (1972). Early human-site-location data was obtained from the Vermont Archeological Inventory (1969 to present) and from Loring (1980). The data is georeferenced in the Vermont State Plane Coordinate System 4400 NAD83. Spreadsheet, graphic, and GIS applications were used to work with the data.[2]

RESULTS

Prior to 20,000 years ago, up to 3-km-thick glacial ice covered New England. During the next 4,000 to 5,000 years, this glacial mass stagnated and underwent a stochastic process of melting punctuated by relatively brief periods of glacial readvancement (e.g., the Shelburne stade). Melting at the glacier's surface first exposed the ridges and mountain peaks, with the valleys remaining below the ice. As melting continued and the valley ice began to retreat, ice- and till-impounded ephemeral lakes of meltwater formed.

By ca. cal. 15,700 B.P. (Chapman 1937), a sequence of glacial lake stages collectively known as glacial Lake Vermont began to form within the Champlain Valley. Each lake stage is defined by relict beach terraces at successively lower elevations that formed during periods of equilibrium in the glacial ice regime as the overall glacial ice mass retreated northward. At least three separate stages of this lake have been defined: Quaker Springs, Coveville, and Fort Anne (Figure 8.3). Although no studies have been undertaken to determine empirical ages for these stages,[iii] glacial Lake Vermont existed from ca. cal. 15,700 until 14,000 B.P. when the impounding ice retreated north of the St. Lawrence Valley and the meltwater drained (McDonald 1968; Wagner 1972).

Meltwater from thousands of other glacial lakes flowed into the oceans, which resulted in the steady rise of sea levels relative to the land. By about cal. 14,000 B.P. the rising sea levels filled the St. Lawrence, Great Lakes, and Champlain Basins to form a large estuary known as the Champlain Sea. Four phases of the Champlain Sea have been defined based on identified relict beach terraces: Champlain Sea Maximum, Pre-Port Kent, Port Kent, and Burlington phases (Figure 8.4). Radiocarbon analysis of shell deposits associated with the Port Kent phase (Wagner 1972) has yielded an age of 11,300 B.P., or 13,200 cal. years ago. Sediment studies (Chase 1972) suggest that the Champlain Sea ended ca. 10,200 B.P. (ca. cal. 12,000 B.P.)

The Champlain Sea phases were located at successively lower elevations relative to the land. Although sea levels rose with glacial meltwater, isostatic rebound of the land (no longer compressed under the weight of nearly 3 km of ice) elevated the Champlain Valley relative to eustatic sea levels. By ca. cal. 12,000 B.P. isostatic rebound separated the St. Lawrence, Great Lakes, and Champlain Basins from the ocean, and these systems returned to freshwater regimes. Isostatic rebound, con-

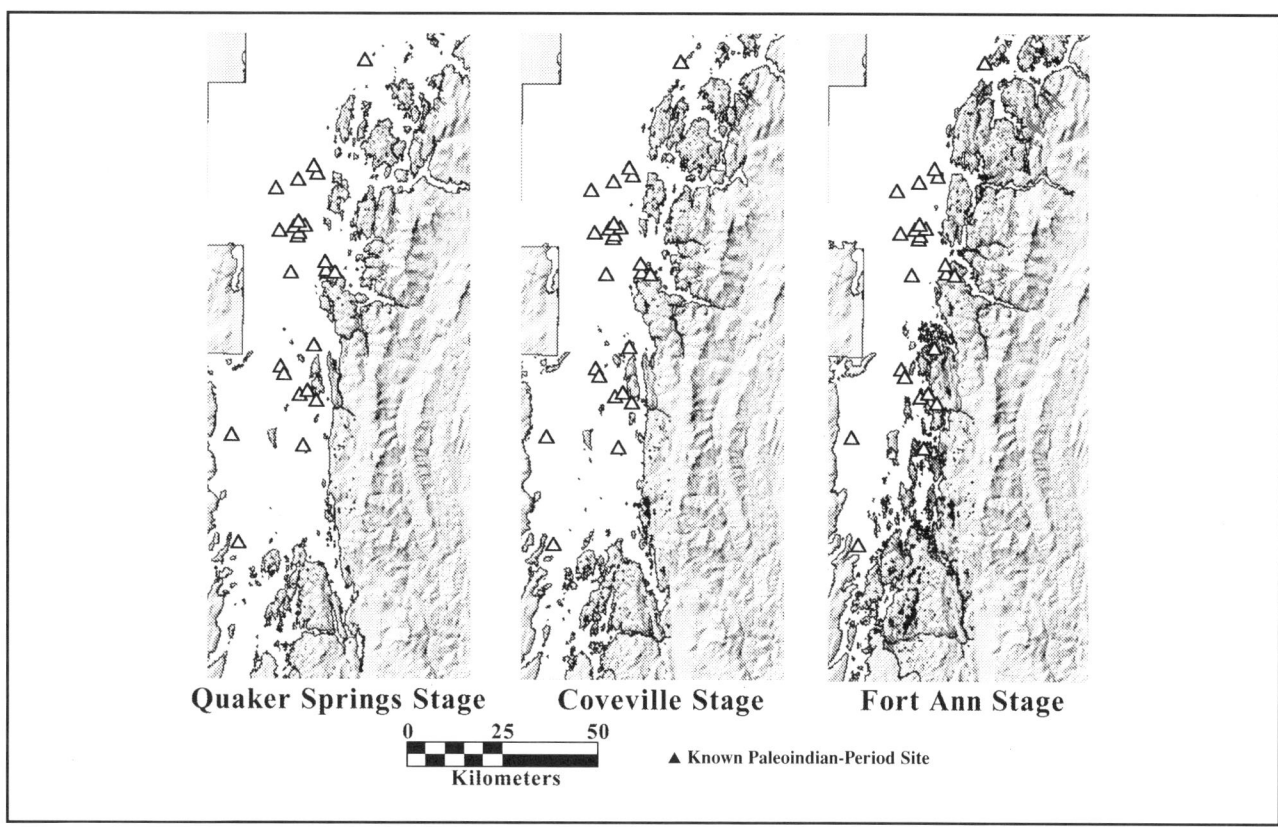

Figure 8.3. Three defined stages of Lake Vermont.

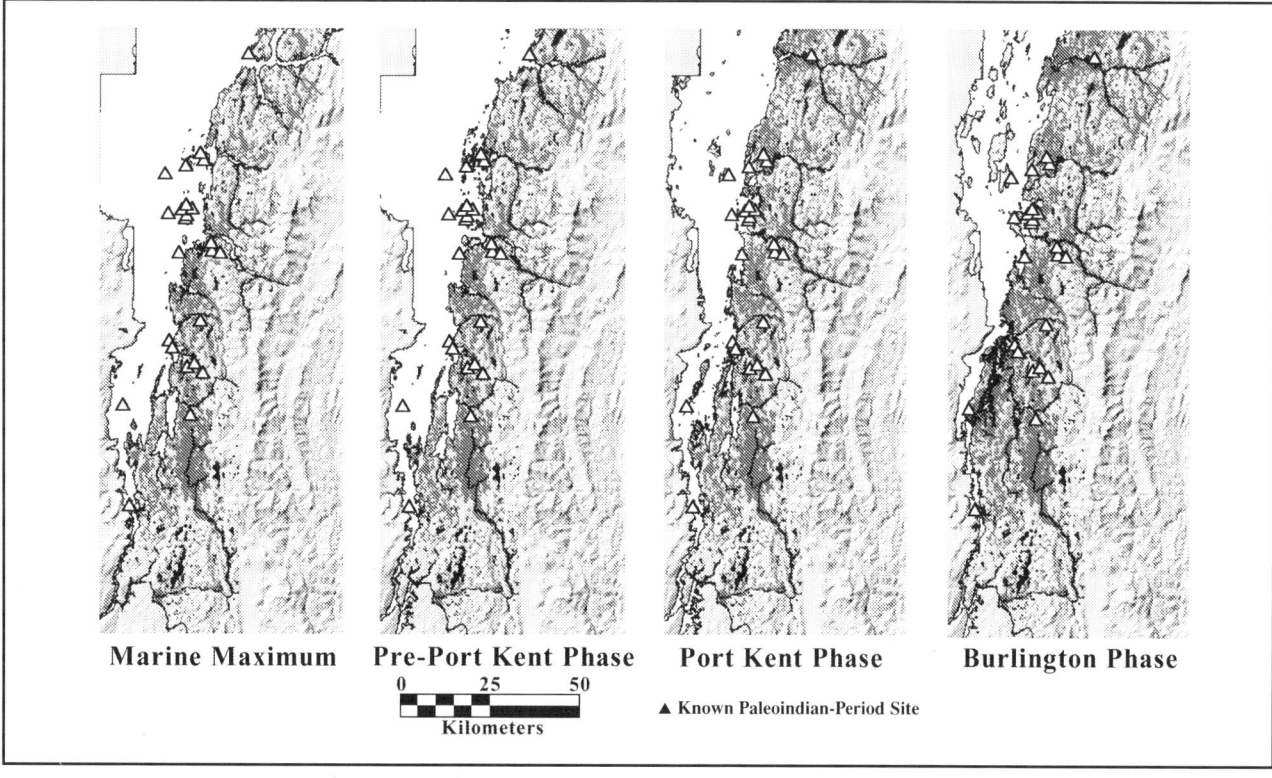

Figure 8.4. Four defined stages of Champlain Sea.

Chapter 8 *Behavioral Continuity on a Changing Landscape* **109**

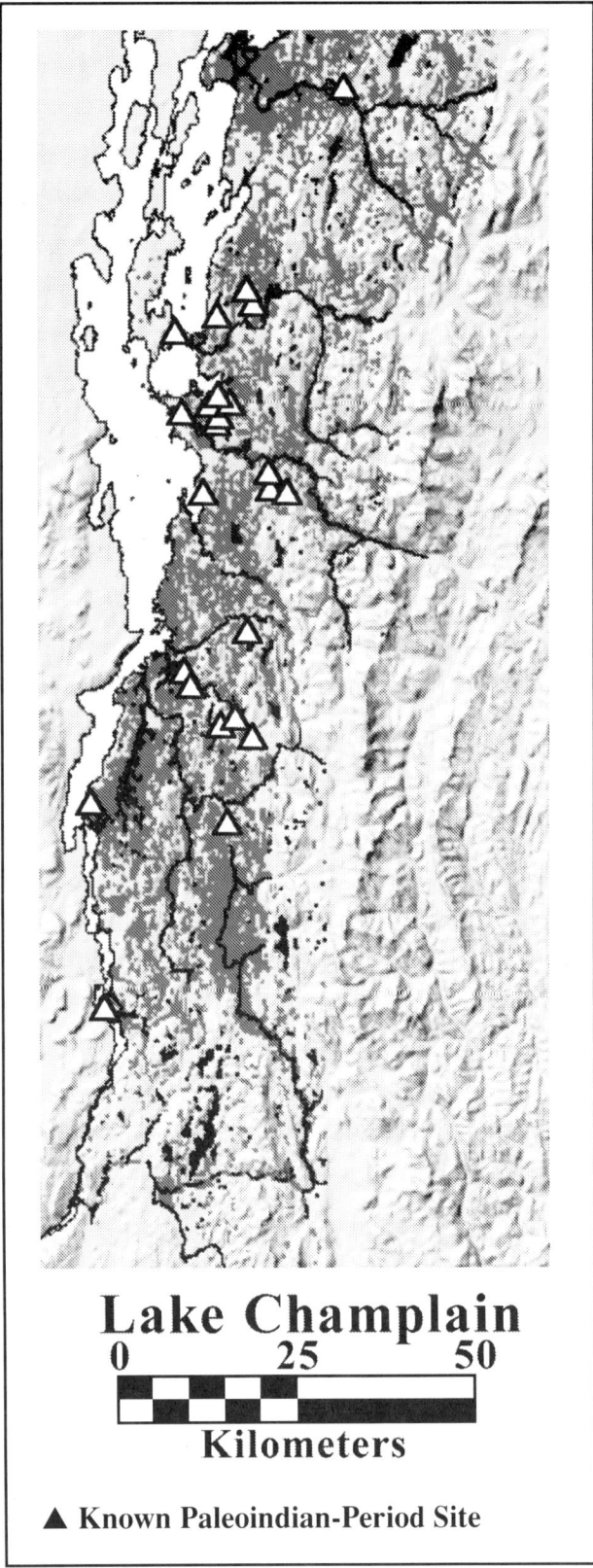

Figure 8.5. Lake Champlain today.

tinuing to the present day, has slowly raised the outlet of freshwater Lake Champlain, with a consequent rise of lake levels relative to the land (Figure 8.5).

A mosaic of changing environments emerged from the inundated conditions associated with evolving glacial lakes, saltwater estuaries, and freshwater lakes. Meltwater from glacial ice impounding adjacent valleys carried gravels, sands, silts, and clay sediments that settled in lake basins and mantled bedrock and tills. Rivers eroded unconsolidated glacial outwash, ice contact features, such as kames and eskers, and former glacial lakes sediments. The rivers transported continuous loads of these eroded sands, silts, and clays to the saltwater estuary and later freshwater bays. As the levels of the glacial lakes and saltwater estuary dropped with the retreating ice, the newly exposed sediment deposits evolved into a mosaic of soils that coevolved with microbial and vegetative communities.

These emergent landscapes did not present a flat and uniform surface. Small differences in initial topographic conditions created divergence in soil seriation (Philips 1999). The newly emergent landscape, best described as undulating, followed the topography of underlying bedrock and glacial till deposits. Numerous small lakes and ponds would have remained separated by elevation from the major lakes and estuaries within this undulating landscape, and surrounding soils would have supported vegetative communities appropriate to the climatic regime of the time period and topographic position.

Some of these lakes and ponds probably lasted only a few decades, while others, although greatly reduced in size, remain today. As drainage systems evolved on the emergent landscapes, most of the residual ponds and lakes drained or underwent eutrification due to vegetation and eroding deposits from upstream. These residual ponds and lakes evolved into marshes, woodland wetlands, and eventually into moderately to poorly drained forest lands.

The U.S. Department of Agriculture soil taxonomic system is based on the recognition that individual soil series are the result of the five interdependent factors of parent material, climate, relief, biota, and time (Jenny 1941, 1961). The coevolution of soils and biological communities is fundamental to the soil classification system, resulting in a direct correlation between soils of similar historic or processual genesis and forest communities. This correlation allows for the reconstruction of pre-European contact forest communities based on defined soil series (Frink 1996). Variability in the defined soil series has been shown to be extremely sensitive to initial conditions (Philips 1995). The evolution of soils with similar textures and mineralogy (parent material) will vary with initial conditions (e.g., beginning as emergent dryland soils vs. evolving from lakes and ponds through marsh and woodland wetlands before becoming emergent dryland soils) and result in distinct soil series. This inherent genetic thread in the defined soil taxonomic units may be used to locate former residual lakes and ponds and former forest communities. As shown in Figure 8.6, a large portion (11 percent) of the emergent landscape would have remained as post-glacial lakes and ponds when Native Americans first arrived in Vermont's Champlain Valley.

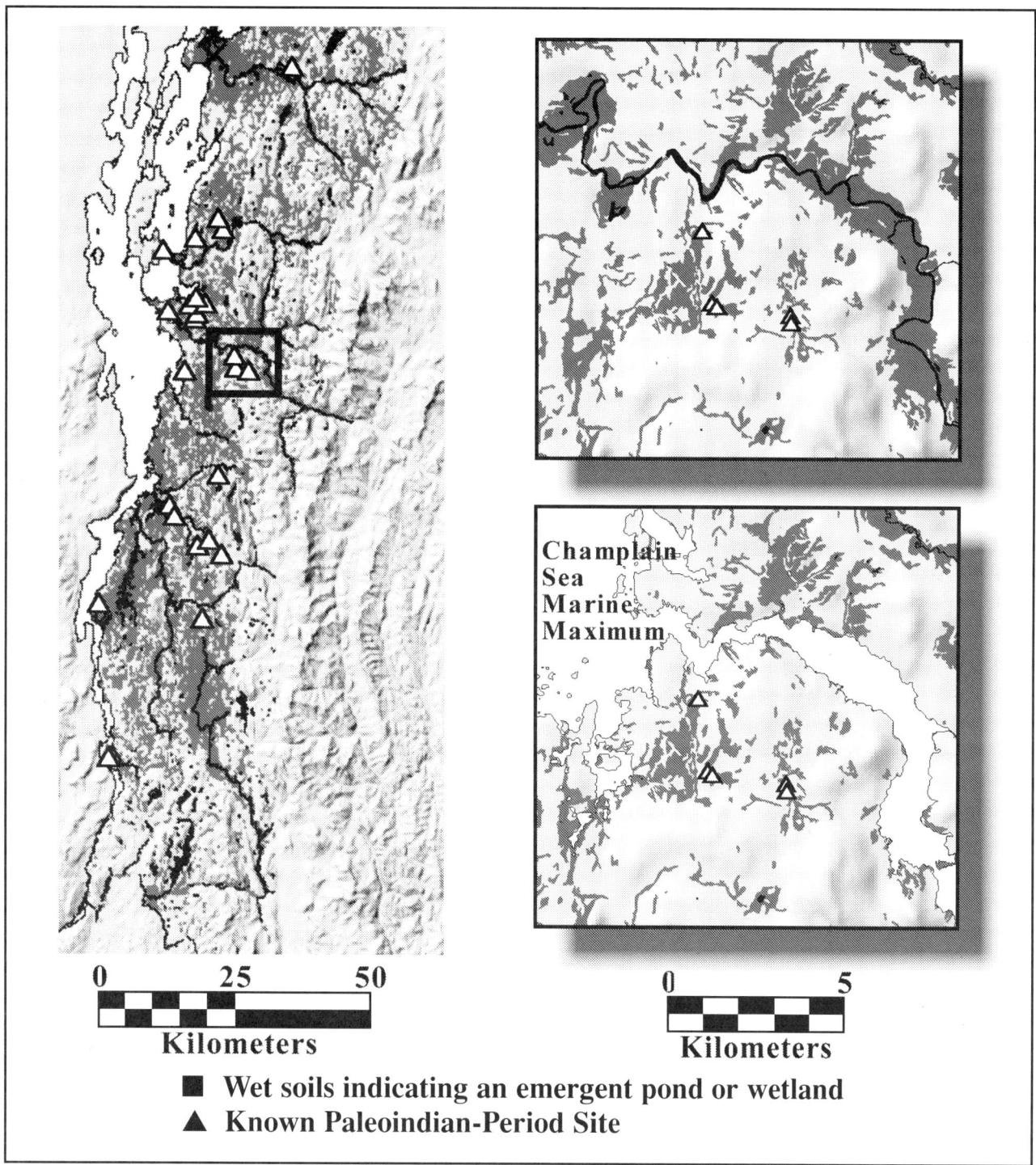

Figure 8.6. Enlarged view of emergent ponds and wetlands as indicated by soils.

Paleoindian-Period Site Locations and Champlain Sea Margins

Figure 8.4 includes Paleoindian-Period site locations in comparison with each of the four Champlain Sea phases. Most (21 of 29, 72 percent) locations were submerged at the time of the Champlain Sea Maximum (ca. cal. 13,800 B.P.). The majority of the locations remained submerged during the Pre-Port Kent Phase ca. cal. 13,500 B.P. (17 of 29, 59 percent), and the Port Kent Phase ca. cal. 13,200 B.P. (16 of 29, 55 percent). However, during the Burlington Phase (ca. cal. 12,500 B.P.), all known Paleoindian-Period site locations were above sea level.

Although the data does not appear to support the hypothesis that Paleoindian-Period sites are associated with the Champlain Sea Maximum, a significant percentage of these sites (24 percent) is within 61 m of the shoreline landforms associated with the four Champlain Sea phases. However, given that the sites would have postdated the demise of the Champlain Sea (post cal. 12,500 B.P.), this relation probably suggests that Native Americans selected sites based on the landforms and the associated soils and vegetative communities that evolved on them, rather than potential resources in the saltwater estuary.

The Landscape at the Time of First Settlement: Freshwater Lakes, Ponds, and the Champlain Sea

As argued above, temporal data throughout the Northeast suggests that early Native American settlement in the region, including the Champlain Valley, may not have occurred until after the climatic change at the end of the Younger Dryas (cal. 11,600 B.P.). When early Native Americans arrived in the Champlain Valley, the weather patterns, seasonality, and forest environments with their associated floral and faunal resources would have approximated those that existed throughout the Holocene. We have also introduced data on the unique landscape characteristics of emergent landforms and shown that 11 percent of the emergent lands would have consisted of freshwater lakes and ponds in various stages of evolution toward marshes and woodland wetlands. Over 40 percent of the Paleoindian-Period site locations are located along the boundaries of residual freshwater lakes and ponds on this reconstructed emergent landform (Figure 8.6).

Continuity in Native American Site Settlement and Procurement Patterns

Frink (1996) presents site distribution data for known Native American sites in Chittenden County, Vermont, within reconstructed forest environments. This data indicates that Native Americans distinctly preferred certain environmental communities in their selection of site locations. Frink hypothesizes that these environmental communities were selected due to their conspicuous seasonal high biomass, which afforded Native Americans with a wide range and large quantity of exploitable resources. A distinct preference (43 percent of known sites) is shown for locations adjacent to freshwater marshes associated with lakes and ponds. Given that the sites themselves are not located in the marshes, the adjacent forest communities occupying the dry land areas are also considered, resulting in 47 forest communities associated with the 29 sites (Table 8.2). When Paleoindian-Period sites in Vermont's Champlain Valley are plotted against similarly reconstructed environmental communities, a similar site preference emerges (Table 8.3).

Table 8.2. Native American Site Components Associated with Defined Forest Communities for Chittenden County, Vermont

Ecological Environment	n	%
Northern Hardwoods – White Pine, Oak Dominant	52	6.2
Northern Hardwoods – White Pine, Maple, Ash, and Beech	40	4.8
Northern Hardwoods – White Pine, Oak, Ash, and Hickory	24	2.9
Northern Hardwoods –Hemlock – Spruce	18	2.1
Pine – Hemlock – Oak	93	11.0
Bottomland Hardwoods	125	14.8
Spruce – Alpine	—	—
Freshwater Marshes	365	43.3
Perpetually Juvenile: Winter Deer Yards	126	14.9
Total	843	100.0

Note. Data from Frink 1996.

Table 8.3. Paleoindian-Period Site Components Associated with Defined Forest Communities in the Champlain Valley

Ecological Environment	n	%
Northern Hardwoods – White Pine, Oak Dominant	8	17.0
Northern Hardwoods – White Pine, Maple, Ash, and Beech	2	4.3
Northern Hardwoods – White Pine, Oak, Ash, and Hickory	6	11.8
Northern Hardwoods –Hemlock – Spruce	—	—
Pine – Hemlock – Oak	11	3.4
Bottomland Hardwoods[a]		
Spruce – Alpine	—	—
Freshwater Marshes	20	42.6
Perpetually Juvenile: Winter Deer Yards[b]		
Total	47	100.1

[a] The dynamic geomorphic nature of the river systems where Bottomland Hardwoods forest communities are located has likely affected this data. Sites dating back to the Paleoindian Period have likely been eroded away by the meandering rivers as they downcut through the glacial lake and Champlain Sea deposits, or have been buried under several meters of alluvium where the rivers have been aggrading.

[b] Forest communities identified as "Perpetually Juvenile" are evolved from the postglacial ponds and lakes. Although some of the identified postglacial ponds and lakes may have already undergone the transformation to Perpetually Juvenile forest communities, there has been insufficient studies to determine which, if any, had done so.

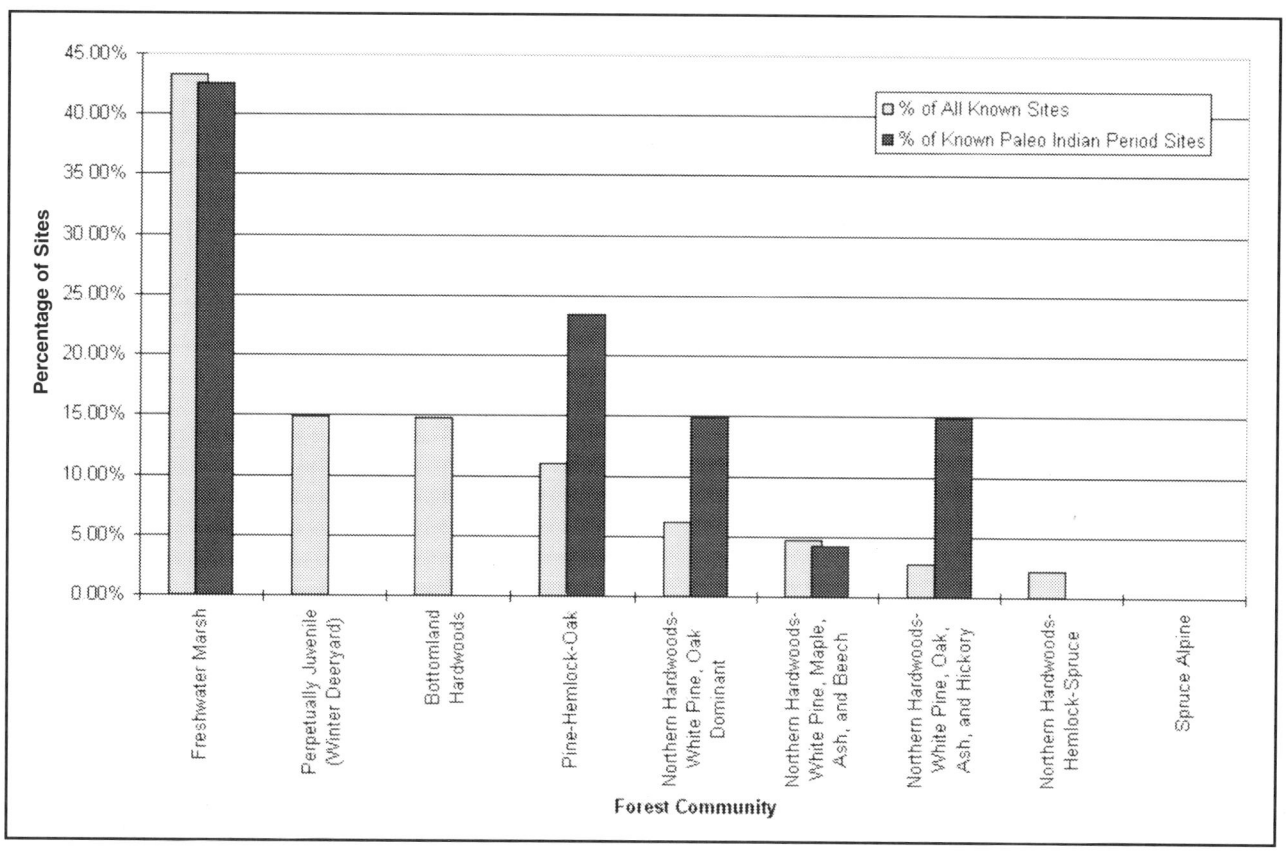

Figure 8.7. Comparison by association with reconstructed forest communities of Paleoindian-Period sites in the Champlain Valley with all known sites in Chittendon County, Vermont.

Chapter 8 *Behavioral Continuity on a Changing Landscape*

Discounting the Bottomland Hardwoods and the Perpetually Juvenile forest communities, a preference for site location relative to forest communities during the Paleoindian Period qualitatively parallels all Native American sites throughout the Holocene. Table 8.3 indicates a distinct preference for locations adjacent to freshwater marshes, followed by the Pine Hemlock Oak and the Northern Hardwoods–White Pine with Oak dominant forest communities. Site locations in the Northern Hardwoods–White pine, Oak Ash Hickory dominant, and the Maple, Beach, Ash dominant communities are also represented, although to a lesser extent. Variability in the quantitative values between these two studies is likely due to the small number of known Paleoindian-Period sites, and the larger geographic sample area of the Champlain Valley versus Chittenden County (Figure 8.7).

CONCLUSION

As archaeological site information is generally incomplete due to postdepositional changes, conclusions drawn from the data must remain hypotheses. As hypotheses, they must be constantly tested against new data both from within archaeology and from other disciplines. This chapter tests one such hypothesis that describes site locational strategies for the Paleoindian Period in the Champlain Valley of Vermont. The settlement model that associates Paleoindian-Period sites with Champlain Sea shorelines fails to explain newly discovered sites not located on shoreline features. This incongruity has been explained by suggesting that the settlement and procurement patterns of behavior for early Native American settlers may have differed significantly from those of later populations.

The time-specific geomorphology and environmental data presented in this chapter demonstrate a continuity in settlement and procurement behaviors throughout the Holocene Period. Environmental conditions during this early settlement period were not as severe as once thought, and early Native American settlement appears to postdate the Champlain Sea. Early Native Americans arrived to a postemergent landscape dominated by a mosaic of freshwater marshes and maturing forest environments. Their site choices paralleled those exhibited by later Native Americans. Thus predictive models for Paleoindian-Period sites in the Champlain Valley should be based on time-specific landscape characteristics and not upon a static reconstructed marine limit or the present environment.

END NOTES

1. Dates are reported in the literature in many different formats. To assist the readers, all dates are given with both their explicit format and a common format of calibrated years (cal. yr) before present (B.P.) defined as 1950 A.D.

2. The following programs were used to work with the data:
   - ArcView copyright Environmental Systems Research Institute (ESRI)
   - CorpsCon copyright U.S. Army Corps of Engineers
   - Microsoft Excel copyright Microsoft Corp.
   - Corel PhotoPaint copyright Corel Corp.
   - GlacialShapeFile copyright Archaeology Consulting Team, Inc. (ACT)

   The elevation and approximate northing of the shoreline features for each geologic stage was entered into an Excel worksheet, the data was plotted as an X, Y scatterplot, and a linear regression equation was calculated for each data set. The USGS DEM data was downloaded in ASCII X, Y, Z (easting, northing, elevation) format. The GlacialShapeFile program was used to "regress" the given DEM elevations to the isostatic tilt derived from the Excel worksheet, to identify submerged areas for each geologic stage, and trace the outlines of those areas in an Environmental Systems Research Institute (ESRI) compatible shapefile format. ArcView was used to assemble and display the maps and shapefiles, and the resulting view was exported to PhotoPaint for conversion to graphics format files.

3. An OCR date of 13,735 +/- 412 cal. yr B.P. (ACT No. 3598) measuring the post-emergent pedogenics of soil deposits from the Fort Ann stage were obtained from behind Pinewood Manor, Essex Junction (Frink and Hathaway 1999).

REFERENCES CITED

Allen, J.R.M., Brandt, U., Brauer, A., Hubberten, H.W., Huntley, B., Keller, J., Kraml, M., Mackensen, A., Mingram, J., Negendank, J., Nowaczk, N., Obehansli, H., Watts, W.A., Wulf, S., and Zolitschka, B. (1999). Rapid environmental changes in southern Europe during the last glacial period. *Nature* **400**:740–743.

Alley, R.B., Meese, D.A., Shuman, C.A., Gow, A J., Taylor, K.C., Gootes, P.M., White, J.W.C., Ram, M., Waddington, E.D., Mayewski, P.A., and Ziellinski, G.A. (1993). Abrupt increase in Greenland snow accumulation at the end of the Younger Dryas event. *Nature.* **362**:527–529.

Anderson, D.G., and Gillam, J.C. (2000). Paleoindian colonization of the Americas: Implications from an examination of physiography, demography, and artifact distribution. *American Antiquity* **65**:43–66.

Chapman, D.H. (1937). Late-glacial and postglacial history of the Champlain Valley. *American Journal of Science* **34**:89–124.

Chappellaz, J., Barnola, J. M., Raynuad, D., Korotkevich, Y. S., and Lorius, C. (1990). Ice-core record of atmospheric methane over the past 160,000 years. *Nature* **345**:127–131.

Chase, J.S. (1972). Operation UP-SAILS: Sub-bottom profiling in Lake Champlain. Unpublished master's thesis, University of Vermont, Burlington.

Cronin, T. (1977). Late-Wisconsin marine environments of the Champlain Valley (New York, Quebec). *Quaternary Research* **7**:238–253.

Curran, M.L. (1984). The Whipple site and Paleoindian tool assemblage variation: A comparison of intrasite structuring. *Archaeology of Eastern North America* **12**:5–40.

Curran, M.L. (1996) Paleoindians in the Northeast: The problem of dating fluted point sites. *The Review of Archaeology* **17**:2–5.

Dent, R.J. (1999). Poster presentation in the *North American Paleoindian and Archaic Session.* Annual meeting of the Society for American Archaeology, Chicago.

Eisenberg, L. (1978). *Paleo-Indian Settlement Patterns in the Hudson and Delaware River Drainages.* Occasional Publications in Northeastern Anthropology No. 4. Bethlehem, Conn.

Frink, D.S. (1996). Asking more than where: Developing a site contextual model based on reconstructing past environments. *North American Archaeologist* **17**:307–336.

Frink, D.S., Perreault, K., and Baker, C.M. (1996). *Phase I Site Identification, Phase II Site Evaluation, and Phase III Data Recovery Archaeological Studies at the Proposed L&M Park Development, Area C South Burlington, Chittenden County, Vermont.* Archaeology Consulting Team, Essex Junction, Vt.

Frink, D.S., and Hathaway, A.D. (1999). Pedomorphic understandings of river systems' behavior through space and time: Key to managing cultural resources. Paper presented at the 64th Annual Meeting of the Society for American Archaeology, Chicago.

Gaudreau, D.C. (1988). The distribution of late Quaternary forest regions in the Northeast: pollen data, physiography, and the prehistoric record. In *Holocene Human Ecology in Northeastern North America,* edited by G.P. Nicholas, pp. 215–256. Plenum Press, New York.

Gramly, R.M. (1982). *The Vail Site: A Paleo-Indian Encampment in Maine.* Vol. 30. Bulletin of the Buffalo Society of Natural Sciences. Buffalo, N.Y.

Haviland, W.A., and Power, M.W. (1981). *The Original Vermonters: Native Inhabitants, Past and Present.* University Press of New England, Hanover, N.H.

Jenny, H. (1941). *Factors of Soil Formation: A System of Quantitative Pedology.* McGraw-Hill, New York.

Jenny, H. (1961). Derivation of state factor equations of soils and ecosystems. *Soil Science Society of America Proceedings* **25**:385–388.

Loring, S. (1980). Paleo-Indian Hunters and the Champlain Sea: A presumed association. *Man in the Northeast* **19**:15–41.

MacDonald, G.F. (1985). *Debert, A Paleo-Indian Site in Central Nova Scotia.* Persimmon Press, Buffalo, N.Y., in collaboration with the National Museum of Man, National Museums of Canada.

Mayewski, P.A., Meeker, L.D., Whitlow, S., Twickler, M.S., Morrison, M.C., Alley, R.B., Bloomfield, P., and Taylor, K. (1993). The atmosphere during the Younger Dryas. *Science* **261**:195–197.

McCulloch, R.D., and Bentley, M.J. (1997). Late Glacial Ice Advances in the Strait of Magellan. *Quaternary Science Review* **17**:775–787.

McDonald, B.C. (1968). Deglaciation and differential postglacial rebound in the Appalachian region of southeastern Quebec. *Journal of Geology.* **76**:664–677.

McNett, C., Jr., et al. (1977). The Shawnee-Minisink site. In *Amerinds and Their Paleoenvironments in Northeastern North America,* edited by W.S. Newman and B. Salwen. New York Academy of Sciences Annals **288**:282–296.

McWeeney, L.J. (1994). Archaeological settlement patterns and vegetation dynamics in southern New England in the Late Quaternary. Unpublished doctoral dissertation, Dept. of Anthropology, Yale University, New Haven, Conn.

McWeeney, L.J. (1995). Searching for the Younger Dryas Period and Paleoindians in southern New England. Paper presented at the 1995 annual meeting of the Eastern States Archaeological Federation, Wilmington, Del.

Moeller, R.W. (1980). *6 LF 21: A Paleo-Indian Site In Western Connecticut.* Occasional Paper No. 2. American Indian Archaeological Institute. Washington, Conn.

Nicholas, G.P. (1987). Rethinking the Early Archaic. *Archaeology of Eastern North America* **15**:99–124.

Nicholas, G.P. (1988). Ecological leveling: The archaeology and environmental dynamics of early post glacial land use. In *Holocene Human Ecology in Northeastern North America,* edited by G.P. Nicholas, pp. 257–296. Plenum Press, New York.

Philips, J.D. (1995). Self-organization and landscape evolution. *Progress in Physical Geography* **19**:309–321.

Philips, J.D. (1999). Divergence, convergence, and self-organization in landscapes. *Annals of the Association of American Geographers* **89**:466–488.

Rabassa, J., Heusser, C., and Stuckenrath, R. (1996). New data on Holocene sea transgressions in the Beagle Channel: Tierra del Fuego. *Quaternary of South America and Antarctic Peninsula* **4**:91–309.

Schuster, P.F., White, D.E., Naftz, D.L., and Cecil, D. (2000). Chronological refinement of an ice core record at Upper Fremont Glacier in south central North America. *Journal of Geophysical Researce* **105**(D4):4657–4666.

Spiess, A.E., and Wilson, D.B. (1987). *Michaud, a Paleoindian site in the New England-Maritimes region.* Occasional Publications in Maine Archaeology No. 6, Augusta.

Stewart, D., and MacClintock, P. (1969). *The Surficial Geology and Pleistocene History of Vermont*. Bulletin No. 19. Vermont Geological Survey, Montpelier.

Taylor, K.C., Lamorey, G.W., Doyle, G.A., Alley, R.B., Grootes, P.M., Mayewski, P.A., White, J.W.C., and Barlow, L.K. (1993). The 'flickering switch' of late Pleistocene climate change. *Nature* **361**:432–436.

Thomas, P. (1990). Draft state historic preservation plan: Prehistory theme. Prepared by the Consulting Archeology Program, University of Vermont, for the Division for Historic Preservation, Montpelier. Copies on file at the Division for Historic Preservation, Montpelier, Vt.

Vermont Division for Historic Preservation (VDHP). (1969–present). *VAI, Vermont Archeological Inventory*. Vermont Division for Historic Preservation. Montpelier.

Vermont Geographical Information System (VGIS). (1998a). *Hillshading Generated from 1:250000 U.S. Geological Survey Digital Elevation Models (USGS DEMs)*. Vermont Center for Geographic Information. Available from: http://geo-vt.uvm.edu.

Vermont Geographical Information System (VGIS). (1998b). *U.S. Geological Survey Digital Elevation Model Data (USGS DEM)*. Vermont Center for Geographic Information. Available from: http://geo-vt.uvm.edu.

Wagner, W.P. (1972). Ice margins and water levels in northwestern Vermont. *New England Intercollegiate Geological Conference Guidebook*, **64**:319–331.

SECTION III
ALLUVIAL SETTINGS

CHAPTER 9

POSTGLACIAL DEVELOPMENT OF THE PENOBSCOT RIVER VALLEY: IMPLICATIONS FOR GEOARCHAEOLOGY

Alice R. Kelley and David Sanger

Rapid geological changes occurred in the Penobscot River valley after glaciation. River systems and drainage patterns developed following deglaciation and subsequent marine inundation. These geological processes, as well as other environmental forces, influenced human decisions relating to resources, shelter, and travel. Regional and local geology influenced resource procurement and site formation and preservation. These processes shaped, and potentially biased, the archaeological record. For this reason, an understanding of the postglacial geological and environmental development of the area is vital to reconstructing prehistoric lifeways.

This study represents the compilation of more than 10 years of archaeological investigation in the Penobscot Valley, combined with geological studies relating to postglacial isostatic adjustment and relative sea level changes conducted during the same time. Archaeological excavations offered a window into the alluvial stratigraphy of the area by providing the opportunity to examine sedimentary sequences in a variety of locations. Geological insights supplied the key to interpreting the sediments and recognizing the dynamic, postglacial landscape of the region.

The Penobscot River drainage comprises 24,306 km^2 of central Maine (Figure 9.1). On the basis of drainage area and average discharge (465 m^3/sec), it is the largest river in the state (U.S. Army Corps of Enginners 1990). The river is more than 160 km long, from the headwaters of either of two major tributaries, the East and West Branches, to its mouth in Penobscot Bay. The upper reaches of the river flow through rugged, mountainous terrain. Numerous islands and rapids characterize the central portion. The lower portion of the river, from Bangor to Penobscot Bay, is tidally influenced. Moosehead Lake, with an area of 303 km^2, is Maine's largest lake and is located immediately to the south of the present day watershed of the West Branch of the Penobscot River. Currently Moosehead Lake forms the source of the Kennebec River, Maine's second largest river.

Although this study addresses the geology of the entire region, archaeological examples are drawn from sites located on the main stem of the river between Old Town and Bangor, Maine (Figure 9.1). Much of the work has been done as part of cultural resource management efforts related to the relicensing of existing dams in the region and the proposed construction of new hydroelectric facilities.

ARCHAEOLOGICAL SITES

The Early Paleoindian Period (11,000–10,000 B.P.) represents the earliest identified human occupation in Maine; the human presence is characterized by distinctive fluted points made of cryptocrystalline quartz (Spiess et al. 1998). Fluted points have not been found in the Penobscot drainage. However, parallel-flaked points associated with the Late Paleoindian Period (10,000–9500 B.P.) have been recovered at the Blackman Stream (Sanger et al. 1992) and Eddington Bend sites (Petersen and Sanger 1987; Sanger et al., this volume). Early (9500–7500 B.P.) and Middle Archaic (7000–6000 B.P.) sites within the Penobscot drainage have been found in thick, stratified alluvial settings adjacent to riverbanks (e.g., Petersen et al. 1986; Putnam 1994; Petersen 1991; Robinson and Petersen 1992; Sanger et al. 1992). Late Archaic-Period (6000–3000 B.P.) sites are more numerous than older sites, and occur throughout the drainage basin. Ceramic-Period (Woodland) (3000–350 B.P.) sites are widely distributed in the Penobscot Basin and are the most numerous sites in the Old Town to Bangor section of the river (Mack et al. 2002).

BEDROCK GEOLOGY

The bedrock geology is important to the geoarchaeology of the region for two reasons: (1) Rock types and the geological structures that bring these formations to the earth's surface shape the character of the landscape. (2) Specific geological units provide lithic resources for the manufacture of tools. Topographic highs and the pres-

---

Geoarchaeology of Landscapes in the Glaciated Northeast edited by David L. Cremeens and John P. Hart. New York State Museum Bulletin 497. © 2003 by the University of the State of New York, The State Education Department, Albany, New York. All rights reserved.

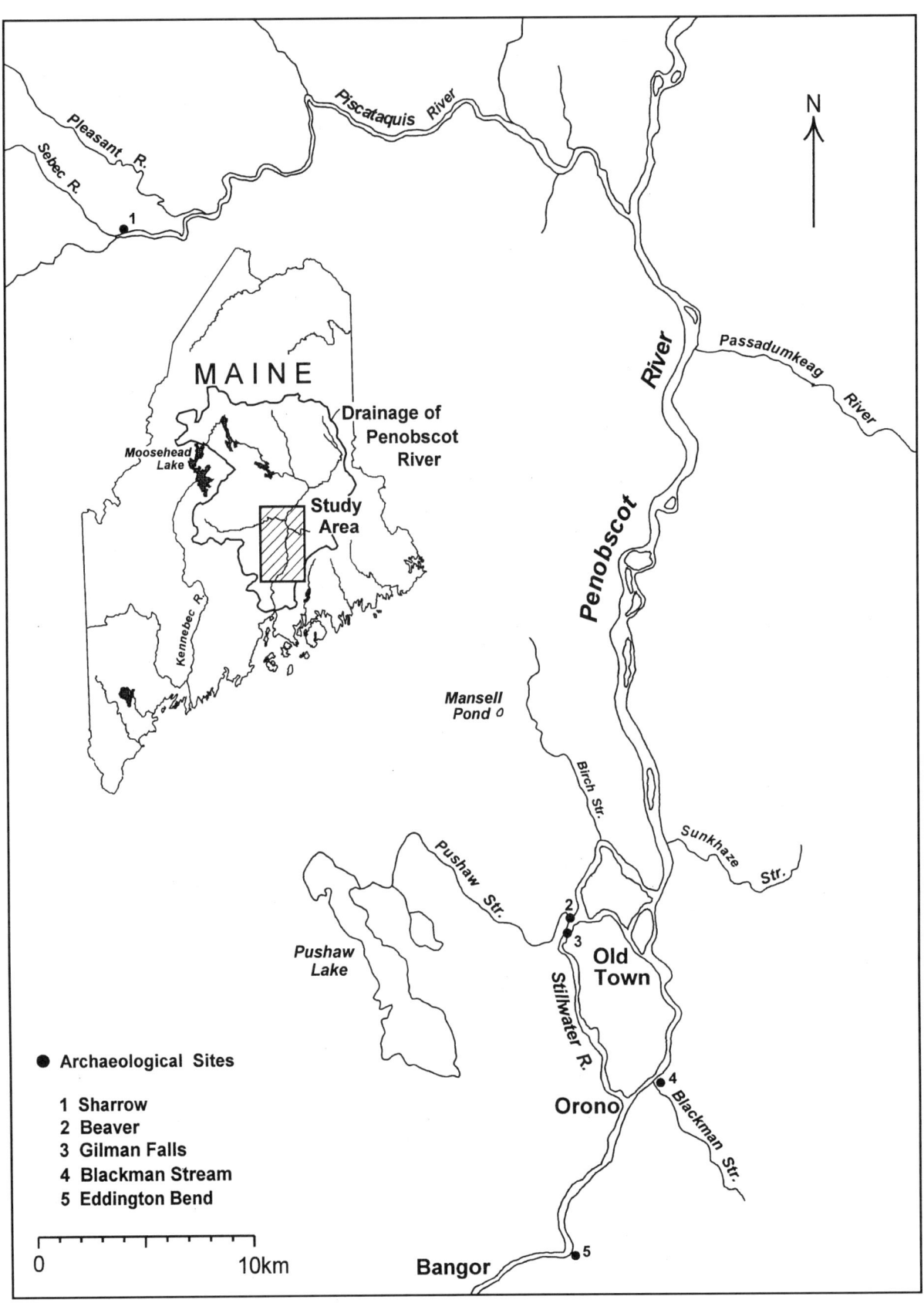

Figure 9.1. Central Maine with locations of major rivers, lakes, and archaeological sites mentioned in text.

ence of rapids and waterfalls are typically related with erosion-resistant rock types. Low-relief areas are associated with easily eroded formations.

Metamorphic and igneous rocks underlie the Penobscot River drainage (Figure 9.2). The Katahdin Pluton, a Devonian-age igneous body composed of granitic intrusives and associated volcanic rocks, dominates the mountainous upper portion of the drainage. Minor low-grade, lower Paleozoic metasedimentary facies are also found in the region. The erosion-resistant rocks found in the upper portions of the drainage are responsible for the rugged topography of the region; lakes are located in more easily eroded lows. The broad, rolling landscape of the central Penobscot drainage is underlain by easily eroded lower Paleozoic, low-grade, fine-grained metasedimentary rocks. Exposures of more resistant facies

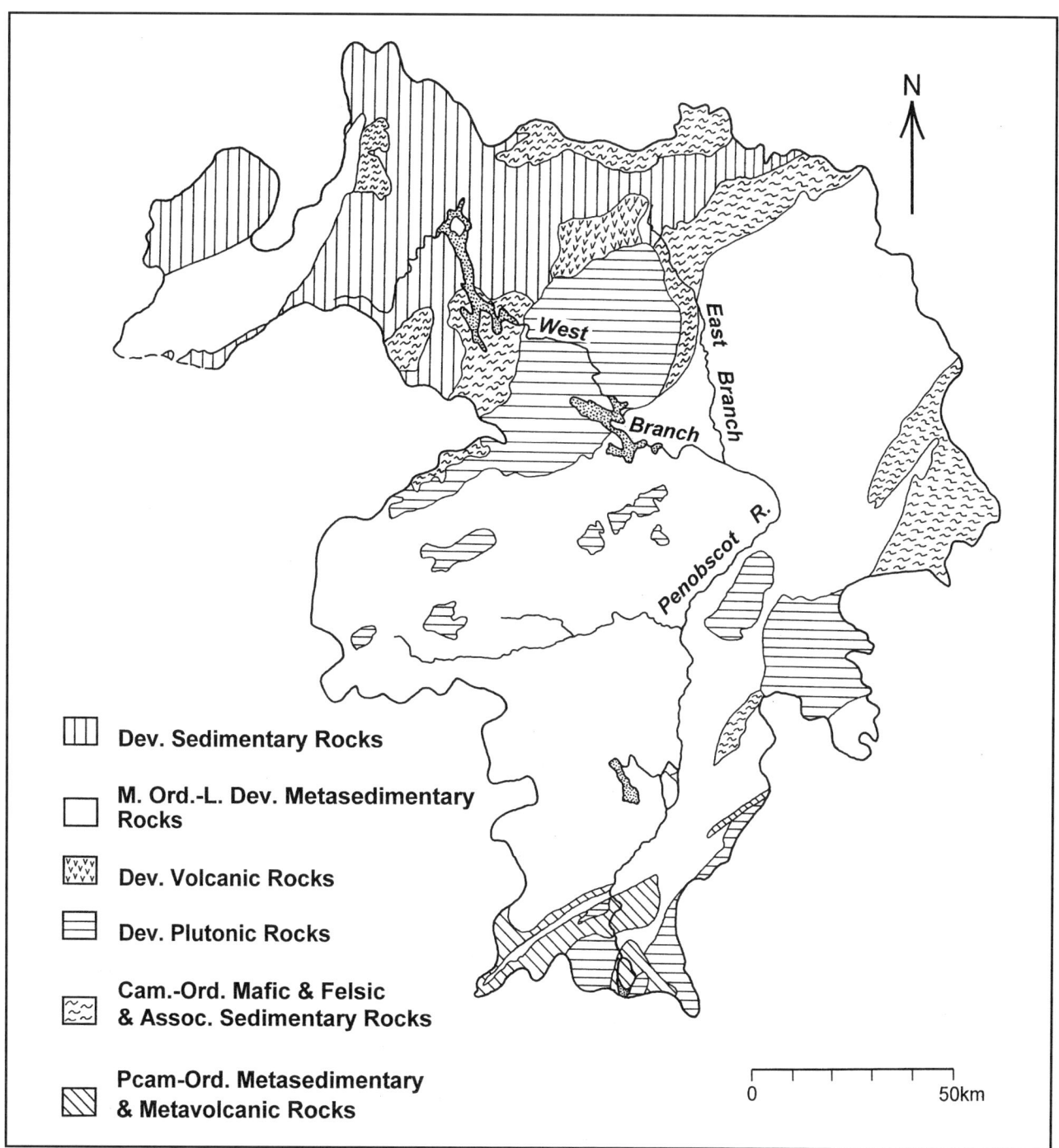

Figure 9.2. Generalized bedrock geological map of the Penobscot River drainage (modified from Osberg et al. 1985).

form the rapids and waterfalls that typify this portion of the river. The bedrock of the area surrounding the mouth of the river includes small granitic plutons, volcanic rocks, and high- to medium-grade metamorphic rocks, all of lower Paleozoic age. This diversity of rock types forms a region of moderate relief with rounded granitic mountains. (Osberg et al. 1985).

The bedrock of the region also provides lithic resources for the manufacture of tools. Several rock types within the Penobscot drainage have been used as a source of raw materials for artifacts. These include Wassataquoik Chert and Traveler/Kineo Rhyolite (Doyle, 1995) from the upper portions of the drainage basin and low-grade metamorphic granofels and phyllite that outcrop in the central portion of the valley (Sanger et al., this volume; Sanger et al. 2001). These rocks were available either at outcrops and quarries or as clasts removed from the numerous glacial and fluvial deposits within the region.

SURFICIAL GEOLOGY

The surficial geology of the Penobscot River drainage (Figure 9.3) is tied to the Quaternary history of the region. At its maximum, 18,000–20,000 B.P., the Laurentide ice sheet extended across New England to its terminus in southern New England and the Middle Atlantic States (Denton and Hughes 1981). By 14,000 B.P. the margin of the melting Laurentide ice sheet reached the Maine coast and disappeared from the central and lower Penobscot Valley by 13,000 B.P. (Dorion 1997). Because the weight of the ice depressed the land's surface, marine waters followed the retreating ice front. This marine transgression created a broad embayment extending up the Penobscot Valley (Thompson and Borns 1985). Glaciomarine deltas and other coarse-grained deposits mark the position of the most landward extent of the marine invasion at approximately 100 m above present sea level.

Marine conditions persisted in the lower Penobscot Valley until approximately 12,000 B.P., when isostatic adjustment brought the land surface above sea level. By ca. 10,000 B.P., the shoreline reached a lowstand of 60 m below present sea level (Belknap et al. 1987). At this time, ice was present in two forms: stagnant ice melting in place and an active, localized ice cap in the headwaters region of the Penobscot River until shortly after 10,000 B.P. (Thompson and Borns 1985). After the marine regression, surface drainage was established, and the present-day Penobscot River began to take shape. The extremely low relative sea level formed a steep gradient for the growing drainage patterns in the region. Rivers formed in a highly erosional regimen due to this steep gradient and the large amounts of available meltwater. In some places the river reoccupied older, pre-Wisconsinan valleys (Calkin 1960); in other places the river has excavated a new path, as indicated by the presence of rapids and waterfalls.

The surficial geology of the Penobscot varies widely with location (Figure 9.3). In the northern, high-relief areas, till of varying thickness blankets the landscape, except at the highest elevations where exposures of bare rock are common. To the south of the Katahdin massif, the till is shaped into ribbed moraines that form a series of hummocky, subparallel ridges. In other portions of the drainage, till is present as drift of varying thickness. Glacial outwash deposits occur in the valleys of the headwaters region (Thompson and Borns 1985). Coarse-grained glaciomarine deposits, including glaciomarine deltas, mark the most landward extent of the Late Pleistocene marine transgression. Large, well-defined esker segments occur in the region to the south of Mt. Katahdin and persist throughout the downstream portion of the Penobscot drainage (Thompson and Borns 1985). The most distinctive surficial deposit of the central and lower Penobscot drainage is the Presumpscot Formation (Bloom 1960). This fine-grained, glaciomarine deposit mantles the lowlands south of the marine limit (Thompson and Borns 1985) (Figure 9.3). Erosion of this unit gives the landscape a rolling, gullied topography. The silts and clays of the formation floor the swamps and bogs that characterize the region.

The surficial deposits within the immediate Penobscot Valley are highly variable, as well. Pleistocene-age deposits of ice contact, glaciofluvial, and glaciomarine sediments are widespead. Both erosional and depositional terraces of apparent Pleistocene age occur within the river valley. Flights of terraces are particularly well developed in the region between Bangor and Old Town. Bedrock exposures are common in the upper portions of the river and to the south of Old Town. In many locations these bedrock outcrops are associated with rapids. Holocene-age sediments are formed from the reworking of preexisting deposits. These deposits are restricted to the immediate riverbanks and form the modern floodplain and depositional terraces. Sediment thickness in these deposits varies from no accumulation in areas of exposed bedrock to sedimentary sequences in excess of 2 m.

The last major geological event in the Penobscot drainage was the construction of dams throughout the region, beginning in the late 18th century and continuing into the 20th (U.S. Army Corps of Engineers 1990). The upper portion of the drainage of the East Branch of the Penobscot River was diverted from the Allagash River by the construction of a dam and canal in the mid-19th century (Barrows and Babb 1912). These changes

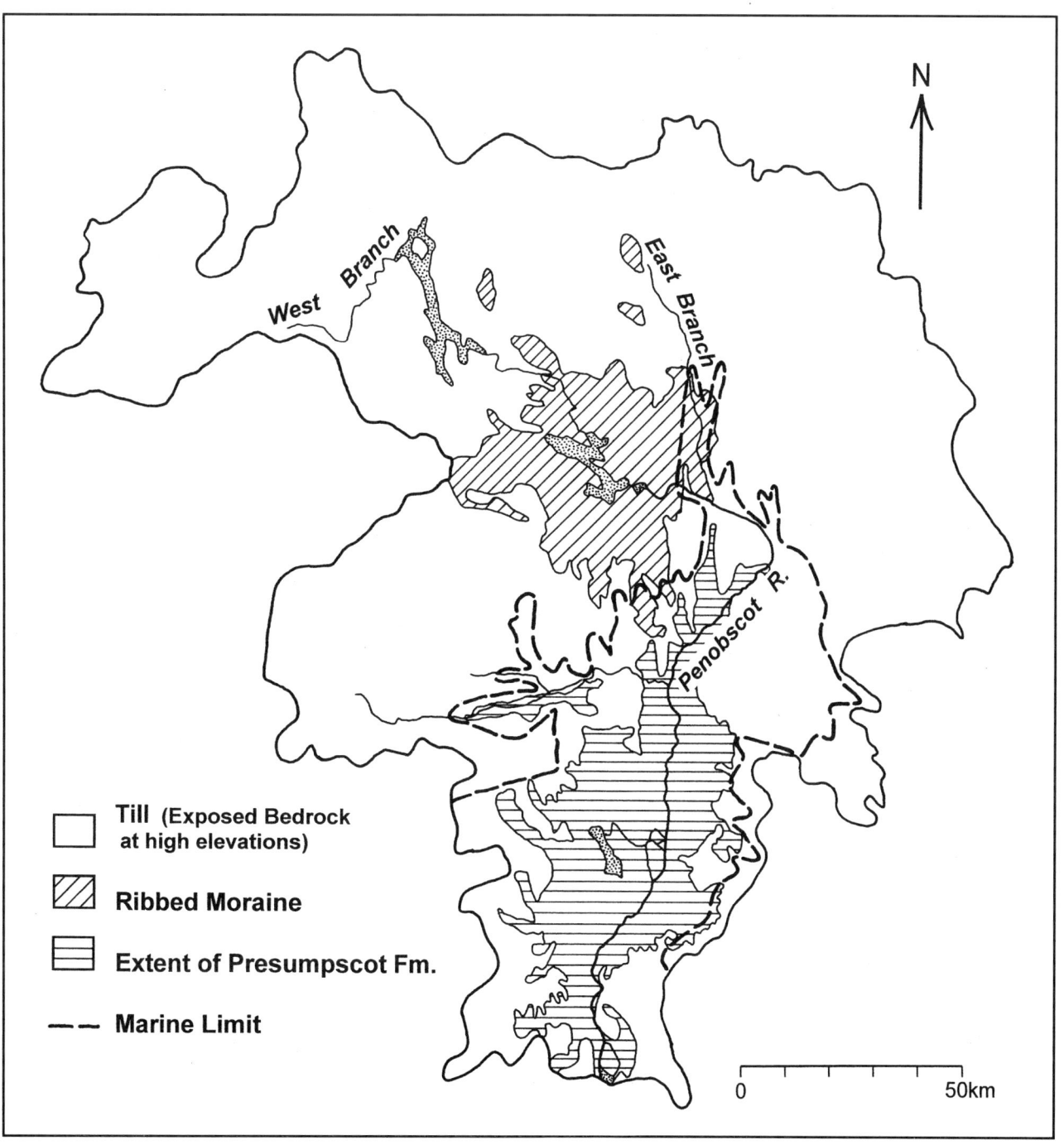

Figure 9.3. Generalized surficial geological map of the Penobscot River drainage (modified from Thompson and Borns 1985).

altered the appearance of the river by drowning rapids in the headponds of dams and raising water levels. Bank erosion of shorelines within impoundments is caused by erosion by raised water levels and spring ice movement. Although the dams on the Penobscot are "run of river" structures built to provide water storage for hydroelectric power generation and log drives in the past, they affect water volumes throughout the year. Water is impounded during the dry summer months, keeping water levels artificially high during these periods. Spring flooding occurs at higher than predam elevations and, when combined with the action of water-driven ice, promotes erosion of riverbanks.

ISOSTATIC ADJUSTMENT OF THE PENOBSCOT REGION

Isostatic adjustments associated with deglaciation have been noted in a variety of locations throughout the world and have been the subject of both field and theoretical studies (Clark 1994; Peltier 1994). Models describe the depression of the earth's lithosphere by an overriding ice sheet, which pushes down on the earth's crust, and causes the earth's mantle to be pushed up into a bulge adjacent to the ice front. As the ice retreats, the weight is removed, and the displaced material flows back into place. Due to the viscosity of the mantle material, this forebulge relaxes slowly and can maintain its shape as it moves inland for some distance.

Comparisons of the eustatic, or worldwide, sea-level rise with the relative sea-level curve for the Maine coast led Barnhardt et al. (1995) to recognize evidence for the migration of a forebulge of approximately 25 m in amplitude moving across the Maine coast at 10,500 B.P. (uncalibrated). Work by Dionne (1988) in the upper St. Lawrence estuary of Quebec suggested the migration of a forebulge from the Maine coast to Quebec City between 10,500 and 7000 B.P.

Balco et al. (1998) tested this hypothesis by examining Moosehead Lake for evidence of tilting. Moosehead Lake, Maine's largest lake, lies between the Maine coast and Quebec and has a northwest-southeast orientation (Figure 9.1), ideally suited to record evidence of isostatic change related to the passage of a migrating forebulge. Examination of historic and calculated prehistoric climatic information, combined with topographic analysis of the lake basin and outlets, suggested that the lake level varies less than a meter due to changes in precipitation (Balco et al. 1998).

Currently the lake forms the headwaters of the Kennebec River, which drains through two outlets at the southwestern end of the lake. The northernmost portion of Moosehead Lake is situated less than 5 km from the West Branch of the Penobscot and is separated from that drainage by a low divide of approximately 4.5 m. Using seismic reflection and side scan sonar, supplemented by coring, Balco et al. (1998) identified lower-than-present shorelines in the southeastern portion of the lake and a higher, abandoned, northern outlet that drained Moosehead Lake into the West Branch of the Penobscot River.

Using this evidence, Balco et al. (1998) postulated that by 10,000 B.P., following deglaciation of the region, the Moosehead Lake region tilted to the northwest, toward the receding ice sheet, at an incline of 0.8 m/km. After 10,000 B.P. lake levels at the southern end of Moosehead rose rapidly due to isostatic adjustment, and the outlet into the West Branch of the Penobscot River was abandoned. The southern outlet to the Kennebec River became active, indicating a lessening of the regional isostatic tilt. Geophysical evidence suggests possible backtilting causing a lowstand of the lake to the north and a highstand to the south at approximately 8750 B.P., followed by an adjustment bringing the land surface to its current elevation.

This change in outlets had a profound effect on the rivers involved. The drainage area of the Penobscot River decreased by 15 percent, whereas that of the Kennebec increased by 22 percent, causing a change in the discharge of each river. During the same period, the migration of the forebulge through the region potentially altered drainage patterns throughout the Penobscot region. This phenomenon was especially significant in the lower Penobscot drainage, where the catchments of lakes and tributary streams are separated by low divides. The resulting alterations may have caused the formation of lakes and the creation of new outlets and streams.

Isostatic depression also affected the gradient of the early Penobscot River. Between 13,000 and 10,500 B.P., relative sea level along the Maine coast dropped to 60 m below present, and then rose to within 5 m of current sea level by 5000 B.P. (Barnhardt et al. 1995). This lower-than-present sea level created a steep gradient for the developing river system, leading to a period of rapid downcutting, followed by decreasing, though still steeper than present, gradient. During this time bedrock sills that trend across the course of the river from Bangor to Old Town were excavated by fluvial erosion, creating local base levels. This process dramatically changed the sedimentation patterns in this region by forming a series of bedrock-dammed reaches of the river.

GEOLOGICAL OBSERVATIONS AT ARCHAEOLOGICAL SITES

Archaeological excavations provide a window into the regional stratigraphy lacking in many geologically oriented studies. Without this insight, details of the sedimentary sequence of events would remain hidden in the geological record. Examination of excavations at numerous sites throughout the lower Penobscot Valley revealed several recurring patterns: (1) Thick, stratified late Pleistocene- to Holocene-age sedimentary sequences were encountered at sites located near the confluence of the river and tributary streams, and were associated with historically recognized rapids. (2) Excavations in these regions showed an abrupt change from a base composed of coarse sediments to a thick overlying sequence of consistently fine-grained material. (3) Dark brown to red horizons, associated with cultural material, occurred at a number of sites.

Sedimentary Sequences

Sites with thick, fine-grained sedimentary sequences could be divided into two groups: (1) those located on Pleistocene-age terrace and outwash deposits that formed prior to pre-European occupation, and (2) those on late Pleistocene- to Holocene-age terrace and floodplain deposits at tributary stream mouths that experienced flooding and sediment accumulation at times during the period of occupation.

Although sites located on Pleistocene-age terrace and outwash deposits were associated with thick sedimentary sequences, all pre-European human activities occurred at approximately the same ground-surface level, yielding unstratified sites with mixed components. Sites positioned on the late Pleistocene- to Holocene-age terraces and floodplains developed through time. As each successive layer of flood-deposited sediment sealed underlying layers, it also created a new surface available for occupation. These thick, fine-grained sedimentary sequences, in excess of 2 m, were encountered at tributary stream mouths and were associated with predam or existing rapids. Sequences were composed of well-stratified Late Pleistocene to Holocene-age sediments with several distinct archaeological components, generally ranging in age from the Archaic to Ceramic Periods.

Sanger (1979) proposed a model linking occupation sites at such locations to seasonally available anadromous fish resources, such as sturgeon, salmon, and eels. Later work by Almquist-Jacobson and Sanger (1999) suggests that these locations also served as gateways to the rich faunal and floral resources of the wetlands associated with some of these tributary streams. A shovel test pit study of the Orono and Veazie portions of the Stillwater and Penobscot River banks using a random sampling strategy illustrated the link between specific environments and site location. Although many locations with at least a meter of sediment were found along the riverbanks, positive shovel test results were related to areas near tributary mouths and rapids (Belcher and Sanger 1988).

At each of the deeply stratified sites the deepest sediments were coarse-grained gravel and/or sand that abruptly changed in texture to fine-grained material that continued to the top of the section. The rapid change in sediment texture at the base of the section suggests an abrupt change in competency of the depositing water. Deposition of gravel and coarse sand is associated with rapidly moving water, whereas fine-grained sediment is deposited by slow-moving or standing water. Little evidence of erosion, such as scoured or deflated zones, was noticed in the fine-grained sediments, supporting the interpretation of deposition by slow-moving floodwater.

Although channel avulsion typically results in a sequence of coarse-grained basal sediments overlain by fine-grained sediments, this process is not applied to the Bangor to Old Town reach of the Penobscot River. In this region the river has a generally straight channel and is confined by bedrock and thick deposits of clay-rich till.

The timing of this abrupt change in sedimentation pattern is dated by using diagnostic artifacts and charcoal associated with artifacts. At Blackman Stream, for example, a parallel-flaked projectile point and associated flakes were recovered from fine-grained sediments 217 cm below the surface and 25 cm above a gravel layer, providing an Early Holocene minimum age for the coarse-to-fine sediment transition (Figure 9.4). The oldest radiocarbon dates from Blackman Stream, approximately 1 m above the gravel horizon, dated between 7300 and 8400 B.P. (Sanger et al.1992; Sanger et al., this volume). At the Gilman Falls site, a minimum date of 7300 B.P. is provided by charcoal in the lowest archaeological component (Sanger 1996; Sanger et al. 2001).

These dates place the abrupt sedimentation change to a time when a number of geologically significant events were occurring. Changes in river volume and base levels were being driven by ice recession, isostatic adjustment, and local geology. The last remnants of the Maine portion of the Laurentide ice sheet were melting in the headwaters of the Penobscot River (Dorion et al. 2001). Greater-than-present discharges and sediment loads associated with meltwater runoff, combined with a steeper-than-present gradient due to lower sea levels probably gave the early Penobscot River a broad, braided channel that occupied the entire width of the available valley. High discharge rates caused rapid downcutting through the till and overlying glaciomarine sediments in the valley. This rapid erosion of the valley ended with the excavation of resistant bedrock sills oriented perpendicular to the river's course in several locations. Each sill formed a local base level and abruptly changed the sedimentary environment upstream. Instead of the anticipated gradually fining upward sequence associated with diminishing flow, fine-grained sediment accumulated in the slower moving water ponded behind each sill (Sanger et al. 2001).

As noted, the isostatically controlled change in the outlet of Moosehead Lake removed the Moosehead Lake and Moose River drainage from the Penobscot watershed, and decreased the amount of water flowing into the main stem of the Penobscot River. This event may have contributed to the rapid change in sedimentation style at the base of these deeply stratified sites. In each case of Gilman Falls and Blackman Stream sites, the minimum dates of the coarse-to-fine transition falls within the time range identified by Balco et al. (1998) for the shift in Moosehead Lake outlets.

Figure 9.4. Plano-like projectile point in fine-grained sediments immediately above gravel layer, Blackman Stream site.

Within the upper fine-grained layers, some geological units fined upward while others appear massive, but in all cases deposits were composed of fine sand, silt, and clay. Strata were recognized on the basis of color and slight textural differences. Contacts between fine-grained units were typically gradational. The location of these sites at the mouth of a tributary stream or rapids suggests that ponding of sediment-laden water behind the rapids and/or backflow of the river into the tributary channel enhanced sedimentation and is responsible for the thick accumulation of material. In the large Penobscot River drainage, individual floods from tributaries reach the main stem quickly, while the combined contribution of all the upstream tributaries causes the main stem flood in the lower portions of the river to be delayed. This delay forces floodwater into the tributary mouths, creating fine-grained slackwater deposition.

Previously unpublished work at the Beaver Site, adjacent to Gilman Falls, has produced a different line of evidence for changing deposition patterns. The Beaver site is located at the confluence of Pushaw Stream and the Stillwater River. At this location the oldest cultural component is Early Archaic (ca. 8000 B.P.), located in the center of the peninsula, 10 m from the present-day shoreline. This stratum is underlain by coarse sediments composed of fine gravel and sand and is covered by approximately 1 m of fine sand and silt. These coarse deposits represent a gravel bar associated with higher-than-present stream velocity; the finer sediments represent deposition by ponded flood water. Sequentially younger occupations, ranging to Late Ceramic Period in age, are found closer to the present-day shoreline. This is interpreted to represent lateral accumulation of sediments through time and provides additional evidence of the abrupt change in the style of deposition.

Buried Soil Horizons

Buried, dark brown to red brown genetic soil horizons have been encountered at a variety of sites within the lower Penobscot Valley. These strata are evident at both the Blackman Stream and Gilman Falls sites and can be dated at these locations through the use of diagnostic artifacts and radiocarbon dating of associated charcoal. In each case the horizon occurs in fine-grained alluvial sediments and is identified by a distinct change in color from the surrounding material. The buried soils are not associated with O, A, or E horizons. The A and O layers have been removed by fluvial erosion at the beginning of the event that buried the soil, or they have been removed due to the acidic nature of the soils in this region. E horizons may have been originally absent, removed by erosion, or mixed with over- and underlying soils by bioturbation. Preservation of faunal or floral remains, unless calcined or carbonized, is extremely poor in the acidic forest soils found at sites in the Penobscot Valley.

Blackman Stream

At Blackman Stream three separate horizons are identified and interpreted as spodic horizons on the basis of color, grain size, and lateral extent (Buol et al. 1997) (Figure 9.5). The stratigraphic section at the Blackman Stream site consists of fine-grained alluvial sediments composed of sand, silt, and clay deposited on coarser fluvial gravels or a fluvially formed lag deposit on a till surface. The surface of the site is covered with a thin 0 to 5-cm organic (O) horizon over a 20 to 30-cm-thick plow layer (Ap). This horizon overlies a 10 to15-cm red brown spodic (Bs) horizon. This horizon is indicative of the process of soil formation occurring at the site since the sediment was deposited. In some portions of the site, a 10-cm-thick, gray white, leached or elluviated (E) horizon is found between the Ap and Bs horizons. A 25 to 30-cm-thick BC horizon (altered parent material) of fine- to medium-grained yellow sand underlies the Bs horizon. The BC horizon is the transition from the Bs horizon to the 5 to10-cm-thick olive yellow C horizon. The C horizon is composed of medium to fine sand that is considered to be the parent material.

Beneath this solum sequence of horizons, at approximately 1 m below the surface, is another dark, reddish brown spodic horizon. This horizon is approximately 10 cm thick and is composed of silt to fine sand. Because it is buried by the overlying material, this horizon is designated as Bsb1 (where b denotes buried). This horizon is associated with large felsite flakes, ground stone artifacts, chipped stone tools, and charcoal dated from 8360 ± 150 to 7400 ± 140 B.P. (Sanger et al. 1992). In the portions of the site most distant from the riverbank and tributary stream, the Bsb1 horizon directly overlies bedded sand and gravel. In other areas it overlies a 20 to 30-cm-thick BCb1 horizon composed of yellow sand over a 10-cm-thick Cb1 horizon of olive yellow sand. This sequence, in turn, overlies another thin, spodic or Bsb2 horizon developed in fine-grained sediments immediately above another Cb2 horizon of bedded sand and gravel. A parallel flaked point, usually associated with a Late Pleistocene age was associated with this Bsb2 horizon. This diagnostic artifact provides an approximate date of 9000 B.P. for this horizon.

This Early Holocene date is significant in that it suggests that, at least in the Blackman Stream reach of the Penobscot River, excavation of the bedrock sill that forms the local base level and the onset of fine-grained sedimentation occurred by ca. 9000 B.P.

Gilman Falls

At the Gilman Falls site, bright to faint red, laterally extensive buried horizons were encountered at several depths throughout the excavation and are interpreted to represent a trisequel spodisol (Soil Survey Staff 1999) formed over the past 8,000 years (Osher, personal communication 2002) (Figure 9.6). Although each horizon was not continuous across the entire site, each could be distinguished on the basis of stratigraphic position and archaeological content. These spodic horizons and their associated artifacts were used to define three vertically distinct cultural zones in the excavation. Each zone consists of a buried spodic horizon that is part of the genetic soil sequence and may include BC and C horizons.

The sedimentary parent material in which these soils formed is composed of 80–150 cm of fine-grained sediments with a clay loam to sandy clay loam texture (Soil Survey Staff 1999). Sandy sediments are found within the lower 50 cm of the sequence. Lenses of gravel (5 by 10 cm in diameter) occur in some horizons. Because of their limited extent and widely scattered appearance in the section, they are interpreted to represent ice-rafted material. The sandy clay loam sediments are deposited on bedded gravel and sand or bedrock. Where the bottom of the section is exposed, glacially polished, and striated bedrock underlies the stratified gravel and sand.

The surface of the site is covered with a 0 to 5 cm-thick organic (O) horizon over a 5 to 25 cm dark brown mineral and organic-rich (A) horizon. A thin, 0 to 10-cm-thick gray white E horizon directly underlies the A horizon in various portions of the site. The A and E horizons overlies a 5 to 20-cm-thick light red to orange red Bs horizon composed of fine sand, silt, and clay (clay loam) that represents the modern soil development. The Bs horizon overlies a 10-20 cm yellow brown BC horizon developed on a 10 to 20-cm-thick C horizon of olive brown sand, silt, and clay (clay loam). The Bs and BC horizons were designated as Cultural Zone 1 (Sanger

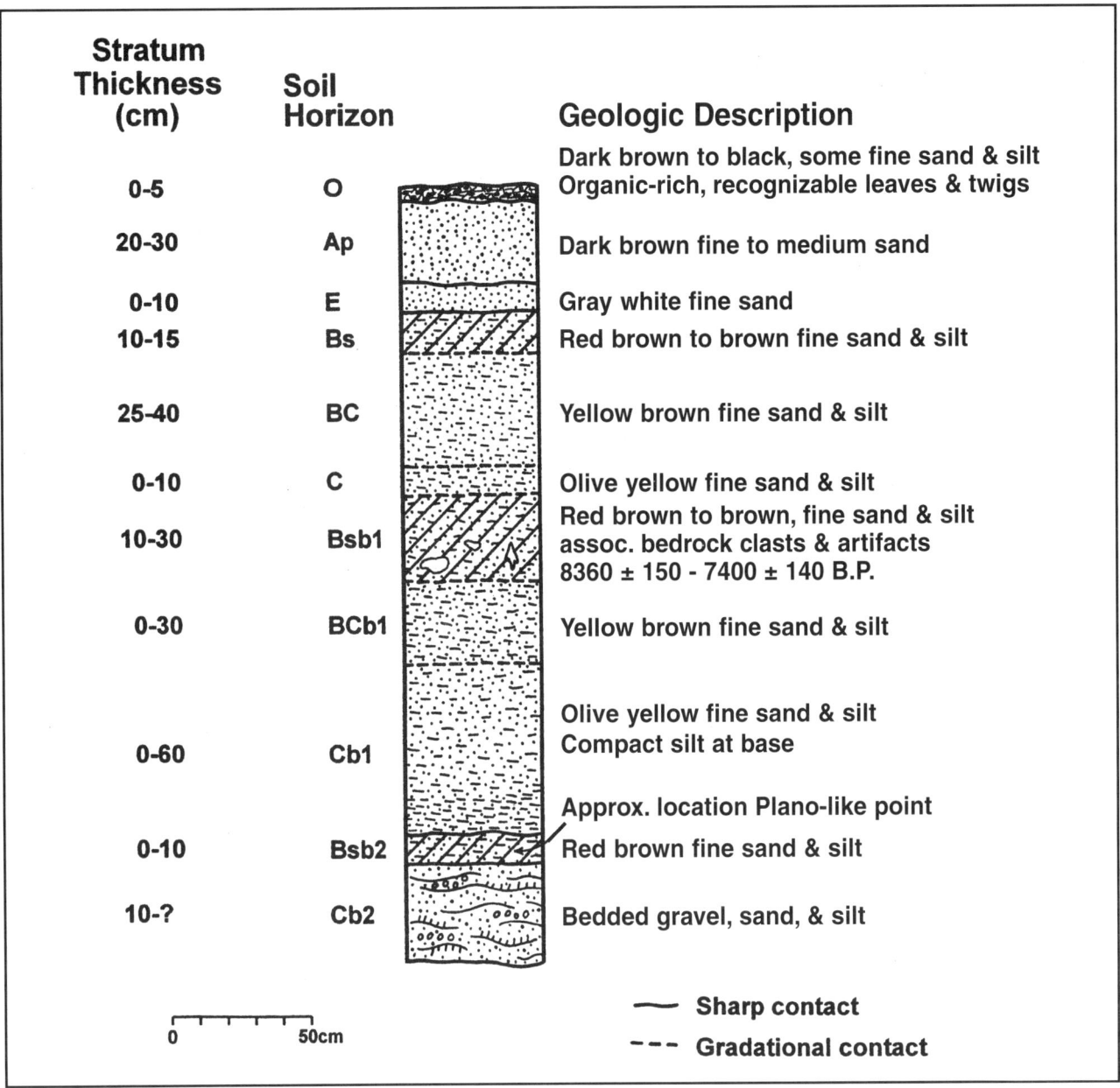

Figure 9.5. Idealized composite stratigraphic section at Blackman Stream site.

1996; Sanger et al. 1994). Artifacts recovered from Zone 1 suggest occupation during Ceramic Periods 2 to 6 (ca. 2250–450 B.P.) (Petersen and Sanger 1991). Evidence of disturbance of this zone by tree throws and rodent burrows was extensive.

A Bsb1 horizon is discontinuous across the site, but where present it occurs at approximately 60–70 cm below the surface. In some locations, it appears as a brightly colored, thin (3–10 cm) layer of red brown sand, silt, and clay with distinct boundaries, while in others it is broader (10–20 cm), more subdued in color, and has gradational boundaries. The Bsb1 horizon overlies a fine-grained, 10-cm-thick, yellow brown BCb1 horizon. This horizon overlies 10–20 cm of olive brown parent material (Cb1 horizon) composed of alluvial sand, silt, and clay (clay loam). The Bsb1 and associated BCb1 horizons were identified as Cultural Zone 2 and assigned a Late Archaic affiliation based on artifact styles and five radiocarbon dates ranging from 3600 to 4500 B.P. (Sanger 1996).

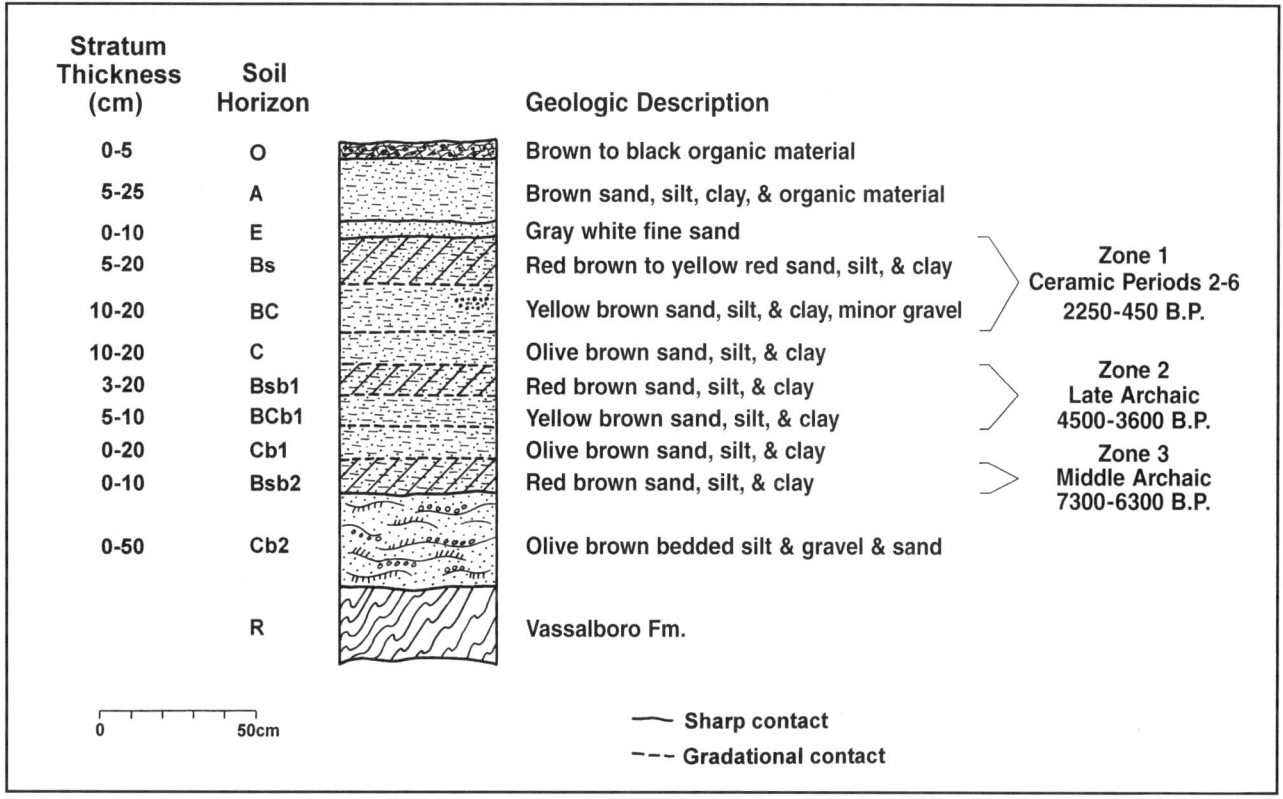

Figure 9.6. Idealized composite stratigraphic section at Gilman Falls site.

The Bsb2 horizon that defines Cultural Zone 3 occurs 90–100 cm below the surface and underlies the Cb1 horizon. Where present, it is a distinct, 5 to10-cm-thick, red-brown horizon composed of fine sand, silt, and clay with distinct boundaries and developed immediately above bedded sand and gravel layers (Cb2 horizon, sandy clay loam). It is associated with numerous artifacts, charcoal, and large (up to 25 kg) rock clasts that match the lithology of the underlying bedrock. Where the horizon is not present, artifacts and rock clasts of similar styles and lithologies to those with the Bsb2 horizon are found at the same stratigraphic levels in the unaltered parent material. Six charcoal samples from Zone 3, the lowest horizon, returned radiocarbon dates ranging from 7300 to 6300 B.P. (Middle Archaic). On the basis of number of artifacts and radiocarbon dates, this occupation at this zone seems to have been the most extensive and may be related to the production of specialized, ceremonial objects (rods) from the lithic resources at the site (Sanger 1996; Sanger et al., this volume). Sediment samples from this horizon showed greatly increased levels of inorganic phosphorus, supporting the interpretation of extended occupation at the site (Sanger 1996; Sanger et al. 2001).

The development of distinct, buried soil horizons at these sites suggests relatively lengthy periods of geological and/or climatic stability of the land surface. The location of these sequences in an active floodplain environment indicates that deposition by flood events was greatly reduced for a time period long enough to allow the development of distinct soil horizons. In the case of the lower two horizons at each site, these periods of decreased sedimentation were followed by renewed alluvial deposition. In the case of the Bsb11 and Bsb2 horizons at Gilman Falls and the Bsb1 horizon at Blackman Stream, radiocarbon dating of associated charcoal suggests a period of time spanning approximately 1,000 years.

Examination of the paleohydrology of the region by Almquist et al. (2001) and Sanger et al. (this volume, Figure 10.5) at Mansell Pond (Figure 9.1) notes two periods of dryer-than-present climates, associated with lower regional water balance. Each of these time periods, 7500–5000 B.P. and 4500–3500 B.P., roughly correlate with the development of the buried spodic horizons seen at Gilman Falls. The lower precipitation rates suggested by the lower regional water balance at these times lowered the level of the Penobscot River and the incidence of flooding to a point that allowed distinct soil horizons to form. The Bsb1 horizon at Blackman Stream may represent the onset of the earlier dry period noted by Almquist et al. (2001), whereas the Bsb2 horizon at Blackman Stream is potentially associated with isostatically influenced volume changes in the Penobscot River.

SITE FORMATION AND PRESERVATION

Site Formation

The first step in site formation is the selection of an area by people as a location for their activities. This selection is based on a number of factors, some cultural and others physical. Idealogical reasons for site selection may not be decipherable in the archaeological record, but the proximity of sites to specific resources, and their location in particular physical settings, indicates active selection rather than random settlement patterns. Landforms or environmental settings inappropriate to a group's travel, shelter, and resource needs are rejected. In interior Maine, Late Paleoindian- to European-age contact occupation sites are generally located along major waterways, near the confluence of streams and rivers, particularly near waterfalls and rapids, and on areas of gentle slopes (Sanger 1979).

The geological location of the site plays an important role in formation and preservation. Bedrock and surficial geology determine the topography and environmental setting of a region. The nature of the bedrock determines the ruggedness of the terrain through the presence or absence of erosion-resistant rock types. The composition of the bedrock also determines the type of sediment produced by erosion. Surficial processes shape landforms and influence the distribution of sediment.

Early Paleoindian sites have been found in southern Maine, New Brunswick, and Nova Scotia, but artifacts from this time period are not known from the Penobscot Valley (Spiess et al. 1998). This apparent absence may have a geoarchaeological explanation. The metasedimentary rocks underlying much of the central and lower Penobscot drainage are prone to erosion and form a region of subdued topography. During the recession of the last ice sheet, this low-relief portion of the Penobscot Basin was inundated by the ocean due to postglacial isostatic depression. Oldale (1985) suggests that areas below the marine limit were not attractive to early inhabitants of the region, because the rapidly falling sea level prohibited the development of marshes, the keystone of the coastal food chain. He suggests that any occupants of this region would be forced to rely on terrestrial resources. Paleoindian sites have been discovered below the marine limit in the western portion of Maine (Spiess et al. 1998), but, as suggested by Oldale (1985), none appears to have a marine focus.

The extensive, sandy outwash deposits frequently linked with Paleoindian sites (Spiess et al. 1988) are extremely limited in the Penobscot drainage. Glacial erosion of widespread metasedimentary rocks and limited exposures of granites produced clay-rich till. Sand-rich deposits are limited to a few well-developed eskers and their associated small deltas. Aeolian deposits are recognized within the Penobscot Valley (Kelley et al. 2000; Thompson and Borns 1985) but are limited in extent and have not been systematically investigated by archaeologists.

The low relief of the central Penobscot Valley, large quantities of glacial meltwater, postglacial forebulge migration, and a mantle of fine-grained glaciomarine sediments combined to create an area of extensive wetlands in the central Penobscot Valley. Broad regions with low drainage divides allowed substantial changes in stream patterns during isostatic adjustment, creating abandoned drainage channels and large bodies of open water trapped by the underlying nonporous clay and silt. Almquist-Jacobson and Sanger (1999) note the presence of a large, Early Holocene lake in an area north of Old Town (Figure 9.1) now occupied by bogs. Many of the extensive bogs in the region may have had a similar history. Although potentially rich with wetland flora and fauna, extensive wetlands would have presented a potential travel barrier to migratory hunter-gatherers.

Later, other settings provided both the setting and environment attractive to people, combined with local geological processes that acted to preserve a record of occupation. Some sites, such as Eddington Bend, are located on preexisting landforms that are not undergoing rapid, large-scale geological change. In this setting, the processes that serve to cover and preserve the site are limited. Human activities during site occupation may add protective strata, whereas geological processes such as the accumulation of organic material and colluviation may cover the site. In areas of very fine grained sediment, human occupation may disturb existing vegetation and allow the development of an aeolian cover. If such occupation surfaces are not quickly separated from the surface, they are prone to disturbance by plants (in the form of tree throws and root growth), animals (burrowing and caching), and people (plowing and other land-disturbing activities). However, in areas without extensive European activity, Early- to Mid-Holocene-age deposits may be relatively intact due to the light use of the landscape at that time. More recent sites found closer to the modern surface are in danger of disturbance as a result of European settlement and agricultural practices.

Sites located in alluvial settings have a higher preservation potential, provided that current velocity is low enough to allow the deposition, rather than erosion of sediment. This accumulation of sediment protects the layers beneath and increases the thickness of the archaeological record, potentially allowing definition of distinct periods of use. If the site experiences erosion during flood events, a deflation surface is created that compresses the archaeological section, and may contribute to the mixing of stratigraphic layers. Channel migration can abandon a portion of riverbank, separating sites from active fluvial processes, thus preserving a record of occupation.

Several of the sites in the Penobscot Valley are ideally positioned to preserve the local archaeological record. These sites, such as Blackman Stream and Gilman Falls, are located where tributary streams enter the main stem of the river and are associated with bedrock sills that extend across the river creating rapids or waterfalls. These resistant bedrock obstacles form local base levels in the river and decrease the local river gradient, decreasing local water velocity. During flood events these bedrock thresholds act as dams, ponding water and decreasing the velocity of flow, allowing sediment to be deposited. At the same time, the presence of the tributary streams allows river water to flow upstream into the tributary mouth, adding to the ponding effect and enhancing the deposition of fine-grained sediment. The repeated flooding of these areas builds a thick (1–3 m) sedimentary sequence that preserves stream-side sites located in these areas.

Farther north within the Penobscot drainage, Putnam (1994) described a 3-m-thick accumulation of stratified, fine-grained sediments at the Sharrow site (Petersen 1991), located near the confluence of the Sebec and Piscataquis Rivers. The damming of water by a Late Pleistocene-age moraine acts as a local base level and formed this multicomponent site. Radiocarbon dates from cultural features at the site spanned the Holocene, ranging from ca. 9500–1510 B.P. Three other sites have been recognized in the confluence area, upstream from the Sharrow site. Sedimentation at these sites was also related to the local base level formed by the moraine, other glacial deposits, and the confluence effect of the smaller Sebec River entering the larger Piscataquis.

Variations in water volume, either isostatically driven or due to climatic changes, affect both the formation and preservation of the sedimentary record. At times of lower water, less accumulation of sediment takes place. If the rate of accumulation is slow enough, distinct soil horizons can develop. These horizons, combined with the artifacts and anthropogenic changes, create distinct layers in the stratigraphic sequence. Due to the low sedimentation rates required to form such features, the sedimentary record for these horizons is compressed, potentially combining material from a wide time range, such as the 1,000-yr time span of Zone 3 in approximately 25–30 cm at the Gilman Falls site (Sanger 1996, Sanger et al. 2001).

In situations of more frequent flooding and deposition, archaeological remains of shorter time periods are separated by alluvial sediments, and soil-formation processes do not have sufficient time to form distinct horizons. In this case the sediments appear uniform and massive and individual flood events are rarely recognized. There is little to distinguish separate events, because the geological processes that form each layer are the same, and the source of material remains constant.

Site Destruction

Although geological processes preserve archaeological sites, the same processes can alter or remove sites. Biotic activity, by plants, trees, animals, or humans, also serves to modify the sedimentary record. Fluvial erosion acts to remove all or a portion of sites, particularly those on the modern floodplains and within the fluctuation zone of impoundments. For example, in the lower Piscataquis River valley, stratified Archaic- to Ceramic-Period sites occur where the river channel has not migrated significantly. Upstream, where the presence of abandoned channels is evident, only Ceramic-Period sites are found in the active floodplains (Sanger and Newsom 2000). This absence of older sites along the modern river may be the result of floodplain erosion, or older sites may occur along abandoned meanders now distant from the geologically active portion of the valley.

The Penobscot River drainage has experienced a series of human modifications in addition to geological events. The construction of dams has led to the raising of water levels and modification of river flow. These changes created a landscape far different in appearance from that used by native occupants of the region, as well as a set of geological processes different from those active at the time of site formation. Construction of past settlement patterns must take these modern, as well as past, geological processes into consideration.

LIFE IN A DYNAMIC ENVIRONMENT

Geological studies conducted in conjunction with archaeological studies in the Penobscot drainage illustrate that the Early to Mid-Holocene was a time of rapid geological changes that affected the landscape and the humans of the region. The rapid change from marine to terrestrial conditions led to dramatic landscape alterations as drainage systems began to take shape. While the well-known transition from tundra to forest vegetation (Davis and Jacobson 1985) was taking place, other broad-ranging geological factors, such as isostatic adjustment and the development of river drainage patterns, were shaping the landscape.

An abundance of water from melting glaciers and relatively moist climatic conditions, combined with steep, isostatically controlled gradients produced highly erosive rivers. In some locations, these rivers occupied pre-existing valleys, but in others the courses were deranged and produced waterfalls and rapids. The blanket of glaciomarine clay, draped across the central and lower Penobscot Valley, created an effective seal and formed the basis for numerous lakes, many of which became local wetlands (Almquist-Jacobson and Sanger 1999). Localized changes in drainage patterns were influenced by the migration of a postglacial forebulge. The most dramatic result of this phenomenon was the shift in drainage outlet of Moosehead Lake from the Penobscot River drainage to that of the Kennebec, affecting sedimentation patterns in each watershed. Smaller-scale changes probably occurred in other portions of the drainage, particularly in the low-relief areas of the central Penobscot Valley where the watersheds of lakes and streams are separated by low divides.

These geological changes occurred with environmental alterations that affected the types and locations of resources available to the human inhabitants of the region. Climatic changes affected the amount of water in rivers and wetlands, influencing geological processes, such as soil formation and sediment accumulation, as well as influencing travel routes and the selection of occupation sites. The rapidly changing environments played a part in shaping lifestyles and resource procurement strategies that are recorded in the archaeological sequence of the area.

CONCLUSIONS

In summary, the Early to Mid-Holocene history of the Penobscot Valley is one of dynamic change, geologically, environmentally, and in terms of human occupation. The geological framework of the region was shaped by a series of dramatic physical changes. A rapidly altering landscape formed through glacial retreat, isostatic adjustment, the establishment of drainage patterns, and variations in climate. All of these factors influenced the human occupants of the region through the availability of resources, travel routes, and occupation sites that changed through time as the landscape continued to evolve. As geological processes active at this time created reorganized drainage patterns and habitats, people living in the region reacted. The results of these reactions have shaped the archaeological record as we see it today.

Acknowledgments

Thhe authors would like to acknowledge the contributions and assistance of Dr. Laurie Osher, soils scientist and assistant professor of soil and water quality, University of Maine, in the descriptions and discussions of buried soils. We would also like to thank Stephen Bicknell, who prepared the figures with his usual grace and expertise.

REFERENCES CITED

Almquist, H., Diffenbacher-Krall, A., Brown, R., and Sanger, D. (2001). An 8,000-yr Holocene record of lake levels at Mansell Pond, Central Maine, USA. *The Holocene* **11**:189–201.

Almquist-Jacobson, H., and Sanger, D. (1999). Paleogeographic changes in wetland and upland environments in the Milford Drainage Basin of Central Maine, in relation to Holocene human settlement history. In *Current Northeast Paleoethnobotany*, edited by J.P. Hart, pp. 177–190. New York State Museum Bulletin 494. The University of the State of New York, Albany.

Balco, G., Belknap, D.F., and Kelley, J.T. (1998). Glacioisostasy and lake-level change at Moosehead Lake, Maine. *Quaternary Research* **49**:157–170.

Barnhardt, W.A., Gehrels, W.R., Belknap, D.F., and Kelley, J.T. (1995). Late Quaternary sea-level change in the western Gulf of Maine: Evidence for a migrating glacial forebulge: *Geology* **23**:317–320.

Barrows, H.K., and Babb, C.C. (1912). *Water Resources of the Penobscot River Basin, Maine.* U.S. Geological Survey, Water Supply Paper No. 279.

Belcher, W.R., and Sanger, D. (1988). *An Assesment of the Cultural Resources in the Veazie and Orono Headponds: Results of the 1987 Phase I Survey.* Report prepared for Bangor Hydroelectric Company, Bangor, Maine.

Belknap, D.F., Anderson, B.G., Anderson, R.S., Anderson, W.A., Borns, H.W., Jr., Jacobson, G.L., Kelley, J.T., Shipp, R.C., Smith, D.C., Stuckenrath, R., Jr., Thompson, W.B., Tyler, D.A. (1987). Late Quaternary sea-level changes in Maine. In *Sea-level Fluctuations and Coastal Evolution*, edited by D. Nummendahl, O.H. Pilkey, Jr., and J.D. Howard, pp. 71–85. SEPM Special Publication No. 41.

Bloom, A. (1960). *Late Pleistocene Changes in Sea Level in Southwestern Maine.* Maine Geological Survey, Augusta.

Buol, S.W., Hole, F.D., McCracken, R.J., and Southard, R.J. (1997). *Soil Genesis and Classification* (4th ed.). Iowa State University Press, Ames.

Calkin, W.S. (1960). The Pre-Wisconsin drainage in the Orono and Bangor quadrangles. Unpublished master's thesis, Dept. of Geological Sciences, University of Maine at Orono.

Clark. J.A., Hendriks, M., Timmermans, T.J., Struck, C., and Hilverda, K.J. (1994). Glacial isostatic deformation of the Great Lakes region. *Geological Society of America Bulletin* **106**:19–31

Davis, R.B., and Jacobson, G.C., Jr. (1985). Late glacial and early Holocene landscapes in northern New England and adjacent areas of Canada. *Quaternary Research* **23**:341–358.

Denton, G., and Hughes, T. (1981). *The Last Great Ice Sheets*. Wiley-Interscience, New York.

Dionne, J.C. (1988). Holocene relative sea-level fluctuations in the St. Laurence estuary, Quebec. *Quaternary Research* **29**:233–244.

Dorion, C. (1997). An updated high resolution chronology of deglaciation and accompanying marine transgression in Maine. Unpublished master's thesis, University of Maine, Orono.

Dorion, C.C., Blanco G.A., Kaplan, M.R., Kreutz, M.R., K.J. Kreutz, Wright, J.D., and Borns, H.W., Jr. (2001). Stratigraphy, paleoceanography, chronology, and environment during deglaciation of eastern Maine. In *Deglacial History and Relative Sea-Level Changes, Northern New England and Adjacent Canada*, edited by T.K. Weddel and M.J. Retelle, pp. 215–242. Special Paper—Geological Society of America 351, Boulder, Colo.

Doyle, R.G. (1995). Analysis of lithic artifacts. In *Diversity and Complexity in Prehistoric Maritime Societies: A Gulf of Maine Perspective*, edited by B.J. Borque, pp. 297–316. Plenum Press, New York.

Kelley, A.R., Dorion, C.C., Balco, G., Dieffenbacher-Krall, A., Garrett, P., Locke, D., and Tolman, A. (2000). Late Pleistocene/Holocene geological evolution of the Central Penobscot River Valley: Surficial geology, geoarchaeology, and water supply. Trip B-6. In *Guidebook for Field Trips in Coastal and East Central Maine*, pp. 168–186. New England Intercollegiate Geological Conference, 92 Meeting, Orono.

Mack, K., Sanger, D., and Kelley, A.R. (2002). *The Bob Site: A Late Archaic and Ceramic Period Sire on Pushaw Stream, Maine*. The Maine Archaeological Society Occasional Publications in Archaeology No. 12. Augusta.

Oldale, R. (1985). Rapid postglacial changes in the western Gulf of Maine and Paleoindian environment. *American Antiquity* **50**:145–150.

Osberg, P.H., Hussey, A.M. II, and Boone, G. (1985). *Bedrock Geologic Map of Maine*, scale 1:500,000. Maine Geological Survey.

Peltier, W. R. (1994). Ice Age paleotopography. *Science* **265**:195–201.

Petersen, J.B. (1991). *Archaeological Testing at the Sharrow Site: A Deeply Stratified Early to Late Holocene Cultural Sequence in Central Maine*. Occasional Papers in Maine Archaeology, No. 8, Maine Historic Preservation Commission, Augusta.

Petersen, J.B., Hamilton, N.D., Putnam, D., Spiess, A.E., Stuckenrath, R., Thayer, C.A., and Wolford, J.A. (1986). The Piscataquis archaeologic project: A late Pleistocene occupational sequence in northern New England. *Archaeology of Eastern North America* **14**:1–18.

Petersen, J.B., and Sanger, D. (1987). *Archaeologic Phase II testing at Eddington Bend Site (74-8), Penobscot County, Maine*. Report submitted to Bangor Hydroelectric Company by the University of Maine, Orono.

Petersen, J.B., and Sanger, D. (1991). An aboriginal ceramic sequence for Maine and the Maritime provinces. In *Prehistoric Archaeology in the Maritime Provinces: Past and Present Research*, edited by M. Deal and S. Blair, pp. 113–170. Council of Maritime Premiers, Fredericton, New Brunswick.

Putnam, D. (1994). Vertical accretion of flood deposits and deeply stratified archaeological site formation in central Maine, USA. *Geoarchaeology* **9**:467–502.

Robinson, B.S., and Petersen, J.B. (1992). Introduction: Archaeological patterning and visibility in northern New England. In *Early Holocene Occupations in Northern New England*, edited by B.S. Robinson, J.B. Petersen, and A. Robinson, pp. 1–11. Maine Historic Preservation Commission, Augusta.

Sanger, D. (1979). *Discovering Maine's Archeological Heritage*, Maine Historic Preservation Commission and the Maine Archaeological Society, Augusta.

Sanger, D. (1996). Gilman Falls site: Implications for the Early and Middle Archaic of the Maritime peninsula. *Canadian Journal of Archaeology* **20**:7–28.

Sanger, D., Belcher, W.R., Fenton, J., and Sweeney, M. (1994). Gilman Falls: A Middle Archaic quarry and workshop in Central Maine. Report on file, Maine Historic Preservation Commission, Augusta.

Sanger, D., Belcher, W.R., and Kellogg, D.C. (1992). Early Holocene occupation at the Blackman Stream site, central Maine. In *Early Holocene Occupations in Northern New England*, edited by B.S. Robinson, J.B. Petersen, and A. Robinson., pp. 149–161. Maine Historic Preservation Commission, Augusta.

Sanger, D., Kelley, A.R., Berry, H.N. IV. (2001). Geoarchaeology at Gilman Falls: An Archaic quarry and manufacturing site in central Maine, U.S.A. *Geoarchaeology* **16**:633–665.

Sanger, D., and Newsom, B. (2000). Middle Archaic in the lower Piscataquis River, and its relationship to the Laurentian tradition in central Maine. *Maine Archaeological Society Bulletin* **40**(1):1–22.

Soil Survey Staff, Soil Conservation Service, U.S. Department of Agriculture. (1999). *Keys to Soil Taxonomy*, Pocahontas Press, Blacksburg, Va.

Spiess, A.E., Bradley, J.W., and Wilson, D. (1998). Paleoindian occupation in the New England-Maritimes region: Beyond cultural ecology. *Archaeology of Eastern North America* **26**:201–264.

Thompson, W.B., and Borns, H.W., Jr. (1985). *Surficial Geologic Map of Maine*, Maine Geological Survey, scale 1:500,000.

U.S. Army Corps of Engineers. (1990). *Water Resources Study: Penobscot River Basin, Maine*. Waltham, Mass.

CHAPTER 10

GEOARCHAEOLOGICAL AND CULTURAL INTERPRETATIONS IN THE LOWER PENOBSCOT VALLEY, MAINE

David Sanger, Alice R. Kelley, and Heather Almquist

Starting in the 1970s, archaeological research sponsored by the University of Maine included in a systematic fashion disciplines other than anthropology. Over the years we have learned much about the cultural record. We have also worked to integrate more effectively the talented work of our colleagues as we attempted to explain aspects of the cultural story. In this chapter we review briefly our understanding of the record and how it has changed, in no slight measure due to insights provided by research efforts of our colleagues.

Our geoarchaeological perspective involves a broad array of disciplines and is perhaps closest to Butzer's (1982) inclusive model, which incorporates the contributions of all physical, earth, and biological sciences that lead to understanding human interactions with past environments. Our emphasis in the Northeast has focused on Maine, both interior and coastal. Except for the final few centuries of the pre-European era, and then only in southwestern Maine, Native people made their living by hunting, fishing, and gathering. As such, they depended totally on what the environment offered and their adaptation to those possibilities. For us, it makes no sense to study the archaeology of this region without close attention to the contemporary ambient environment. To forestall the disparaging epithet "environmental determinists," we feel that people always made choices, based on their background and perceived needs. However, these decisions were limited by what Trigger (1991:556) has called "external constraints," such as environmental potential, as opposed to purely cultural or "internal constraints." Given this perspective, we favor a broad definition of geoarchaeology.

The analytic techniques we employ are guided by the following general model, which we do not claim as original or unique:

- Humans make decisions about specific sites and areas to live in based on perceived needs and levels of satisfaction that can be met in an area given a certain level of culture.

- Geological and ecological processes create the physical settings and therefore the adequacy of any site or area to meet the perceived human needs.

- Past human activities affect the character of the environment, at the levels of habitation sites and surrounding area.

- Ongoing geological, ecological, and cultural (including modern excavation and development) processes continue to modify the environment of sites and areas.

In this chapter we present the background to more than a decade of intensive geoarchaeological investigations, including the cultural history; the impact of geology, both bedrock and Quaternary; the evolution of local peatlands; the evolution of upland vegetation; and critical aspects of water levels. Clearly the habitats available to indigenous people at the arrival of Europeans had changed substantially during the Holocene. Culture-environmental relationships based on relatively recent environments are clearly inappropriate for the Archaic Period, just as they are for the Paleoindian.

We begin our review with a brief summary of early attempts, followed by a more detailed look at research that spans the 1990s. In the interests of brevity, we refer to published accounts for the details.

Radiocarbon dates in the text are not calibrated. Neither are date ranges for regularly used cultural periods, such as Archaic Periods, which are presented in radiocarbon years B.P. Actual radiocarbon dates, when given, have been calibrated using the CALIB 4.3 program (Table 10.1).

AN EARLY ATTEMPT: THE HIRUNDO PROJECT

Test excavations in 1971 revealed a stratified site at the Hirundo Game Preserve on Pushaw Stream, a tributary of the Stillwater River, a branch of the Penobscot

---

Geoarchaeology of Landscapes in the Glaciated Northeast edited by David L. Cremeens and John P. Hart. New York State Museum Bulletin 497. © 2003 by the University of the State of New York, The State Education Department, Albany, New York. All rights reserved.

Table 10.1. Calibrated Radiocarbon Dates (CALIB 4.3) from Sites

Radiocarbon Date ± sigma	Laboratory Number	Calibrated Age(s) B.P.
5389 ± 70	TO-1840	6200
6480 ± 70	TO-1841	7420
7400 ± 140	Beta-21682	8180
7760 ± 130	Beta-22125	8540; 8530; 8520
8360 ± 150	Beta-21681	9430; 9400; 9340; 9330
9175 + 230/−225	A-70580	10,360; 10,340; 10,320; 10,250
9540 ± 80	Beta-101669	11,050; 11,020; 11,000; 10,970; 10,750

River in central Maine (Figure 10.1). From the beginning, a Quaternary geologist, soils scientist, a paleoecologist, and a radiocarbon specialist were involved (Sanger et al. 1977; Sanger and MacKay 1973). Although we were able to develop some data on upland vegetation and an appreciation that the local wetlands had changed, lack of resources greatly hampered the program. Goodwill and volunteered time go only so far. The sediment aspect of the project never developed really useful data, in part because of the nature of the site sediments, fine sand deposited on a boulder till lag deposit shallowly draped over bedrock. Pushaw Stream flood events periodically deposited and then stripped sediments, leaving deflated surfaces. At the time it was thought that, in general, the Penobscot drainage system had been in an erosional cycle since the regression of the Late Pleistocene marine incursion (Kelley and Sanger, this volume). Given this model, we did not think it possible to locate, and therefore did not actively seek, archaeological components deeply buried in alluvium. In hindsight, we erred.

EARLY AND MIDDLE ARCHAIC HIATUS?

In the 1960s archaeologists from the Great Lakes to Maine expressed puzzlement over the apparent absence of Early and Middle Archaic sites (Fitting 1968; Ritchie 1965). Faced with a similar dilemma in Maine, Sanger (1977) suggested four hypotheses, actually "talking points," that singly, or in combination, might account for the supposed absence: (1) data incomplete, (2) coastal sites drowned by a rising sea level, (3) vegetation inadequate to support humans (basically the Ritchie-Fitting model), and (4) changing river gradients and effects on anadromous fisheries. Less than a decade later, the demonstrated presence of intact Early and Middle Archaic sites, deeply buried in alluvium, rendered moot the hiatus speculation, at least for central Maine (Petersen et al. 1986; Petersen and Putnam 1992).

After a brief overview of the cultural history, we discuss the results and implications of our geoarchaeological investigations.

GEOARCHAEOLOGY IN THE PENOBSCOT VALLEY AND THE MILFORD DRAINAGE BASIN

Introduction

During the past 15 years, ongoing research between the cities of Bangor and Old Town (Figure 10.1) in the Penobscot Valley (ca. 20 km) focused on pre-European sites. A geoarchaeological perspective was involved from the beginning, although the level of integration changed with experience. The program's impetus was cultural resource management, and the need to relicense two hydroelectric reservoirs plus a permit application for a new reservoir. Bangor Hydro-Electric Company contracted with the University of Maine to conduct the archaeological portion of the permitting process. Phase I level surveys identified nearly 200 previously unreported sites despite abundant evidence for considerable site destruction due to nearly two centuries of Euro-American activities. Long stretches of Penobscot River bank revealed nothing but disturbed sites or no sites at all. The situation changed dramatically once we left the main stem of the river and explored tributary streams and areas that had suffered relatively less Euro-American disturbance. Although the vast majority of sites date to the Ceramic Period (Woodland) (3000–350. B.P.), older occupations also occurred. Phase II and Phase III excavations enabled us to examine a carefully selected sample of sites in accordance with guidelines established by the Maine Historic Preservation Commission (Spiess 1990).

Two sites contain evidence for Late Paleoindian (10,000–9000 B.P.?) presence, in the form of parallel-flaked points, while a number of others produced early (9000?–7500 B.P.) and Middle (7500–6000 B.P.) Archaic assemblages. Buried under alluvium ranging up to 3 m

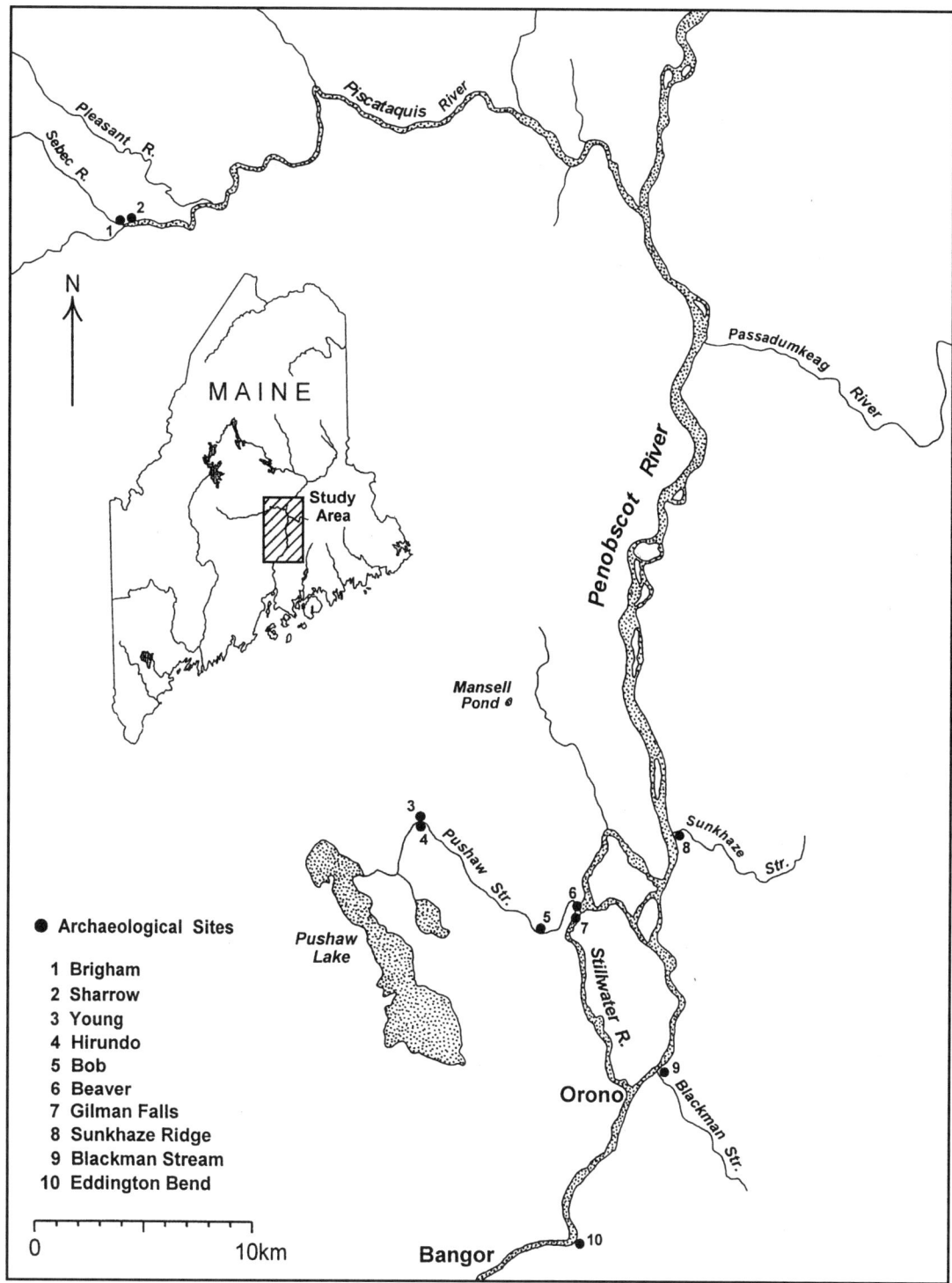

Figure 10.1. Central Maine with locations of major waterways and archaeological sites.

in depth (Sanger 1996; Sanger et al. 1992), previous surveys had failed to explore deeply enough. The task of description and evaluation of all excavated site sediments became Kelley's responsibility. Built into the scope of work was a detailed examination of the local environments from which aboriginal people had to make a living. For this, the program involved paleoecologist Heather Almquist. Other specialties, including bedrock geology and soils sciences, were incorporated as needed.

The Archaeological Record

The well-known Early Paleoindian culture type, characterized by fluted projectile points, has not been reported anywhere in the Penobscot drainage (see Spiess et al. 1998, for a review). Rather than attribute this to the archaeologists' "luck of the draw," it seems more likely that a geoarchaeological explanation involving preservation of suitable landforms and habitat will provide useful hypotheses to explain the apparent absence. The occurrence of Early Paleoindian sites in the Munsungun Lake area (drained by the Aroostook River), which is in the St. John River watershed, is mostly likely explained by lithic raw material acquisition, the famous Munsungun chert that was widely distributed (Pollock et al. 1999; Spiess et al. 1998).

Late Paleoindian finds in the Penobscot Valley occurred in two sites, Blackman Stream and Eddington Bend. At Blackman Stream, the base of a single parallel-flaked biface (Figure 10.2), plus some flakes, came from a very fine sand and silt stratum above coarser sand and fine gravel at a depth of 2.17 m below surface (Sanger et al. 1992, Figure 3). Unfortunately, no charcoal accompanied the biface, and denial of the reservoir permit has resulted in no additional research. However, the biface occurred 1 m below a buried remnant soil horizon radiocarbon dated by three charcoal samples, 7400 ± 140 B.P. (Beta-21682), 7760 ± 130 B.P. (Beta-22125), and 8360 ± 150 B.P. (Beta-21681). These buried soil remnants, which occur commonly in central Maine alluvial sequences, are dark red or orange-colored lenses lacking any evidence for an A Horizon (for a more extended discussion see Kelley and Sanger, this volume).

The excavation unit that yielded the parallel-side point began as a 4-by-4-m pit. At 1.50 m below surface, we continued excavation in a 2-by-1-m portion of the test unit to a depth of 2.6 m, where we encountered sandy gravel above coarse gravel. Artifacts and flakes associated with Ceramic (Woodland) Period and Late Archaic (6000–3000 B.P.) Period continued until roughly 1.3 m below surface. We then excavated another 1.1 m of culturally sterile sediment until we encountered the Late Paleoindian biface and a flake. The absence of any artifacts for more than 1 m above the biface indicates to us that it and the associated flake did not migrate down through the sediments due to natural causes. How long it took to accumulate the 1 m of sediment over the artifacts is unknown as no dateable materials occurred with the artifacts. Although additional flakes occurred in comparably deep stratigraphic situations in other test units, none is diagnostic. The age of parallel-flaked

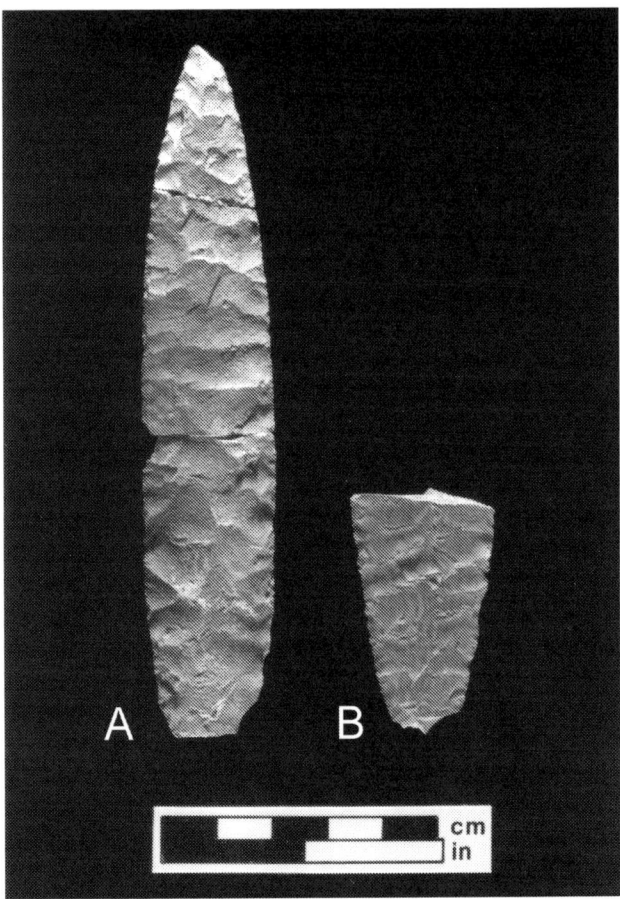

Figure 10.2. Late Paleoindian bifaces from Eddington Bend (a) and Blackman Stream (b).

points in the Maine-Quebec region remains problematic (Dumais 2000; Petersen et al. 2000).

The overall stratigraphic situation suggests that sometime prior to ca. 8500 B.P. Late Paleoindians lived along the banks of the Penobscot River that was already experiencing sediment accumulation. Deposition continued until 8500 BP. For at least 1,500 years, according to our three dated hearths, flood events were intermittent enough to permit a stable land surface and the development of a forest soil. People living in Early and Middle Archaic Periods settled on the surface and made the hearths.

A second biface, also typologically Late Paleoindian, came from the Eddington Bend site, farther downriver, near Bangor. To explore an area slated to be destroyed by a proposed new powerhouse, a hand-excavated trench progressed from the riverbank, across sandy deposits on outwash gravels, and up to a till surface. Right on the till we recovered an almost intact biface. Scatted charcoal returned accelerator mass spectrometry (AMS) dates of 5389 ± 70 B.P. (TO-1840) and 6480 ± 70

B.P. (TO-1841) clearly too young for the biface, but appropriate for other items in the deposit.

Cultures of later occupations left their remains at both sites and apparently lived in an environment that began to approximate more modern drainage conditions, something that tends to characterize the Late Paleoindian settlement pattern in Maine in general (Doyle et al. 1985; Petersen et al. 2000). Known Early and Middle Archaic sites in central Maine occur in two clusters in the Penobscot drainage: the Piscataquis River and in the Bangor to Old Town region. These happen to be the scenes of intensive survey and excavation. Their presence may be more widespread but hidden under deep alluviation or alongside now-abandoned river channels in areas where meandering occurred, as, for example, in the lower Piscataquis River (Sanger and Newsom 2000). Upstream of the area of abandoned channels, near the town of Milo, deeply stratified deposits are found (e.g., Petersen 1991; Petersen et al. 1986).

First recognized at the Brigham site in the early 1980s, and later at the nearby Sharrow site, the deep alluvial sediments that formed at the confluence of the Sebec and Piscataquis Rivers gave archaeologists their first real glimpse into the Early and Middle Archaic of central Maine (Figure 10.1) (Petersen 1991; Petersen at al. 1986; Petersen and Putnam 1992; Putnam 1994). It is safe to say that without the stratigraphic context, plus buried land surfaces, and/or cultural features, which could be dated by radiocarbon, many of the recovered artifacts might not be recognized as humanly modified pieces. Others, clearly artifacts, could be considered Late Archaic in affiliation. These two sites provide a remarkable sequence of human habitations extending back to at least 9,400 years ago. An earlier date of about 10,300 years ago could not be associated with any cultural material. Putnam's (1994) review of the geoarchaeology at the confluence illustrates the stratified cultural and noncultural sequence nearly 3 m deep.

At the confluence of the Stillwater River and Pushaw Stream are sites clustered around a bedrock sill known as Gilman Falls (Figure 10.2) (Sanger 1996; Sanger et al 2001). Although 19th- and 20th-century activities damaged some sites significantly, enough intact deposits remained at the Gilman Falls site and the Beaver site to warrant additional research. Together with sites farther up Pushaw Stream, they provide insights into human activities from 8,000 years ago to the contact period. We use "Milford drainage basin" to refer to a region that includes Pushaw Stream, Pushaw Lake, the surrounding peatlands and Stillwater River, all areas studied in connection with the relicensing of the Milford Dam at Old Town (Figure 10.3).

Near continuous human occupation began almost as soon as early Holocene Pushaw Lake achieved its modern configuration, about 8000 B.P. (Almquist-Jacobson and Sanger 1999, Figure 11.6:3). The only major gap in the cultural record is between the Laurentian Tradition termination at about 4500 B.P. (Mack et al. 2002) and the initial appearance of the Susquehanna Tradition about 3800 B.P. This hiatus in dated occupation sites occurs throughout interior central Maine. The Susquehanna Tradition is well represented at Hirundo (Sanger et al. 1977) and especially across Pushaw Stream at the Young site (Borstel 1982). Midway between Hirundo-Young and the downstream end of Pushaw Stream is the Bob site, another multicomponent site first occupied during Laurentian Tradition times. Stratified above this tradition are a few artifacts of the Susquehanna Tradition, and then a series of Ceramic (Woodland)-Period components (Mack et al. 2002). Other Ceramic-Period components occur at various sites along Pushaw Stream.

Given the opportunity to return to Pushaw Stream, scene of the 1970s Hirundo project, we reincorporated geoarchaeology into the research designs. The remainder of the chapter describes these activities.

Bedrock Sources

Bedrock plays an important role in the regional geoarchaeology. As described in Kelley and Sanger (this volume), bedrock controls the drainage and grade. Bedrock also profoundly influences lithic technological systems, defined as the availability and technological ability to use those rocks.

The Paleoindian Period of the Northeast is well known for extensive use of fine-grained chert and other silicates that often were collected many kilometers away from excavated sites (e.g., Ellis et al. 1998; Pollock et al. 1999; Spiess et al. 1998). A major trait of the Archaic Period in central Maine is the near absence of these lithic types. Instead, people use locally available rocks of three basic lithologies and sources: porphyritic rhyolite, vein quartz, and low-grade metamorphic rocks.

Porphyritic rhyolite, often called "felsite" by local archaeologists, does not outcrop in the study area. However, it is readily available in cobble form in streambeds where it erodes out of ice-contact features. During Early and Middle Archaic times, felsite cobbles served as hammer stones and provided raw material for scrapers and the very occasional biface. A specialized hammer stone form, the ridged hammer stone, was extensively used in the reduction of low-grade metamorphic rocks. Starting with a fist-sized felsite cobble, bifacial flaking produced a ridge that served as the point of impact. Experiments with low-grade metamorphic rocks indicate ridged hammer stones produce better results than the usual rounded hammer stones that tend to crush the edges rather than remove unwanted material (Sanger 1996).

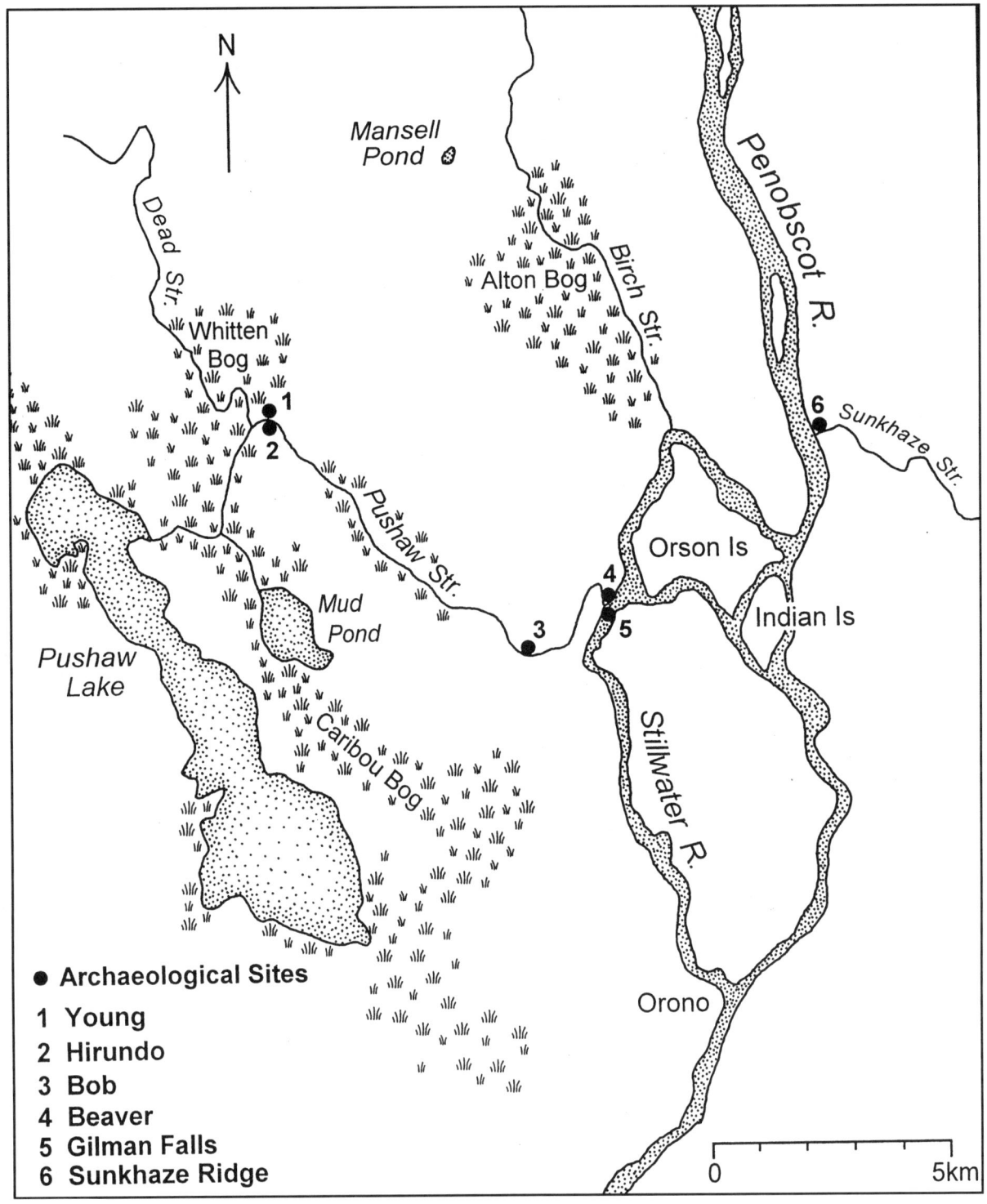

Figure 10.3. Pushaw Stream and the Milford drainage basin.

Geologist Henry Berry IV joined the research team after the recognition that stone artifacts from the Gilman Falls site apparently matched bedrock outcrops around and under the site (Figure 10.4). After field mapping and detailed petrographic analysis of field and archaeological specimens, Berry was able to separate those artifacts

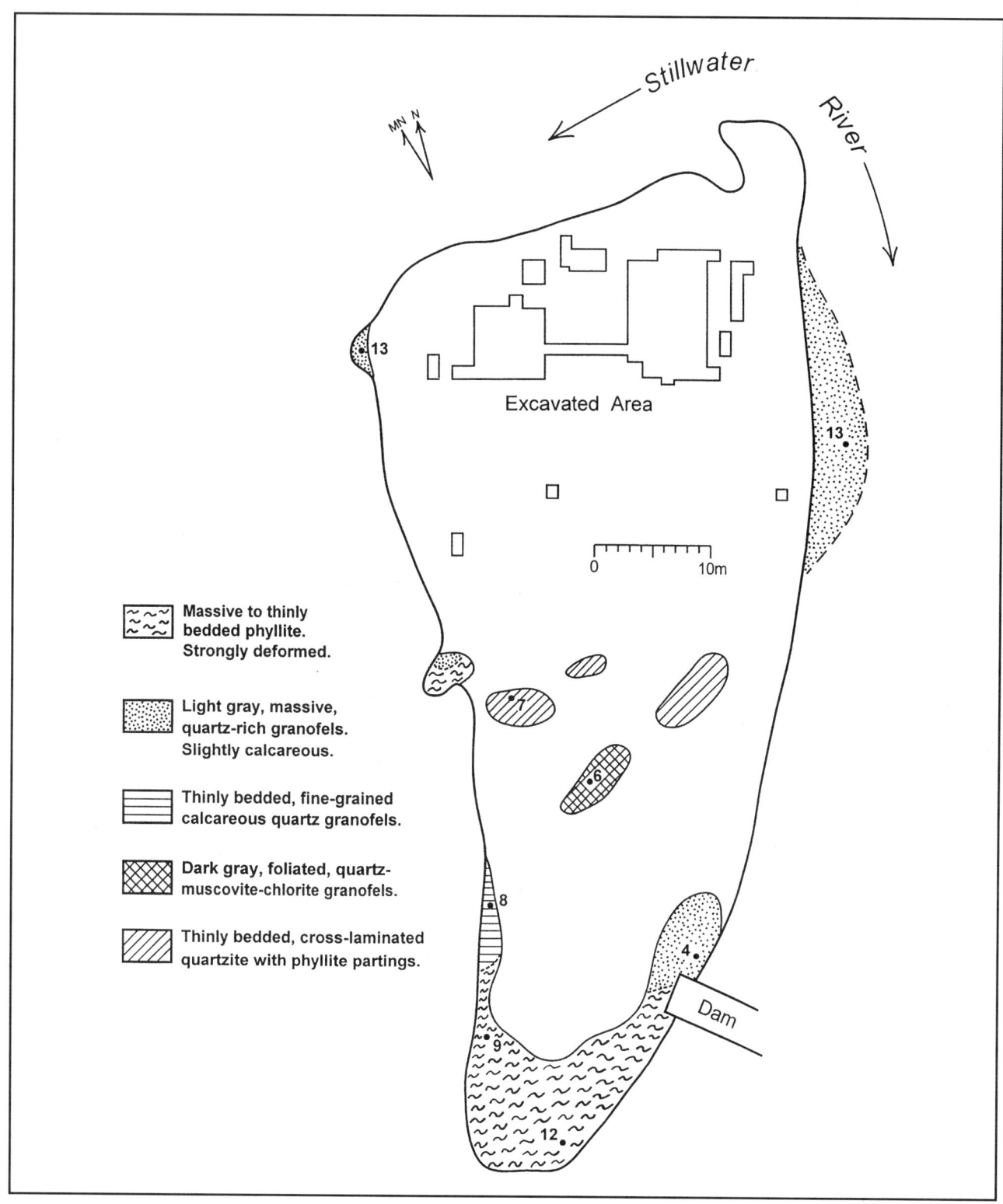

Figure 10.4. Bedrock map of Gilman Falls Island.

that could not have derived from the local bedrock, as opposed to those that might be of local origin.

Local bedrock is dominated by Silurian marine clastic rocks grouped as the Vassalboro Formation, a heterogeneous unit of metamorphosed bedded sandstones and pelites, as shown on the Bedrock Geologic Map of Maine

(Osberg et al. 1985; Sanger et al. 2001). Metamorphism, which is at low grade (chlorite zone) in central Maine, becomes lower north and east, and substantially higher grade south and west (Figure 10.5). This difference is reflected in archaeological collections, not only in the characterization of the rocks but also in the ways in

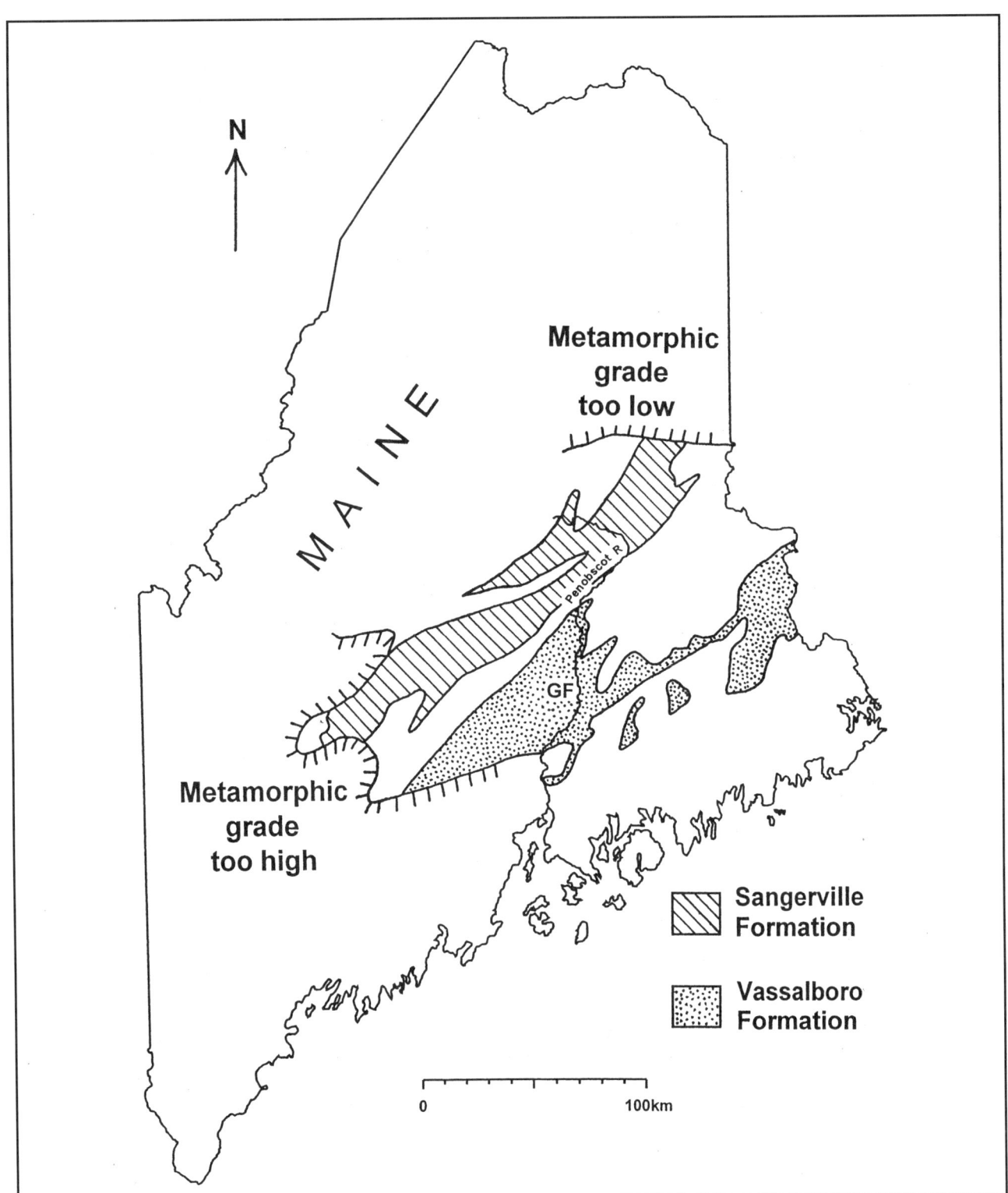

Figure 10.5. Bedrock metamorphic grades in central Maine (modified from Osberg et al. 1985).

which they fracture. For example, Richard Will (personal communication to Sanger, 2000) notes that artifacts and technology similar to Gilman Falls do not appear to exist in archaeological sites from northern Maine, an observation that meets our expectations given the differences in the grade of metamorphism. Even slight differences in grade can be reflected in the artifact production process.

Nearly two-thirds of Middle Archaic artifacts from the Gilman Falls site could easily have derived from either the small island itself or from outcrops within a radius of 200 m. Most are described as phyllite, granofels, and quartzite. Of these, foliated quartz-muscovite granofels and phyllite outcrop under and beside the site. Large chunks of raw material (up to 25 kg) have been recovered from the site, while a specialized ridged hammer stone occurred on a quarried outcrop adjacent to the Gilman Falls site. The rocks provided the source material for the most prominent artifact class at the site, cylindrical stone rods (Sanger et al. 2001).

Although rods occur at habitation sites, they also functioned in the mortuary domain. Robinson (1992) illustrated a number of rods associated with Early and Middle Archaic red ocher burials in New England. The Sunkhaze Ridge cemetery, located by the Penobscot River just 6 km northeast of Gilman Falls, contained a series of rods and full-length grooved gouges. Rods in all stages of manufacture (Figure 10.6) occurred in the Zone 3 (Middle Archaic) occupation at Gilman Falls (radiocarbon dated by 10 assays to ca. 6300–7300 B.P., uncalibrated).

A number of refits enables us to reconstruct the entire rod production sequence from quarry blanks, to flaked rough-outs, to pecked, and finally polished specimens. All 147 fragments broke in manufacture, and none shows any signs of use wear. This can be contrasted with some rods of nonlocal lithology that do exhibit polished facets, interpreted as use wear (Sanger 1996). We hypothesize that the rods in production at Gilman Falls were intended exclusively as grave offerings, a trait of the Moorehead burial tradition (Sanger 1973).

Robinson's (1996) analysis of Moorehead burial tradition grave artifacts indicated that rods dropped out of inventories during the latter stages of the burial tradition. It may be more than just a coincidence that the rod workshop activities at Gilman Falls terminated at a comparable time (ca. 6000 B.P.), after which the site was rarely inhabited.

Paleo-Environments in the Milford Drainage Basin

As noted, when we first began our investigations at Hirundo in 1971, we attempted a reconstruction of the

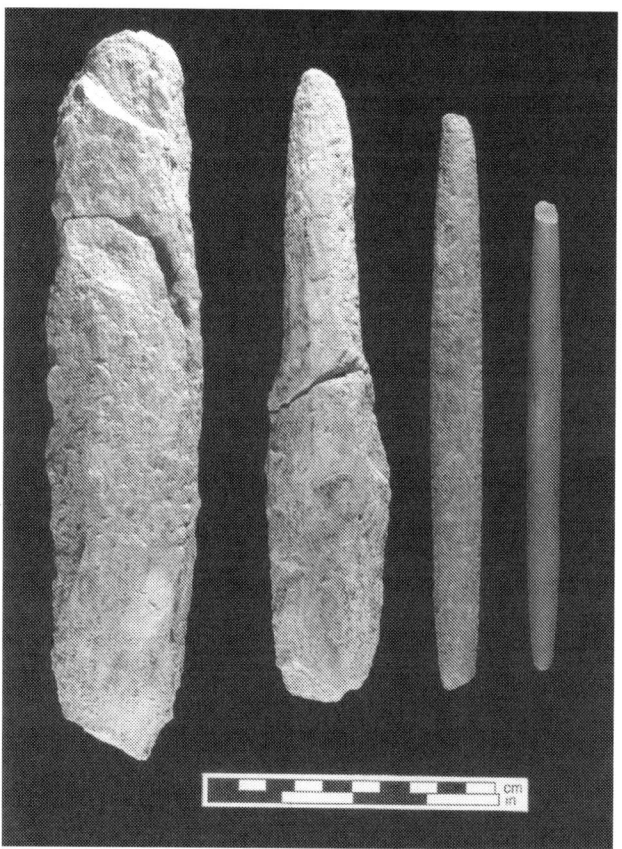

Figure 10.6. Production sequence of rods from early to final stages of manufacture.

paleo-environments (Sanger et al. 1977), including upland vegetation history and evolution of the wetlands. Our limited resources precluded much more than some dates for vegetation changes at Mud Pond (Sanger and MacKay 1973) and a preliminary pollen diagram (presented in Sanger et al. 1977). Over the next decade various students supervised by R.B. Davis produced useful data on the general history of peatland development (Gajewski 1987; Hu and Davis 1993). With the advent of much more extensive funding built into the Cultural Resource Management projects sponsored by Bangor Hydro-Electric Company, paleo-ecologist Almquist carried out a more ambitious program, strongly supported by radiocarbon dates to a level not available in the earlier investigations.

In the interim, the potential importance of wetlands adaptations had come to the fore in southern New England archaeology, largely due to the advocacy of George Nicholas (e.g., Nicholas 1990, 1991, 1998). Pushaw Stream and the Stillwater River (Figure 10.3) provided exceptional access to many hectares of wetlands. Today many of these are *Sphagnum* moss peatlands with relatively low carrying capacity for human use. In the past, however, the situation was quite different.

Three interrelated projects comprise our paleo-environmental research package: upland vegetation, peatlands evolution, and water levels. In each instance, we explicitly considered the need to match scales between the reconstructions of paleo-environments and the anticipated scope of human activities. On the one hand, overly broad environmental reconstructions may mask variability and detail that is potentially important for the understanding of human behavior. On the other hand, reconstruction at the level of the archaeological site ignores the likelihood that people traveled some unknown distance to exploit select features of the environment. The latter aspect is especially likely in a riverine environment where canoes can transport harvests a considerable distance with minimal effort, thus greatly increasing the potential catchment area. We have operated at the "neighborhood scale" (Dincauze 1996). A problem, of course, lies in defining the "neighborhood" limits.

Upland Pollen Record

Mansell Pond, a small (4 ha) kettle-hole pond situated less than 10 km north of Pushaw Stream (Figure 10.3), was selected for a sediment core because of its location and size, and the potential to obtain a very local pollen signal appropriate to the nearby archaeological record. Lacking any surface inlets or outlets, Mansell Pond water depth is about 7 m. The pond contains 8 m of Holocene sediment that has accumulated since deglaciation. Twelve-bulk sediment and AMS radiocarbon dates on *gyttia* constrained the sedimentation rate, beginning at 9175 + 230/-225 B.P., (A-7058). Bearing in mind the archaeological rationale for obtaining a regional, fine-grained vegetation history, the 4.5-m-long core was sampled at 2-cm intervals (Almquist-Jacobson and Sanger 1995).

The Mansell Pond pollen research indicates that upland vegetation during Early Archaic Period (ca. 9200 B.P.) began as open woodland dominated by *Picea* (spruce), *Populus* (poplar), and *Larix* (larch) replaced shortly by a *Pinus strobus* (white pine) Phase-1 forest. By around 8400 B.P., more temperate taxa, especially *Quercus* (oak) and other hardwoods, appeared in the pollen diagram. The *Pinus* forest changed into a *Tsuga canadensis* (hemlock) dominated forest by 7400 B.P., followed once again by another *Pinus* forest (Phase 2) at 6400 B.P. A second *Tsuga* forest emerged by 5700 B.P.

Many changes occurred in the Late Archaic (6000–3000 B.P.). After the well-known and widespread demise of the *Tsuga* forests at 4700 B.P., came the northern hardwood genera, such as *Betula* (birch) and *Fagus* (beech). As Almquist-Jacobson and Sanger (1995) pointed out, the advent of northern hardwoods would have increased potential beaver and muskrat habitat, both important species in the wetlands of central Maine. Many of the same tree species constituted key resources for deer and moose.

Peatlands Evolution

Within the Milford drainage basin there are a number of *Sphagnum* moss peat bogs (Figure 10.3). The North Caribou Bog complex surrounding Pushaw Stream includes several discrete peatlands, of which Caribou and Whitten bogs are probably the most important. A network of 35 coring sites, integrated with previous work on the Caribou Bog (Hu and Davis 1993), produced a series of time-transgressive wetland maps (Almquist-Jacobson and Sanger 1999: Figure 11-6) constrained by 14 radiocarbon dates and 8 sediment ages determined by pollen correlations with the Mansell Pond diagram. It is noteworthy that at 10,000 B.P., Pushaw Lake extended over the Hirundo and Young sites, and even by 8000 B.P. there was a substantial body of water in the vicinity. By 6000 B.P. the landscape began to assume a more modern configuration, although there was still more open water in the region than there was later in the Holocene. However, whereas modern *Sphagnum* peatlands dominate the Hirundo-Young area, at 6000 B.P. the reconstructions reflect an extensive *Typha* (cattail) marsh surrounding Dead Stream, a tributary of Pushaw Stream just minutes from Hirundo-Young by canoe (Figure 10.3). As Nicholas (1991) noted, this is an extremely productive wetland form for humans, because of the mix of its plant and animal species. According to our reconstructions, the extensive Dead Stream *Typha* marsh did not evolve into a *Sphagnum* peatland (now known as the Whitten Bog) until after sometime after 1000 B.P.

The link between Hirundo and Young sites and the nearby wetlands is imperfectly reflected in the site fauna and flora due to the generally acidic soil. However, the calcined remains that have survived indicate the presence of more beaver and muskrat, both aquatic mammals, than deer. For example, of more than 12,000 bone elements recovered from Hirundo (mostly in Ceramic-Period contexts), approximately 93 percent of identified elements represent beaver. As the Hirundo faunal analyst James Knight (1985) pointed out, differential preservation of calcined beaver bone versus calcined deer bone probably cannot explain the substantial variation in numbers of identified elements for each species. In his opinion cultural explanations, such as differential disposal, may have played a role. We hypothesize that although deer were probably available in the forested uplands near the site, the Hirundo residents focused on the wetland environment for a greater part of their subsistence. Large numbers of very small, unidentified fish bones are also present in the faunal collection; but in the absence of positive identifications we cannot determine if these were of riverine or marshland derivation. In a previous report on the Hirundo site, the potential importance of an anadromous-catadromous fishery was stressed (Sanger et al. 1977) because of the bedrock rapids that would have enabled fish traps of various sorts. Although this may be true, at the time we did not fully appreciate the potential for wetlands to provide a stable, even year-round, subsistence.

A small Ceramic Period faunal sample from the Bob site, another Pushaw Stream station, also reflects the wetlands emphasis on beaver, muskrat, and turtle (Mack et al. 2002).

Water Levels

The third portion of our paleo-environmental package involved a study of water levels in Mansell Pond (Almquist et al. 2002). This study assumed key proportions in light of others that modeled east-coast mid-Holocene aridity, perhaps 30 percent less precipitation than today, combined with thermal warming of perhaps 2° C higher than modern conditions (Davis et al. 1980; Harrison 1989; Prentice et al. 1991; Webb et al. 1993). We wanted to see if the general trends identified across the broader region were reflected in a more localized record.

Once again Mansell Pond meets the paleo-ecologists' requirements for an ideal setting from which to recover highly localized information. Mansell Pond, as noted earlier, is small (ca. 4 ha) and lacks inlet or outlet streams. Lined with marine clay from the Late Pleistocene marine incursion (Belknap et al. 1987; Kelley and Sanger, this volume), the water containment is such that the bottom of a gravel pit, just meters away and excavated below modern pond level, remains dry. Eight cores, taken along a littoral to deep water transect, provided organic material to determine sedimentary transitions that reflected changes in water levels.

The oldest date from this study is 9540 ± 80 B.P. (Beta-101669) from core HT-710 taken at the deepest point in Mansell Pond. The date accords well with the earliest date derived from core MS-660 acquired for the pollen study (9175 + 230/−225 B.P.; A-7058). Transitions between *gyttia* and *Sphagnum* peat were dated by 29 radiocarbon determinations on the premise that the former was deposited below the water surface and peat above. As recorded in the 8-core transect, water levels fell between ca. 8000 B.P. and 6000 B.P., and remained low until after 5000 B.P. At its lowest level it may have been about 8 m below modern water levels. Shortly after 5000 B.P. water levels began to rise, coming up most rapidly ca. 3225–2780 B.P., and reaching its highest levels in recent centuries (Figure 10.7).

Linkages among Holocene lake-water levels, climatic regimes, vegetation, wetland evolution, and hunter-gatherer adaptations would appear intuitively transparent. Yet, with the impoverished floral and faunal collections from our sites, only the broadest of statements can be sustained by archaeological evidence. As we have pointed out previously (Almquist-Jacobson and Sanger 1995; Almquist-Jacobson and Sanger 1999), the incidence of aquatic mammals, especially beaver and muskrat, as well as abundant evidence for turtles and small fish remains, is expected and partially confirmed by site fauna (Mack et al. 2002). Pushaw Stream acts as a corridor for anyone wishing to gain access to marshes or Pushaw Lake. It is probably no accident that we have many sites arrayed along its banks. Archaeological surveys of the peatlands have not occurred. Given the raised, blanket bog systems of northern New England (Johnson 1985), we anticipate that sites once located at the fringes of marshes will be buried under the expanding peat. Although the thought of an inundated site with excellent organic preservation is intriguing, there would appear to be very little incentive for Native peoples to actually live by a marsh. If access could be obtained by canoe from a better-drained locality, such as a riverbank, extensive camping by a marsh seems unlikely.

Nicholas (e.g., 1990, 1991, 1998) has emphasized the advantages accrued from living in proximity to wetlands, including high biological productivity and predictability. Yet it is not until about 8000 B.P. that we begin to detect settlements that were poised to take advantage of what the wetlands have to offer. Our study stands alone in northern New England, so it is not possible at this time to detect any trends through comparative analyses. Nevertheless, the evidence indicates that Pushaw Lake and Mud Lake did not shrink to near-modern configurations until the middle of the Holocene. As Pushaw Lake retracted, *Typha* (cattail) marshes appeared, followed by encroaching peat. That intermediate period would appear to represent an optimal period for an intense wetland adaptation. These events appear coincident with the hypothesized mid-Holocene warmer and drier period (e.g., Davis et al. 1980; Harrison 1989; Prentice et al. 1991; Webb et al. 1993).

DISCUSSION

Our research program has documented the highly dynamic nature of Late Pleistocene and Holocene environments, right to the present. This has important implications for models of culture change in the region,

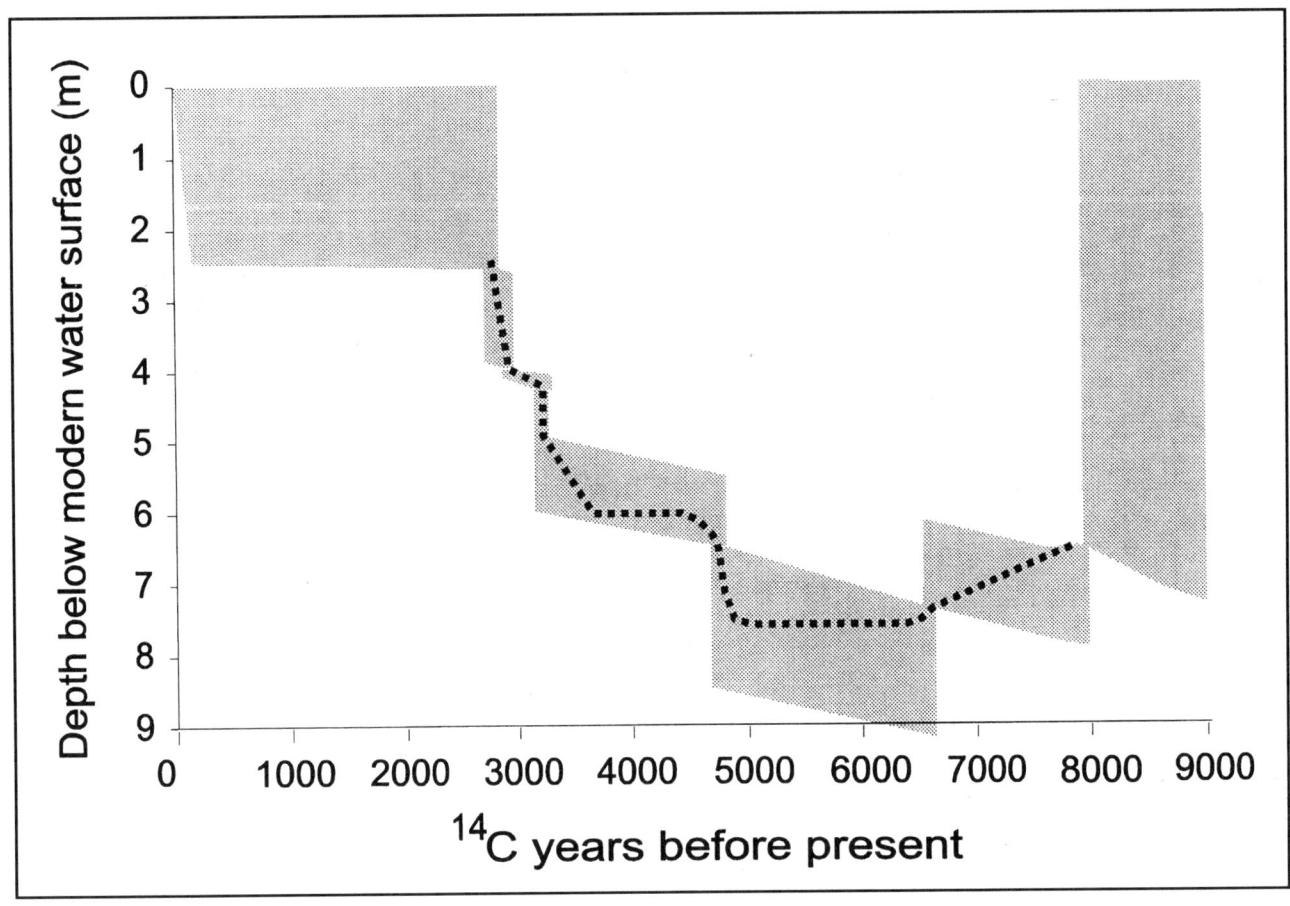

Figure 10.7. Water levels in Mansell Pond.

because it means that highly localized environments must be taken into account. In addition, we argue that the nature of bedrock outcrops in central Maine, plus Quaternary glacial history, has affected the trajectory of culture history. Our attempt at integration of cultural and environmental data recognizes the intersection of broad-scale models with more localized environments.

Paleoindian-Demise Models

As an example of a large-scale cultural model we note the Paleoindian Period in the Northeast, featuring very similar looking assemblages scattered across late Pleistocene landscapes. In the absence of well-preserved faunal assemblages, many archaeologists have linked terminal Pleistocene tundra and spruce parkland environments of the Northeast with a caribou-hunting focus. The few caribou bones in sites that have survived give a certain amount of credence to the model (e.g., Ellis et al. 1998; Spiess et al. 1998). Anderson et al. (1996) pointed out that many Paleoindian researchers have assumed a collector or logistically based adaptation (sensu Binford 1980). However, given the chronological constraints imposed by such a model, plus the need to document all the various site types implied, this appealing model will remain just that, intuitively reasonable but currently undemonstrated until more sites and firmer chronologies are developed.

With the advent of the Early Archaic, Ellis et al. (1998:162) noted a number of changes that appeared to be "pan-eastern." The Early Archaic assemblages of central Maine followed the trend by making use of local lithologies, which in this instance meant eschewing the previously popular but regionally restricted cherts and fine-grained silicates. Ground stone woodworking tools appeared, as did less formally shaped scraping and cutting implements made of low-grade metamorphic rocks. Finally, Ellis et al. (1998:162) noted an impression of decreased residential mobility, something that probably occurred in Maine. Explanation may lie in vegetative and climatic changes that made possible a more broadly based subsistence pattern, one that took advantage of the newly evolved wetlands.

As mapped by the Surficial Geological Map of Maine (Thompson and Borns 1985), large parts of central and eastern Maine and adjacent New Brunswick are dominated today by extensive wetlands, mostly *Sphagnum* moss peatlands. Unfortunately we know relatively little about the evolutionary history of these wetlands. To date, our extensive research efforts in Caribou and Whitten bogs have not been duplicated. By extrapolation, it would seem reasonable to assume that the other peatlands experienced a somewhat similar general history; that is, open-water late glacial lakes and ponds evolved into *Sphagnum* bogs through a process known as lakefill (Johnson 1985). Intermediate in this evolution were swamps and marshes of the kind we model as being highly desirable for hunter-gatherers. Although it seems likely that the chronology will vary depending on a number of local control variables, the overall impression is that while the uplands were assuming a more closed forest configuration (Davis and Jacobsen 1985), the wetlands became highly desirable places, providing certain changes in lifestyle occurred. Foremost among these was abandonment of the putative caribou-hunting pattern and regular visits to localized chert and silicate quarries. Instead, people used local lithics with all the technological constraints imposed by nature of the raw material. It is important to emphasize we do not claim this scenario applied outside of the region of extensive wetlands in central and eastern Maine. Undoubtedly aboriginal peoples in Maine participated generally in what Ellis et al. (1998) have termed a "pan-eastern" technological and settlement pattern shift. It may help to explain, however, why we are finding so many Early and Middle Archaic sites compared with surrounding areas that lacked the draw of such extensive wetlands.

Impacts of the Mid-Holocene Warmer and Dryer Period

Climate modelers have speculated on the extent of apparent increased temperature and decreased precipitation during the late mid-Holocene (e.g., Davis et al. 1980; Harrison 1989; Prentice et al. 1991; Webb et al. 1993). Lowered lake levels in Mansell Pond between 8000 and 6000 B.P. may be linked to this broad-scale climatic change. After about 5000 B.P. when climate reverted to more precipitation and lower temperatures, water levels in Mansell Pond began to rise, more rapidly after 3500 B.P. (Almquist et al. 2001). The apparent increase in regional precipitation is reflected in increased alluvial sedimentation rates witnessed in local sites.

By historic times, Maine's aboriginal peoples were heavily dependent on the waterways for fish habitat and for transportation routes (Speck 1940). We speculate on what lowered water levels might mean, because they were highest just as Europeans arrived.

1. The high water levels in recent centuries likely drowned any Archaic-Period archaeological sites situated on the edge of a lake, unless the site was perched on a high landform.

2. A drop in water levels would have meant that travel along the waterways was hampered or unprofitable, depending on the watercourse. Even the Penobscot River, the main drainage and artery into northern Maine, could be very tedious to canoe during periods of low water. For example, in the fall of 1820, surveyor Joseph Treat (1820) ascended the Penobscot River by canoe, led by Penobscot Indian guides. Low water frustrated the travelers and rocks damaged their birch bark canoes. Many of Maine's rivers may be "run" only at high water in the spring (Cook 1985). Less water in mid-Holocene times would imply even more difficult travel.

3. Lowered water levels and less flow would also have meant decreased fish habitat and access to spawning beds for anadromous fish, while warmer water temperatures could prove lethal to salmon, as recent studies in the Penobscot River have demonstrated. In 1988 more than 70 of 200 trapped and released Atlantic salmon perished as water temperatures exceeded 27° C. Low flow rates in the Piscataquis River, a major salmon-spawning river, inhibited upstream migration (Sanger and Newsom 2000; Shepard 1995).

4. On the positive side, Almquist-Jacobson (1995) and Sanger pointed out that the ca. 5000 B.P. increase in hardwoods—especially *Populus* and *Acer rubrum* (red maple)—following the *Tsuga* decline would have greatly enhanced beaver habitat.

5. Beaver and their ponds create ideal muskrat habitat. Both species were highly prized, as documented in the archaeological site faunal lists and in ethnographies (e.g., Speck 1940). In addition to beaver and muskrat, the ponds provide habitat for a wide variety of fish, amphibian, reptile, bird, and plant remains.

6. Finally, beaver dams in small tributaries will create a series of stepped ponds, each only a meter or so higher than the previous one. These ponds

cover rocks and greatly assist canoe access to otherwise nonnavigable waterways (Cook 1985).

Milford Basin Culture History

The multiyear investigation in the Milford Drainage Basin area has resulted in a cultural sequence that spans all pre-European periods except for the Early (fluted point) Paleoindian. Within the Milford Basin proper, our earliest occupation is dated to just over 8000 B.P. Prior to this period, Pushaw Lake and other regional lakes were considerably larger. As the more modern configuration of wetlands developed, archaeological sites become located along the banks of Pushaw Stream. Native peoples had access to riverine as well as wetland resources, especially those found in the highly productive *Typha* marshes. Part of the adaptation to this environment involved the use of local rocks, such as felsites from the streambeds, vein quartz from local bedrock, and low-grade metamorphic rocks quarried from outcrops of the Vassalboro Formation. The latter's structure quite severely constrained artifact form, so that functional implements often appear misleadingly casual or "expedient."

At one occupation site, Gilman Falls, we have evidence for quarrying local granofels and phyllite, and the manufacture of polished stone rods (Sanger 1996; Sanger et al. 2001). The highly specialized ground stone forms comprised part of a symbolic suite of pecked and ground artifacts placed with the dead in red ocher cemeteries of the Moorehead burial tradition (Robinson 1996, 2001; Sanger 1973).

Shortly after the end of the Middle Archaic Period (ca. 6000 B.P.), a new tool form enters the record in central and eastern Maine. This is the broad side-notched projectile point, known as Otter Creek, that evolved from midcontinent origins (Fank 1998; Ritchie 1965, 1968; Wright 1995). In central and eastern Maine, these points became attached Middle Archaic assemblages that contained slate points, celts, gouges, ulus, and plummets, all associated with the Vergennes Phase of the Laurentian Tradition (Cox 1991; Mack et al. 2002; Petersen 1991; Petersen et al. 1986; Ritchie 1986; Sanger et al. 1977; Sanger and Newsom 2000; Sanget et al. 2001).

CONCLUSION

As we continue to work with our database that resulted from over a decade of interdisciplinary research, the usefulness of a broad geoarchaeological approach becomes ever more apparent. In this study, we have been able to integrate bedrock and sediment analysis into the history of wetland evolution, upland vegetation, water levels, and human adaptation over 8000 years. Although our record is localized, we recognize that each component is part of a larger picture that includes the bedrock geology of eastern North America, glaciation and deglaciation, hydrological regimes, vegetation responses to climate and pathogens, wetland ecosystem evolution, and finally humans. The latter combined localized environmental regimes with broad cultural patterns recognized in much of the Northeast.

Acknowledgments

We are pleased to recognize the help of many students and staff at the University of Maine, the cooperation of various landowners, and local residents. Stephen Bicknell prepared Figures 10.1–10.4 plates and assisted in many other tasks, both field and laboratory. Ann Dieffenbacher-Krall produced Figure 10.5. Research funding came from Bangor Hydro-Electric Company; The National Geographic Society; the Maine Historic Preservation Commission administering funds from the U.S. Department of the Interior; the National Science Foundation (EPSCoR program); and the University of Maine.

REFERENCES CITED

Almquist, H., Dieffenbacher-Krall, A., Brown, R., and Sanger, D. (2001). An 8000-yr Holocene record of lake levels at Mansell Pond, Central Maine, U.S.A. *The Holocene* **11**:189–210.

Almquist-Jacobson, H., and Sanger, D. (1995). Holocene climate and vegetation in the Milford Drainage Basin, Maine, U.S.A., and their implications for human history. *Vegetation History and Archaeobotany* **4**:211–222.

Almquist-Jacobson, H., and Sanger, D. (1999). Paleogeographic changes in wetland and upland environments in the Milford drainage basin of central Maine, in relation to Holocene human settlement history. In *Current Northeast Paleoethnobotany*, edited by J. Hart, pp. 177–190. New York State Museum Bulletin 494, The University of the State of New York, Albany.

Anderson, D.G. (1996). Models of Paleoindian and Early Archaic settlement in the lower Southeast. In *The Paleoindian and Early Archaic Southeast*, edited by D.G. Anderson and K.E. Sassaman, pp. 29–57. The University of Alabama Press, Tuscaloosa.

Anderson, D.G., O'Steen, L.D., and Sassaman, K.E. (1996). Environmental and chronological considerations. In *The Paleoindian and Early Archaic Southeast*, edited by D.G. Anderson and K.E. Sassaman, pp. 3–15. The University of Alabama Press, Tuscaloosa.

Belknap, D.L., Andersen, B.G., Anderson, R.S., Anderson, W.A., Borns, H.W., Jr., Jacobson, G.L., Kelley, J.T., Shipp, R.C., Smith, D.C., Stuckenrath, R.J., Thompson, W.B., and Tyler, D.A. (1987). Late Quaternary sea-level changes in Maine. In *Sea Level Fluctuations and Coastal Evolution*, edited by D. Nummedahl, O.H. Pilkey, and J.D. Howard, pp. 65–79. SEPM Special Publication No. 41. Society for Economic Paleontology and Mineralogy.

Binford, L.R. (1980). Willow smoke and dog's tails: Hunter-gatherer settlement systems and archaeological site formation. *American Antiquity* **45**:4–20.

Borstel, C. (1982). *Archaeological Investigations at the Young Site, Alton, Maine*. Occasional Publications in Maine Archaeology, No. 2. Maine Historic Preservation Commission, Augusta.

Butzer, K.W. (1982). *Archaeology as Human Ecology: Method and Theory for a Contextual Approach*. Cambridge University Press, Cambridge.

Cook, D.S. (1985). *Above the Gravel Bar*. Milo Publishing, Milo, Maine.

Cox, B.J., and Petersen, J.,B. (1997). The Varney Farm (36-57 Me): a Late Paleoindian encampment in western Maine. *The Maine Archaeological Society Bulletin* **37**(2):25–48.

Cox, S.L. (1991). Site 95.20 and the Vergennes Phase in Maine. *Archaeology of Eastern North America* **19**:135–161.

Davis, M.B., Spear, R.W., and Shane, L.C. (1980). Holocene climate of New England. *Quaternary Research* **14**:240–250.

Davis, R.B., and Jacobsen, G.L., Jr. (1985). Late glacial and early Holocene landscapes in northern New England and adjacent areas of Canada. *Quaternary Research* **23**:341–368.

Dincauze, D.F. (1996). Modeling communities and other thankless tasks. In *The Paleoindian and Early Archaic Southeast*, edited by D.G. Anderson and K.E. Sassaman, pp. 421–424. The University of Alabama Press, Tuscaloosa.

Doyle, R.A., Hamilton, N.D., Petersen, J.B., and Sanger, D. (1985). Late Paleo-indian remains and their correlations in Northeastern prehistory. *Archaeology of Eastern North America* **13**:1–34.

Dumais, P. (2000). The La Martre and Mitis Late Paleoindian sites: A reflection on the peopling of southeastern Quebec. *Archaeology of Eastern North America* **28**:81–112.

Ellis, C., Goodyear, A.C., Morse, D.F., and Tankersley, K.B. (1998). Archaeology of the Pleistocene-Holocene transition in eastern North America. *Quaternary International* **49/50**:151–166.

Fitting, J. (1968). Environmental potential and the postglacial readaptation in eastern North America. *American Antiquity* **33**:441–445.

Funk, R.E. (1988). The Laurentian concept: A review. *Archaeology of Eastern North America* **16**:1–41.

Gajewski, K. (1987). Environmental history of Caribou Bog, Penobscot Co., Maine. *Le Naturaliste Canadien* **114**:133–140.

Harrison, S.P. (1989). Lake levels and climate change in eastern North America. *Climate Dynamics* **3**:157–167.

Hu, F.S., and Davis, R.B. (1993). Postglacial development of a Maine bog and environmental implications. *Canadian Journal of Botany* **73**:638–649.

Johnson, C.W. (1985). *Bogs of the Northeast*. University Press of New England, Hanover, N.H.

Knight, J.A. (1985). Differential preservation of calcined bone at the Hirundo Site, Alton. Unpublished master's thesis, Institute for Quaternary Studies, University of Maine, Orono.

Mack, K.E., Kelley, A.R., and Sanger, D. (2002). *The Bob Site: A Multicomponent Archaic and Ceramic Period Site on Pushaw Stream, Maine*. Occasional Publications in Maine Archaeology, No. 12, Maine Archaeological Society and Maine Historic Preservation Commission, Augusta.

Nicholas, G.P. (1990). *The Archaeology of Early Place: Early Postglacial Land Use and Ecology at Robbins Swamp, Northwestern Connecticut*. Doctoral dissertation, University of Massachusetts, Amherst. University Microfilms, Ann Arbor, Mich.

Nicholas, G.P. (1991). Putting wetlands into perspective. *Man in the Northeast* **42**:29–38.

Nicholas, G.P. (1998). Assessing climatic influences on human affairs: Wetlands and the maximum Holocene warming in the Northeast. *Journal of Middle Atlantic Archaeology* **14**:147–160.

Osberg, P.H., Hussey, A.M., and Boone, G.M. (1985). *Bedrock Geologic Map of Maine*. Maine Geological Survey, Augusta.

Petersen, J.B. (1991). *Archaeological Testing at the Sharrow Site: A Deeply Stratified Early to Late Holocene Cultural Sequence in Central Maine*. Occasional Papers in Maine Archaeology, No. 8. Maine Historic Preservation Commission, Augusta.

Petersen, J.B., Bartone, R.N., and Cox, B.J. (2000). The Varney Farm site and the Late Paleoindian Period in northeastern North America. *Archaeology of Eastern North America* **28**:113–140.

Petersen, J.B., Hamilton, N.D., Putnam, D., Spiess, A.E., Stuckenrath, R., Thayer, C.A., and Wolford, J.A. (1986). The Piscataquis archaeological project: A late Pleistocene occupational sequence in northern New England. *Archaeology of Eastern North America* **14**:1–18.

Petersen, J.B., and Putnam, D.E. (1992). Early Holocene occupation in the central Gulf of Maine region. In *Early Holocene Occupation in Northern New England*, edited by B.S. Robinson, J.B. Petersen, and A.K. Robinson, pp. 13–61. Occasional Papers in Maine Archaeology, No. 9. Maine Historic Preservation Commission, Augusta.

Pollock, S.G., Hamilton, N.D., and Bonnichsen, R. (1999). Chert from the Munsungun Lake Formation (Maine) in Palaeoamerican archaeological sites in northeastern North America: Recognition of its occurrence and distribution. *Journal of Archaeological Science* **26**:269–293.

Prentice, I.C., Bartlein, P.J., and Webb, T.I. (1991). Vegetation and climate change in eastern North America since the last glacial maximum. *Ecology* **72**:2038–20513.

Putnam, D.E. (1994). Vertical accretion of flood deposits and deeply stratified archaeological site formation in central Maine, U.S.A. *Geoarchaeology: An International Journal* **9**:467–502.

Ritchie, W.A. (1965). *The Archaeology of New York State*. Natural History Press, New York.

Ritchie, W.A. (1968). The KI Site, the Vergennes Phase, and the Laurentian tradition. *The Bulletin of the New York State Archaeological Association*, No. 42:1–5.

Robinson, B.S. (1992). Early and Middle Archaic Period occupation in the Gulf of Maine region: mortuary and technological patterning, In *Early Holocene Occupation in Northern New England*, edited by B.S. Robinson, J.B. Petersen, and A.K. Robinson, pp. 63–116.Occasional Publications in Maine Archaeology, No. 9, Maine Historic Preservation Commission, Augusta.

Robinson, B. S. (1996). A regional analysis of the Moorehead burial tradition: 8500–3700 BP *Archaeology of Eastern North America* **24**:95–148.

Robinson, B.S. (2001). *Burial Ritual, Groups, and Boundaries on the Gulf of Maine: 8600–3800 B.P.* Doctoral dissertation, Brown University, University Microfilms, Ann Arbor, Mich.

Robinson, B.S., and Petersen, J.B. (1992). Introduction: archaeological patterning and visibility in northern New England. In *Early Holocene Occupation in Northern New England*, edited by B.S. Robinson, J.B. Petersen, and A.K. Robinson, pp. 1–11. Occasional Publications in Maine Archaeology, No 9. Maine Historic Preservation Commission, Augusta.

Sanger, D. (1973). *Cow Point: An Archaic Cemetery in New Brunswick.* Mercury Series 12. Archaeological Survey of Canada, National Museums of Canada, Ottawa, Ont.

Sanger, D. (1977). Some thoughts on the scarcity of archaeological sites in Maine between 10,000 and 5,000 years ago. *Maine Archaeological Society Bulletin* **17**:18–25.

Sanger, D. (1996). Gilman Falls site: Implications for the Early and Middle Archaic of the Maritime Peninsula. *Canadian Journal of Archaeology* **20**:7–28.

Sanger, D., Belcher, W.R., and Kellogg, D.C. (1992). Early Holocene occupation at the Blackman Stream site, central Maine. In *Early Holocene Occupations in Northern New England*, edited by B.S. Robinson, J.B. Petersen, and A.K. Robinson, pp. 149–161, Occasional Publications in Maine Archaeology, No. 9. Maine Historic Preservation Commission, Augusta.

Sanger, D., Davis, R.B., MacKay, R.G., and Borns, H.W., Jr. (1977). The Hirundo archaeological project: An interdisciplinary approach to central Maine prehistory. In *Amerinds and Their Paleoenvironments in Northeastern North America*, edited by W.B. Newman and B. Salwen, pp. 457–471. Annals of the New York Academy of Sciences. Vol. 288. New York Academy of Sciences, New York.

Sanger, D., Kelley, A.R., and Berry, H.N. IV. (2001). Geoarcheology of Gilman Falls: An Archaic quarry and manufacturing site in central Maine, U.S.A. *Geoarchaeology* **16**:633–665..

Sanger, D., and MacKay, R.G. (1973). The Hirundo archaeological project: Preliminary report. *Man in the Northeast* **6**:21–29.

Sanger, D., and Newsom, B. (2000). Middle Archaic in the lower Piscataquis River, and its relationship to the Laurentian tradition in central Maine. *The Maine Archaeological Society Bulletin* **40**:1–22.

Shepard, S.L. (1995). Atlantic salmon spawning migrations in the Penobscot River, Maine: Fishing, flows, and high temperatures. Master's Thesis, University of Maine.

Speck, F.G. (1940). *Penobscot Man.* University of Pennsylvania Press, Philadelphia.

Spiess, A.E. (1990). *Maine's Unwritten Past: State Plan for Prehistoric Archaeology.* Maine Historic Preservation Commission, Augusta.

Spiess, A.E., Bradley, J.W., and Wilson, D. (1998). Paleoindian occupation in the New England-Maritimes region: Beyond cultural ecology. *Archaeology of Eastern North America* **26**:201–264.

Thompson, W.B., and Borns, H.W.J. (1985). *Surficial Geologic Map of Maine.* Maine Geological Survey, Augusta.

Treat, J. (1820). *Journal and Places of Survey.* Maine State Archives, Vol. 14. Maine Land Office Field Notes.

Trigger, B.G. (1991). Distinguished lecture in archaeology: Constraint and freedom—a new synthesis for archaeological explanation. *American Anthropology* **93**:551–569.

Webb, T.I., Bartlein, P.J., Harrison, S.P., and Anderson, K.H. (1993). Vegetation, lake levels, and climate in eastern North America for the past 18,000 years. In *Global Climates Since the Last Glacial Maximum*, edited by H.E.J. Wright, J.E. Kutzback, T.I. Webb, W.F. Ruddiman, F.A. Street-Perrott, and P.J. Bartlein, pp. 415–467. University of Minnesota Press, Minneapolis.

Wright, J.V. (1995). *A History of the Native People of Canada, Vol. I (10,000–1,000 B.C.).* Mercury Series, Archaeological Survey of Canada 152. Canadian Museum of Civilization, Ottawa, Ont.

CHAPTER 11

GEOCHRONOLOGY FROM ARCHAEOLOGY: AN EXAMPLE FROM THE CONNECTICUT RIVER VALLEY

Kathryn Curran

In the Riverside District of Gill, Massachusetts (Figure 11.1) a series of river-cut terraces lies above the modern level of the Connecticut River (Figure 11.2). In the late 1960s and early 1970s, avocational and professional archaeologists found sequentially aged cultural sites associated to these terraces (Figure 11.3). At this location, older prehistoric sites lie on higher terraces, while younger sites are at lower elevations (Thomas 1980:75). The long cultural sequence, from Paleoindian (12,000–10,000 B.P.) through the Contact Period (A.D. 17th century) suggests that this area remained a popular settlement location for thousands of years.

In and around Riverside, a unique series of geological events occurred postglacially. As a result, a terrace set developed at a location conducive to habitation. Because archaeological materials were deposited after the terraces were cut and their surfaces exposed, cultural remains here indicate minimal ages for the alluvial surfaces (Figure 11.3). The movements of the proto-Connecticut River are thus tracked and timed with archaeological materials. In this chapter I use archaeological data to date both the abandonment of an initial course of the Connecticut River through White Ash Swamp (near Greenfield, Mass.) and a pair of now relict waterfalls near the project area. The Riverside terraces can be employed to estimate time of geomorphic processes and are useful in interpreting regional drainage chronologies for glacial Lake Hitchcock as well as proto-Connecticut River development.

Cultural remains of human activity were placed coeval to natural depositional and erosional events. As a result archaeological artifacts and features are an integral part of Riverside stratification. After Late Wisconsinan deglaciation, a new Connecticut River channel section formed and the resource base around Riverside expanded. This rich landscape attracted Native Americans to the region. The changing shape of relict landforms helped to define settlement patterns and site placement styles. Human populations returned over the millennia to the shores of the Connecticut, living in and around Riverside in close proximity to the river and its abundance of resources.

The unwavering popularity of the area (for thousands of years) helped to first augment and then to erode Riverside deposits. Today little visible evidence of individual terrace boundaries remains. Over hundreds of years increased development changed the face of Riverside forever. In 1996 while visiting the area, I found that construction of roads and buildings had smoothed land contour, making terrace edges impossible to clearly identify. However, while examining Richard Jahns' (1966) surficial map of Greenfield, I noted that scarp lines he illustrated in Riverside in 1966 delineated boundaries for a total of five river-cut terraces.

Figure 11.1. Location of Gill, Massachusetts, in southern New England (adapted from Mulholland 1984:449).

Geoarchaeology of Landscapes in the Glaciated Northeast edited by David L. Cremeens and John P. Hart. New York State Museum Bulletin 497. © 2003 by the University of the State of New York, The State Education Department, Albany, New York. All rights reserved.

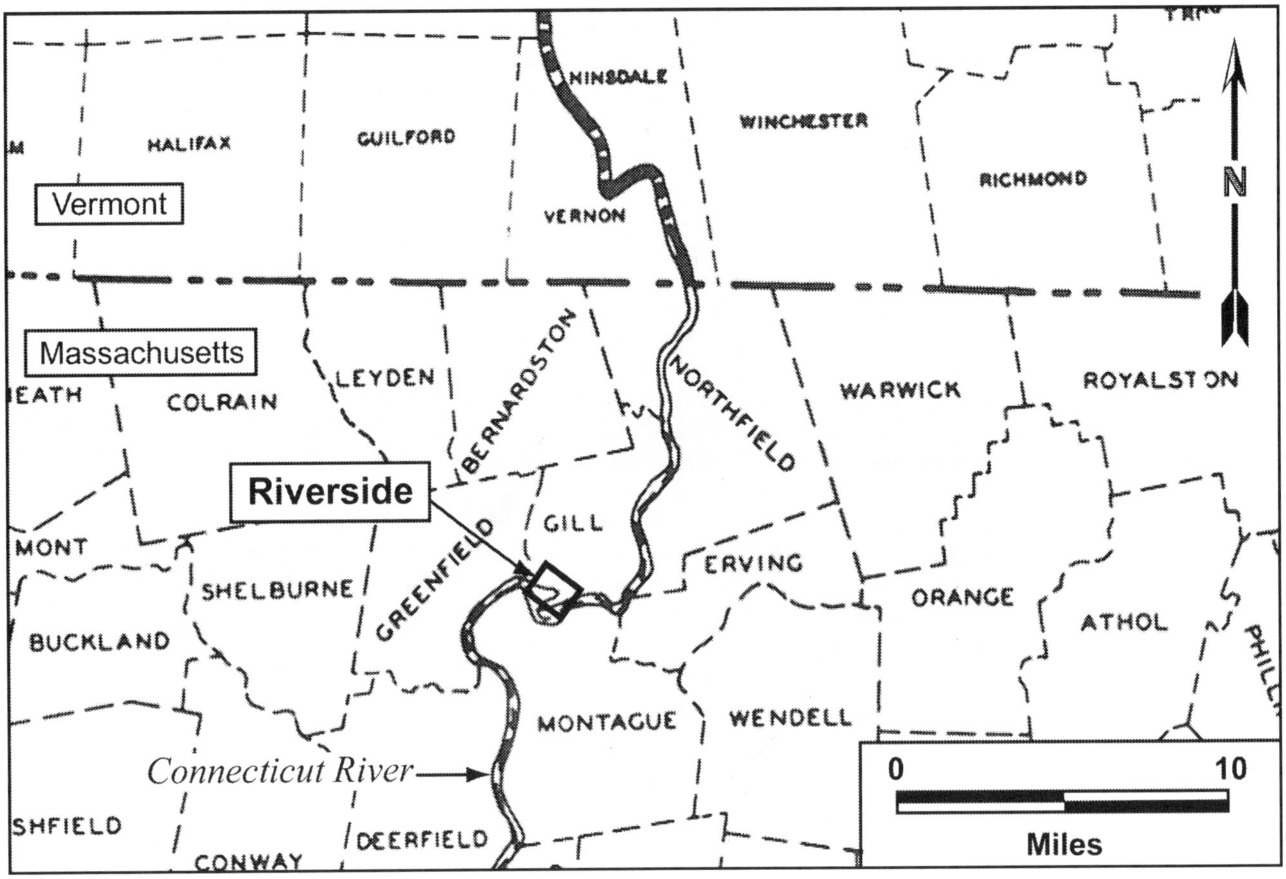

Figure 11.2. Riverside section of Gill, Massachusetts (adapted from Commonwealth of Massachusetts 1988).

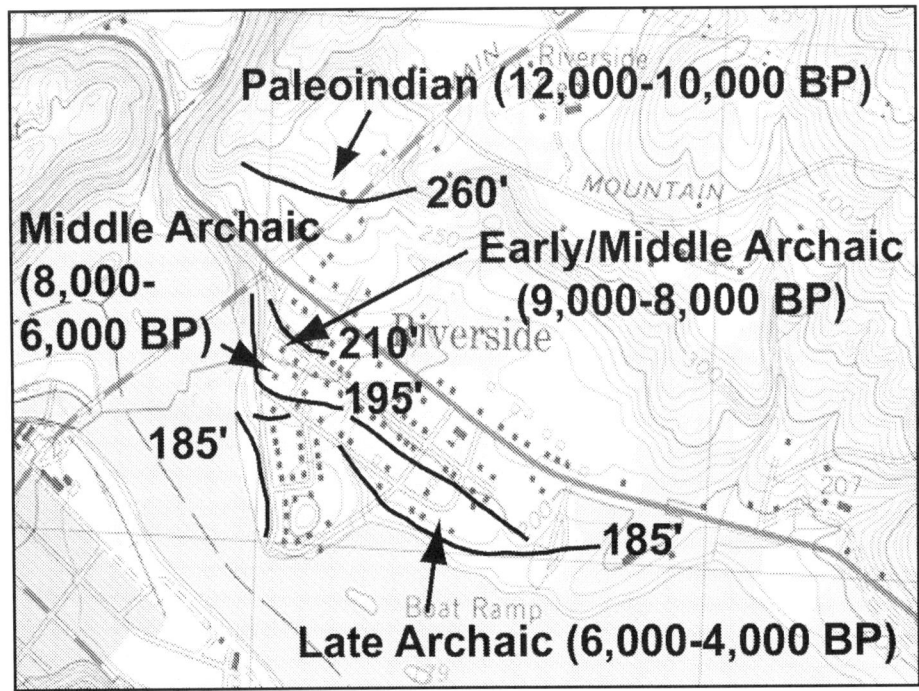

Figure 11.3. Riverside terraces.

Jahns (1966) indicated the southern limit of each terrace by drawing a scarp line to denote the base of a steep visible slope. In all, five concentric terraces run parallel to the flow of the Connecticut River. They decrease in elevation from north to south. Human cultural deposits have been found only on the lowest four of these former riverbanks (Figure 11.3). In 1996 I transferred scarp lines from Jahns' surficial map to a 1979 version of the U.S. Geological Survey topographic Greenfield, Massachusetts, Quadrangle (these scarp lines are depicted on Figure 11.3). Only then did I recognize that archaeological deposits found in and around Riverside neatly correlate with Jahns' terrace boundary outlines. At this location, site data supplements geological information, providing a timed sequence of proto-Connecticut River development.

THE DEVELOPMENT OF TERRACES IN RIVERSIDE

Prior to the last glaciation, the western stretch of the Connecticut between Northfield and Greenfield (Figure 11.2) did not exist. The river may have coursed southward (Figure 11.4), along East Mineral Hill (Jefferson 1898:468). This earlier channel filled with glaciolacustrine deposits as the Montague Plain delta expanded into glacial Lake Hitchcock. The current course of the Connecticut River and western jog occurred only after the lake drained (Brigham-Grette and Wise 1988:234).

The large glacial Lake Hitchcock, 200 mi long at its extremes, formed as the Laurentide ice sheet receded. The ice sheet reached its maximum extent between 21,000 and 18,000 B.P., during the Late Wisconsinan glaciation (Stone and Borns 1986:46). This final glacial advance is marked in New England by evidence of a terminal moraine along Martha's Vineyard, Massachusetts, and Long Island, New York (Figure 11.5). Deglaciation in North America began soon after this geographical margin was reached. The melting ice receded slowly northward, taking 8,000 to 9,000 years for the edge of the ice sheet to be north of the St. Lawrence lowland

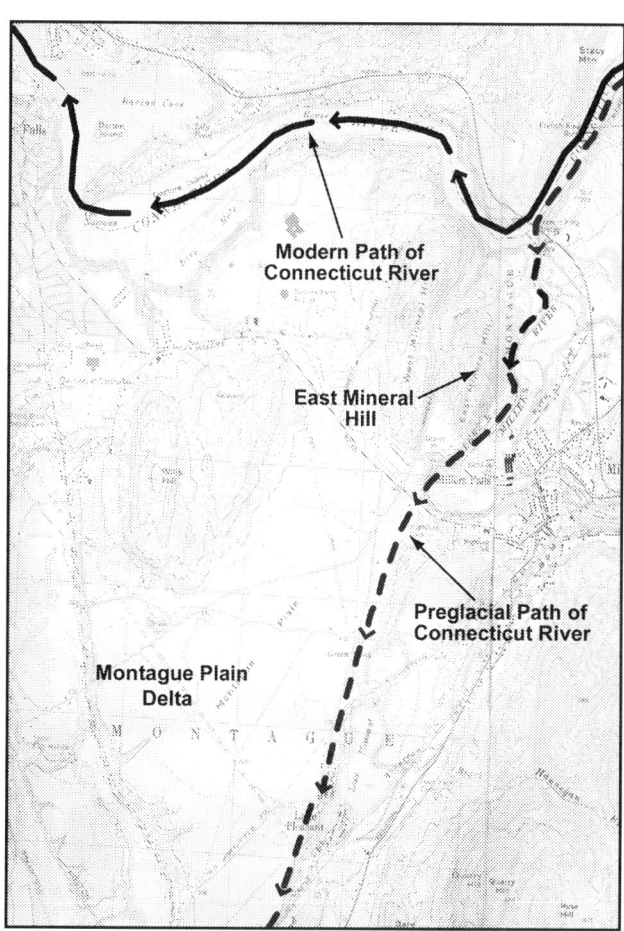

Figure 11.4. Preglacial and modern paths of the Connecticut River.

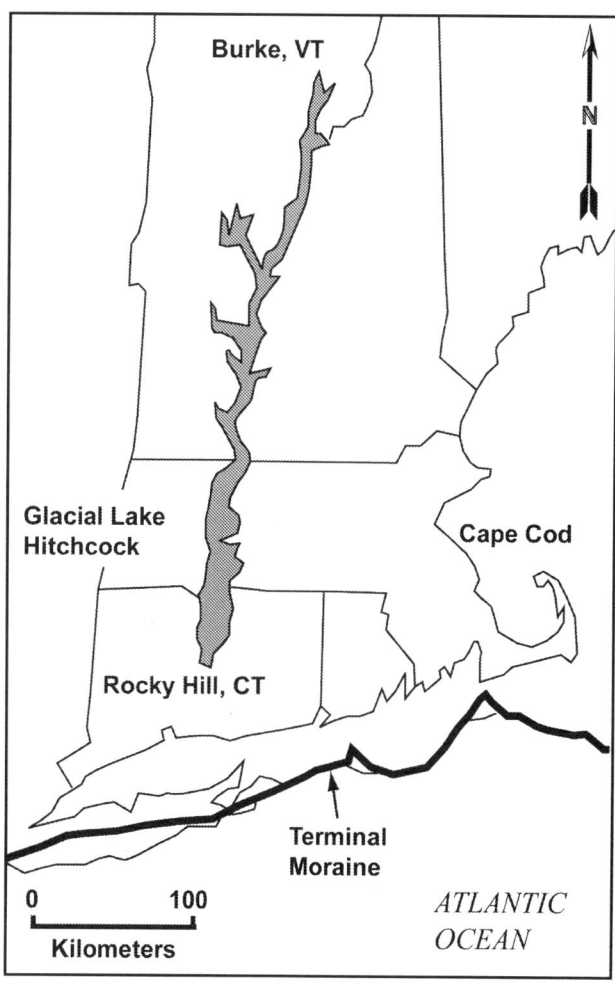

Figure 11.5. Glacial Lake Hitchcock (adapted from Thorson and Schile 1995:752).

(Brigham-Grette and Wise 1988:210).

In central Connecticut, glacial Lake Hitchcock formed behind a sediment dam at Rocky Hill, ca. 16,000–17,000 B.P. Water pooled in a low-lying expanse behind the deltaic complex of the earlier glacial Lake Middletown, which existed at a slightly higher elevation (Koteff et al. 1988:173–175). By coordinating radiocarbon dates (from organic samples) with varve chronologies in Vermont, Ridge and Larsen (1990:897) suggested a lake inception date prior to 15,600 B.P.

The area of standing water, which collected between relict deltas of glacial Lake Middletown and the Laurentide ice sheet, represents the early bounds of glacial Lake Hitchcock. The stagnant ice margin continued to recede northward, being as far north as Chicopee, Massachusetts, by 15,000–16,000 B.P. and to Burke, Vermont, by 14,000 B.P. (Brigham-Grette and Wise 1988:210). Water from the melting ice pooled between bedrock walls in the lower-lying Connecticut River valley. Evidence of glacial Lake Hitchcock extends for more than 200 mi, from Rocky Hill, Connecticut, to Burke, Vermont (Koteff et al. 1988:172). It is important to remember, however, that the entire 200-mi length of lake was not simultaneously active (Figure 11.5).

Varve chronologies imply that portions of the lake remained stable and water level was maintained for 4,000 years (Ridge and Larsen 1990:899). However, the precise dates of inception and drainage are currently in debate. The movement of the ice and the creation of glacial Lake Hitchcock produced a series of ice contact as well as lacustrine deposits in the region (Brigham-Grette and Wise 1988:209). The Montague delta is one of a complex of deltas in the Massachusetts stretch of the former lake, lying between present-day elevations of 300 and 350 ft (Brigham-Grette and Wise 1988:234). The delta formed as meltwater poured down the Miller's River Valley into the glacial lake (Figure 11.6). The coarse sediment load carried by the Miller's River produced a steep-fronted Gilbert-style delta. Foreset and topset beds overran Lake Hitchcock bottom deposits over a large area, including Riverside.

After the sediment dam at Rocky Hill eroded, the pooled lake water drained. Episodic drainage of lake basins occurred from south to north, with portions of southern glacial Lake Hitchcock drained by 14,000 B.P. (Ashley and Stone 1992:305). The section at issue here drained by 13,000 B.P. (Ridge and Larsen 1990:898). In this central portion of the lake, between the Chicopee deltaic complex and the Montague Delta complex, a series of smaller pools and interconnecting river channels resulted (Rittenour 1999:54). By 12,000 B.P., the Connecticut River began to erode sands and gravels

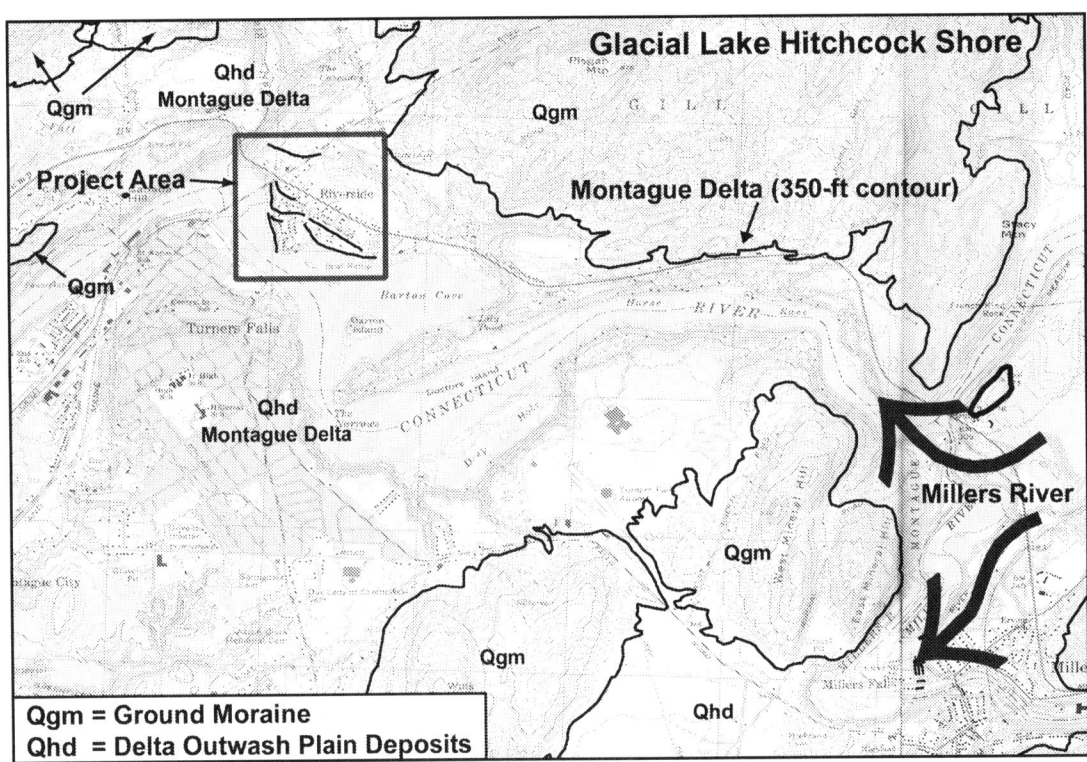

Figure 11.6. Extent of glacial Lake Hitchcock at the 350-ft elevation line.

deposited in its former channel during glaciation and deglaciation (Jahns 1947:29).

The postglacial Connecticut River drainage channel flowing toward Miller's River encountered the Montague Plain delta. The combination of thick, coarse delta deposits and bedrock prevented the river from reoccupying its earlier direct south-trending channel. The river diverted around the sediment body (Brigham-Grette and Wise 1988:234), coursing westward, near the thinner delta margin in Riverside (Figure 11.7).

Similar lake and delta deposits also buried the channel of the south-trending Falls River (Bain and Meyerhoff 1963:59). The Connecticut and Falls Rivers combined just east of Canada Hill (Figure 11.7). At the confluence of these two postglacial rivers, while still at high elevations, the larger and stronger Connecticut temporarily restricted the southerly flow of the smaller Falls River. Neither river could penetrate Montague Delta deposits south of the capture point. The combined fluvial systems flowed northwestward on the delta surface, north of Canada Hill, and then southward cutting the White Ash Channel and continuing into Greenfield (Figure 11.7). Glacial Lake Hitchcock bottom deposits of silt and clay were exposed in this short-lived river channel by the force of the combined flow.

The new westward path of the proto-Connecticut removed Montague delta deposits southeast of Riverside. Over time a ridge of Jurassic-age sandstone and mudstone now known as the Lily Pond Barrier (the southeast wall of Barton's Cove) was exposed (Brigham-Grette and Wise 1988:234). Coeval to the diversion of the Connecticut through White Ash Swamp, the flow of the river was impeded as water began to collect behind this rock peninsula. At this time only a small portion of the Lily Pond Barrier was emergent above water surface (Figure 11.7). No archaeological sites are found along the river shore in Riverside at or above 300-ft (above sea level [ASL]) elevation, implying a prehuman landscape prior to ca. 11,000 B.P.

Eventually, the Connecticut and Falls Rivers cleared Montague Delta deposits creating the present course east of Canada Hill (Figure 11.8). The channel through White Ash Swamp was abandoned. Meanwhile, at the Lily Pond Barrier, the fast-flowing Connecticut deeply incised the bedrock peninsula, cascading over the barrier in two huge waterfalls. At the base of the waterfalls, the sand and gravel carried by the river scoured plunge pools into bedrock.

The smaller southern plunge pool, once known as Poag's Hole, was initiated earlier than the northern plunge pool, named the Lily Pond (Figure 11.8). The plunge pools were paired only briefly. The proto-Connecticut abandoned Poag's Hole early, after the lowering river level dropped beneath the rock crest of the

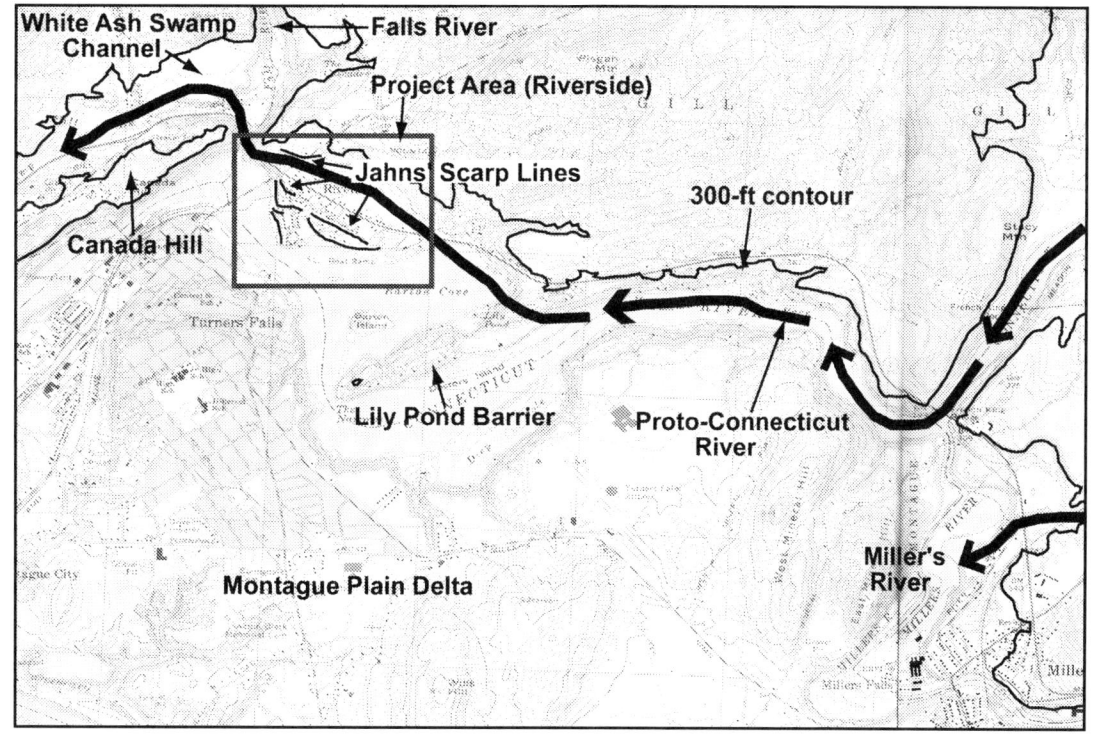

Figure 11.7. The proto-Connecticut River flows west through White Ash Swamp.

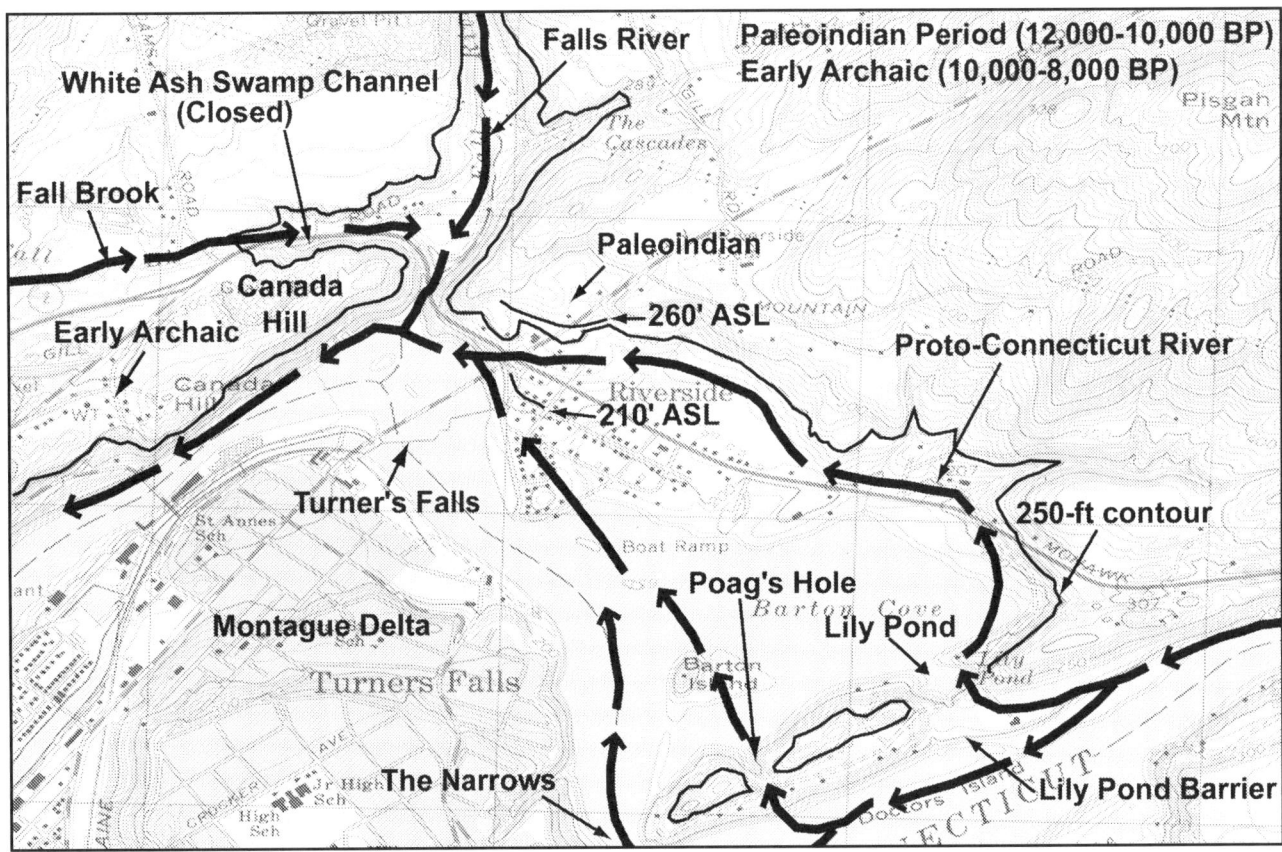

Figure 11.8. The White Ash Swamp channel is closed. The proto-Connecticut spills over the Lily Pond Barrier as two waterfalls.

southern waterfall. The wall behind Poag's Hole is some 20 ft higher in elevation (Jefferson 1898:465) than that behind the Lily Pond.

The waterfall above the Lily Pond remained active for a long time after Poag's Hole spillway ceased to exist. U.S. Geological Survey topographic maps from an 1890 atlas of Massachusetts detail the longer duration of water erosion in line with the Lily Pond waterfall (Figure 11.9). In front of the waterfall, the river carved a semilunate channel westward. This feature is known as Barton's Cove. Water flowing in the lunate channel met bedrock east of Riverside. The westward flow of the proto-Connecticut was blocked and the water diverted to the southwest. The bedrock exposure protected Riverside, forcing the Connecticut away from the triangular peninsula. In time channel downcutting and alluvial deposition formed a series of five successive terraces.

The area between the waterfall and the bedrock exposure was deeply eroded, unlike the delta remnant in front of the Lily Pond barrier (Figure 11.9). When river level later fell to below 160 ft ASL, a new landform was exposed. On the 1890 atlas, this area was called Barton's Field. Barton's Cove existed as an abandoned channel remnant until after the construction of a hydroelectric dam at Turner's Falls in the early 20th century. Pool level rose, inundating both Barton's Cove and Barton's Field. All that remains of this landform today is Barton's Island, the highest elevation point of the former Barton's Field.

Human entrance into southern and central New England is estimated between 11,000 and 12,000 years ago (Thomas 1977:6). The highest Riverside terrace, between 210 ft and 260 ft in elevation, yields sparse evidence of earliest human occupation (Figure 11.8). A single Paleoindian (12,000–10,000 B.P.) projectile point recorded at 260 ft ASL is similar in style to other tools dating to 11,000–10,000 B.P. (Dincauze 1989:182). The artifact was unearthed during construction in the 1970s. Although eyewitness accounts suggest that cultural features were found at the time of excavation, no documented evidence of these features exists. Only the single projectile point remains (D. Dincauze, personal communication 1997).

The fluted projectile point near 260 ft (ASL) in eleva-

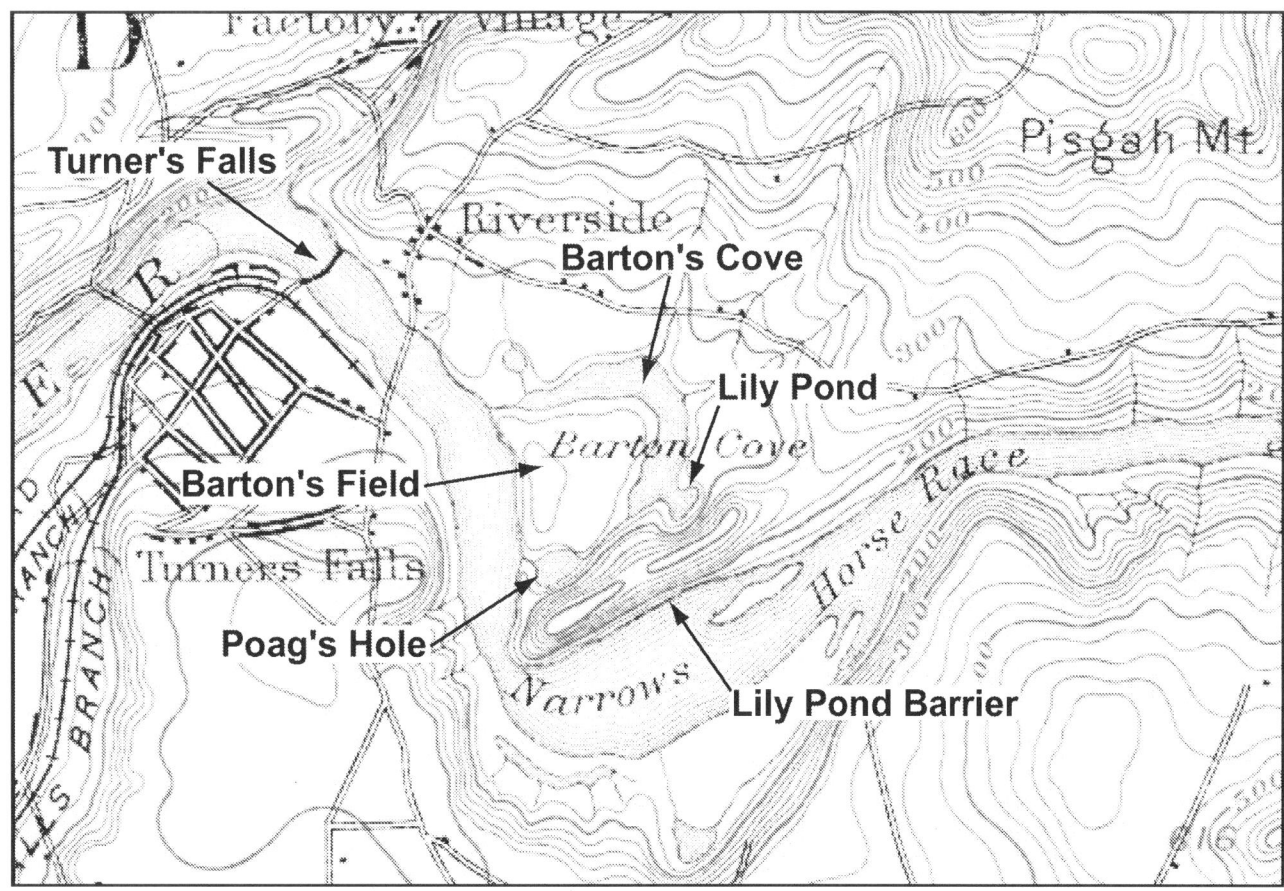

Figure 11.9. Barton's Cove and Barton's Field ca. 1890.

tion implies that the postglacial Connecticut was below 260 ft by 10,000–11,000 B.P. I suggest that the White Ash Swamp detour closed between 10,000 and 11,000 B.P. and that the first local inhabitants lived near an impressive pair of waterfalls at the Lily Pond Barrier. Also important to note is the fact that this site was well placed to allow for the exploitation of abundant local resources. The recently abandoned White Ash channel remained swampy and was a perfect refuge for migratory fowl, reptiles, turtles, and a host of large and small mammals (Thomas 1980:82). Farther upstream, the imposing Lily Pond Barrier likely impeded the movement of anadramous fish moving upstream to spawn, which made Riverside an ideal location for fishing (D. Dincauze, personal communication 1997).

As more of the Montague Plain delta wore away, the river incised the area known as the Narrows (Figure 11.8). For a time, the proto-Connecticut traveled over the rock (as waterfalls) as well as around the Lily Pond peninsula. When the channel eroded out to below 240 ft ASL in elevation, the water was no longer high enough to flow into Poag's Hole. As downcutting of the proto-Connecticut continued, river level fell beneath 200 ft ASL, and water no longer crossed over the Lily Pond Barrier (Figure 11.10). The river incised the Narrows as it flowed around the Lily Pond peninsula.

Farther downstream, the larger Connecticut now dominated the southern-flowing channel of the postglacial Falls River (Figure 11.10). The steep eastern wall of this immature river (Sammartino 1981:39) was eroded by the Connecticut. Over time, weak spots in the bedrock were exploited, and a second set of falls, Turner's Falls, was initiated. A series of channels carved through the rock, creating islands in the remnant bedrock (Stoughton 1978:32–33). The river followed a course similar to that seen today, west over Turner's Falls, remaining east of Canada Hill and then flowing southward.

The present course of the river, east of Canada Hill, was established after the White Ash Swamp channel was abandoned (Jahns 1966:3). The combined force of the Falls and Connecticut Rivers trenched southward,

Chapter 11 *Geochronology from Archaeology* **157**

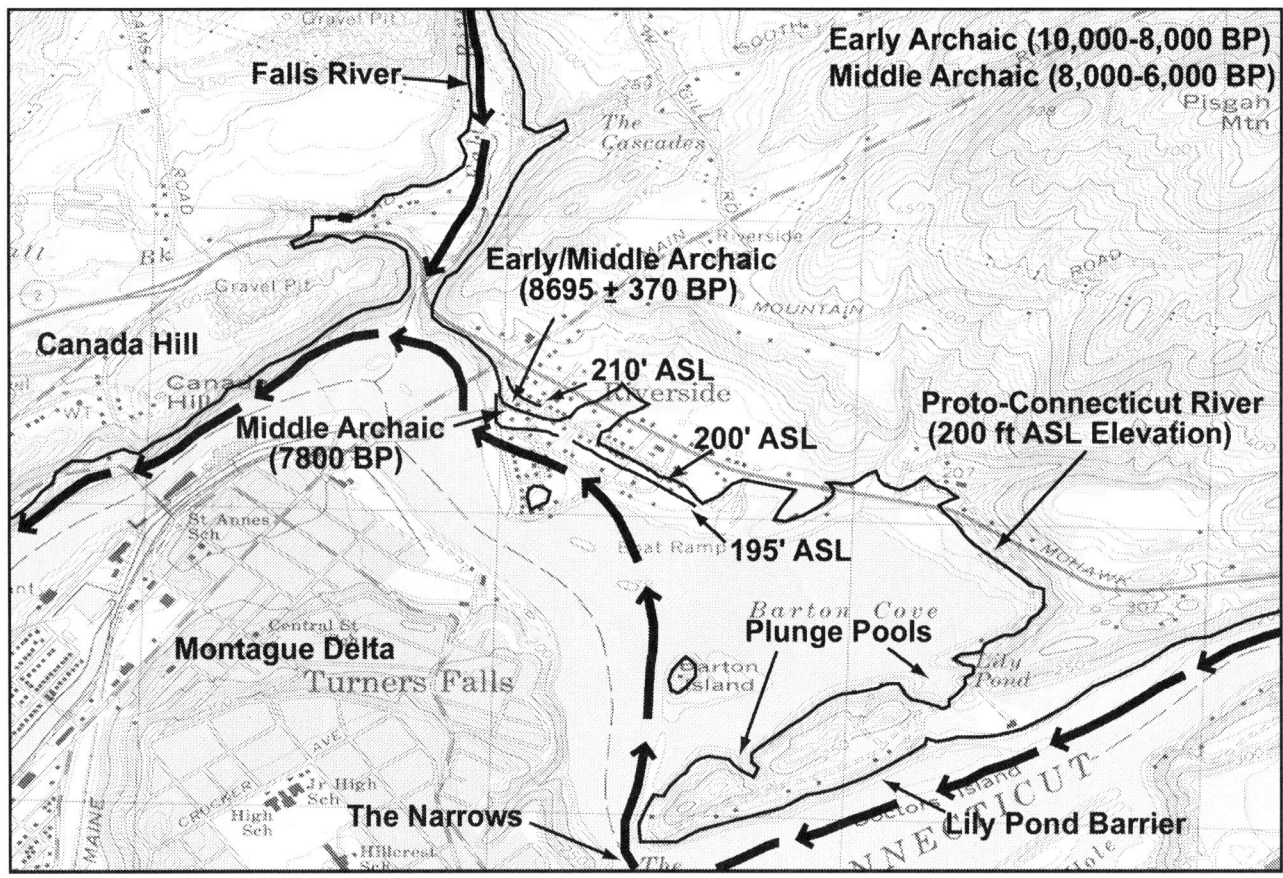

Figure 11.10. The proto-Connecticut River travels around the Lily Pond Barrier through the Narrows. The plunge pools are now closed.

removing portions of the Montague Delta and scouring to bedrock the eastern face of Canada Hill (Figure 11.10). The rivers removed the sandy delta deposits and exposed bedrock from Turner's Falls to the point of confluence with the Deerfield River (farther south).

The proto-Connecticut entrenched and as the channel migrated southward, it cut a series of terraces. As the Connecticut periodically overflowed its banks, these terraces were coated with well-sorted silt, sand, and gravels (Jahns 1966:3). Nearby at Turner's Falls, the waterfalls continued to form until the man-made dam hindered the flow of water and retarded further erosion of the bedrock.

The terrace between 210 and 195 ft ASL in elevation has two important archaeological sites in association (Figure 11.10). At the top of this terrace, between 200 and 210 ft in elevation, is an Early to Middle Archaic archaeological site excavated in 1979 (Curran and Thomas 1979:39). The site yielded an intact Middle Archaic (8000–6000 B.P.) midden, which extended 85–95 cm (34–38 in) below present ground surface. The oldest type of point recovered from this location was a Neville point, dating to ca. 7800 B.P.

A radiocarbon sample taken during excavation (P. Thomas, personal communication, 1997) indicates that the site was occupied by 8685 ± 370 B.P. (cal. 10,640–8920 B.P., p = 0.965, GX-6995; ^{13}C-corrected and calibrated at 2-σ with the program CALIB 4.3, Method B [Stuiver et al. 1998]). This radiocarbon age fits the cultural evidence well. Although a distinctive scraper found during the excavation suggested an Early Archaic presence, there were no projectile points from this era to fully support the conjecture (Curran and Thomas 1979:37–39). The settlement was inhabited between the end of the Early Archaic (10,000–8000 B.P.) and the beginning of the Middle Archaic (8000–6000 B.P.).

Toward the tip of this same terrace is another Middle Archaic site, between 195 and 200 ft ASL in elevation (Figure 11.10). Known as WMECO (an acronym for Western Massachusetts Electric Company), this archaeological site was excavated in 1972. Middle Archaic projectile points, Stark (7000 B.P.) and Neville (7800 B.P.), were found in an aeolian silt mantle atop alluvium and channel lag (Thomas 1980:73).

It is interesting to note that this second and younger of these Middle Archaic sites is lower in elevation but on the same terrace. On the basis of information combined from these two sites, I speculate that human populations occupied this terrace as early as 9000 B.P. Water level dropped and a steep scarp near 195 ft was carved sometime after 7800 B.P. (Figure 11.10). Because the Lily Pond barrier was abandoned at 200 ft ASL in elevation, I suggest that the plunge pools were active from 11,000 B.P. to ca. 8000 B.P. Human populations during Paleoindian (12,000–10,000 B.P.), Early Archaic (10,000–8000 B.P.) and perhaps even Middle Archaic (8000–6000 B.P.) times witnessed dramatic waterfalls at the Lily Pond Barrier.

Both archaeological sites on the terrace at 195 to 200 ft ASL in elevation are characterized as Middle Archaic middens (Curran and Thomas 1979:23, Thomas 1980:75). The soil here and across much of Riverside is black, very fine sandy loam that extends to a depth of approximately 3 ft. The color reflects constant reuse over time and the addition of plentiful organic materials to the soil. The texture indicates a high content of very fine sediments; probably wind transported (Jahns 1966:3) onto Riverside terraces. These thick aeolian deposits necessitated a sediment source to the west or northwest.

Wind transport of silt was a continual event in Riverside, beginning during the Middle Archaic Period (8000–6000 B.P.). The movement of silt coincides with colder weather in the region, dating to ca. 8200 B.P. (Dincauze 1989:1). As for a sediment source, the best location seems to be either the White Ash Swamp area or Canada Hill. The now-dry and exposed lake bottom in White Ash Swamp was a likely place for silt from 10,000 B.P. and later. The fine silts transported to Riverside were interspersed with intensive occupations by Native Americans. Over time, the soil became black as organic matter infused with wind-blown sediments, forming a thick midden.

Beneath the aeolian silt mantle are water-laid sediments. Alluvium was deposited as the proto-Connecticut entrenched and migrated southward. Deeper still, channel lag was recorded at WMECO at the base of the excavated deposits (Thomas 1980:75). River cobbles here suggest that the channel thalweg once sat at this location when the river coursed at a higher elevation.

The Lily Pond waterfalls and later Turner's Falls were important to subsistence and archaeological site placement. Primarily the waterfall sets proved a hindrance to the movement of shad, which seasonally swam upstream to spawn. The site was close enough to the waterfalls for easy gathering of fish in spring and summer and was adjacent to fresh water. Evidence from the WMECO excavation suggests that a diet of reptiles and turtles as well as small and large mammals supplemented the shad (Thomas 1980:87). Several ecological zones were now present around Riverside. Resources from riverine, deciduous forest, open grassland and swampy environments were abundant during the Middle Archaic (8000-6000 B.P.) and in close proximity to the terraces (Thomas 1980:87).

Further downcutting by the Connecticut soon carved the two lowest terraces (Figure 11.11). Water level dropped below 160 ft ASL in elevation. These surfaces both date to the same extended cultural period with sites of the Late Archaic Period (6000-3000 B.P.) to present. Between 185 and 195 ft ASL in elevation a midden deposit found just above the 190-ft scarp line yielded a radiocarbon date (P. Thomas, personal communication 1997) of 5,530 ± 180 B.P. (cal. 6671–5928 B.P., p = 0.992, GX-7007; ^{13}C-corrected and calibrated at 2-σ with the program CALIB 4.3, Method B [Stuiver and et al. 1998]). Cultural affiliation cannot corroborate radiocarbon evidence because no identifiable tools were recovered during the excavation (Curran and Thomas 1979:26). Based on the radiocarbon age alone, this terrace is dated to the beginning of the Late Archaic (6000–3000 B.P.).

The proto-Connecticut dropped below 160 ft ASL, exposing the lowest terrace between 160 and 185 ft in elevation (Figure 11.11). When river level fell to 160 ft, the local inhabitants lived on Barton's Field, an area now inundated by the dammed river. Sites lying between 160 and 185 ft ASL also date between the Late Archaic (6000–3000 B.P.) and the present. By contrast, the site above the 190 ft ASL scarp dates to the very beginning of this era. Based on the archaeological evidence, it seems that the final recession of the proto-Connecticut began sometime after 5530 ± 180 B.P. (cal. 6671–5928 B.P., p = 0.992, GX-7007; ^{13}C-corrected and calibrated at 2-σ with the program CALIB 4.3, Method B [Stuiver et al. 1998]). The water remained at this low 160-ft-ASL level behind the bedrock sill of Turner's Falls until the 19th century, when successive dam construction raised the Connecticut River to its present 180-ft level.

During the Late Archaic (6000–3000 B.P.) and through the Contact Period (A.D. 17th century), reliance on shad is well documented in both the archaeological and later in historical records. Floral and faunal recovery from Late Archaic sites suggests a diet comparable to Middle Archaic (8000–6000 B.P.) layers. The local inhabitants relied heavily on fish and reptile from the adjacent river and nearby swamps. They supplemented their diet with birds as well as small and large mammals from the surrounding forest (Thomas 1980:86–87).

Into the 17th century, Native Americans seasonally established a village at Riverside. Prior to the Contact Period (A.D. 17th century), these falls were known as Peskeomskut. The literal translation references the place

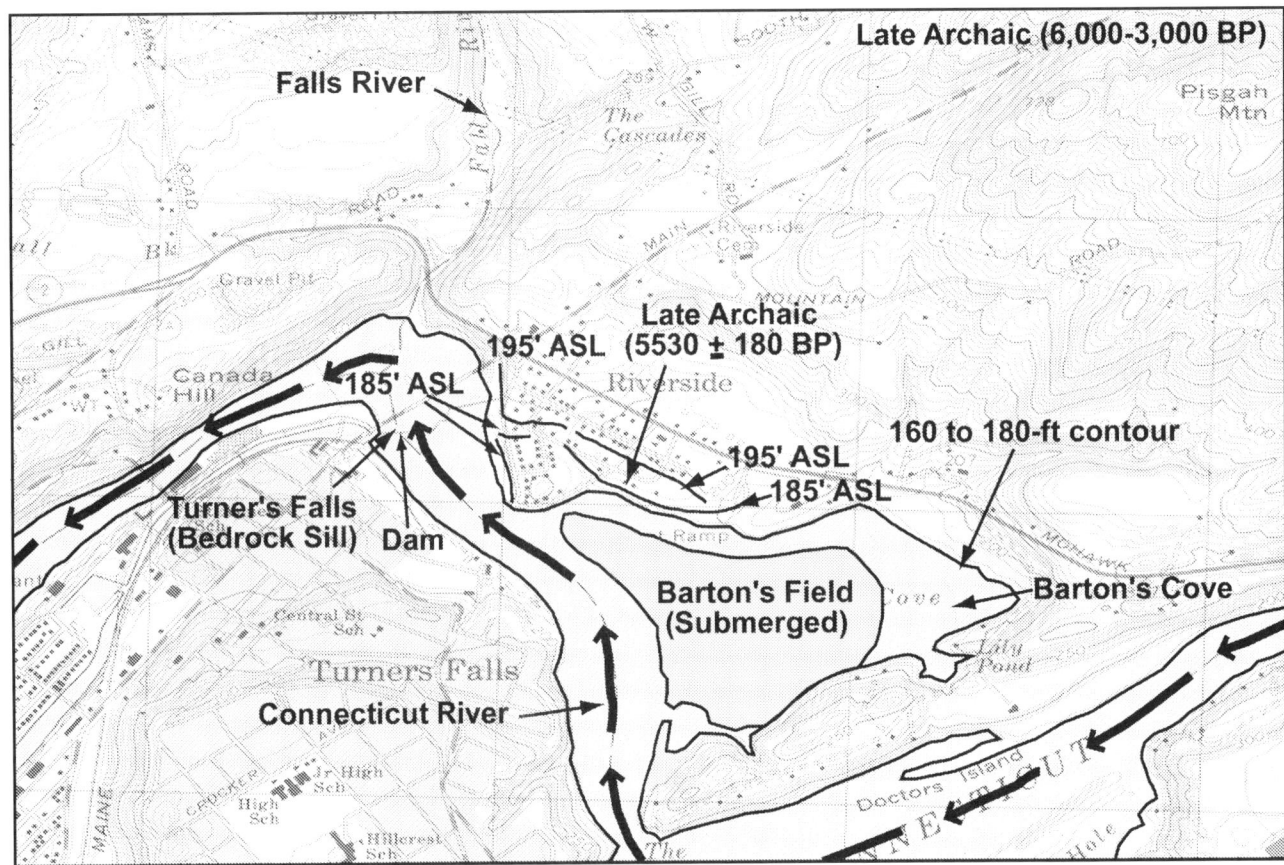

Figure 11.11. The Connecticut River establishes a course similar to that seen today.

along the river where the flow of water is divided by a cleft in the rock (Turners Falls Reporter 1875:1). The village was annually inhabited to coincide with the spawning of shad. Fish were collected in bulk during the spring and summer and then processed, dried, and stored for the winter (Stoughton 1976:6). Into the late 19th century, Europeans used Riverside in similar fashion, as a staging ground to gather and process shad at Turner's Falls.

SUMMARY

By combining geological and archaeological information, timing of proto-Connecticut River development is fine tuned. Data from the terraces suggests that a relict channel through White Ash Swamp existed between 12,000–11,000 B.P. Coeval to the White Ash Swamp detour, a pair of waterfalls were initiated on the Lily Pond Barrier. These falls remained active until 8000–9,000 B.P., when the river slipped around the Lily Pond Barrier through the Narrows.

As the proto-Connecticut scoured the east face of Canada Hill, Peskeomskut (later Turner's Falls) formed as the sand and gravel carried by the water eroded Triassic- and Jurassic-aged sandstone, conglomerate, and mudstone. For the next 8,000 years, the proto-Connecticut entrenched and migrated south until a low 160-ft-ASL water level in the 19th century. During this period the river shaped Riverside terraces. Alluvial and later aeolian deposits coated the region as people seasonally returned to occupy this terrace set. Turner's Falls dam construction in the early 20th century raised the pool level, inundating the lowest in elevation of the five Riverside terraces.

Archaeological evidence from Riverside terraces allows an accurate chronology of postglacial downcutting by the Connecticut River to be established. Radiocarbon and known cultural chronologies date the use of these terraces by Native Americans and thus their exposure as surfaces. Archaeological sequences, recorded in terms of human social history and experience, discretely time events in the Holocene. At Riverside, placement of the terraces alongside many and varied resources led to continual site use and a unique overlap between human and geological histories. In time we expect to be able to better define the postglacial chronology of river cutting in this region, including smaller lakes and interconnected river systems not previously identified.

Acknowledgments

I would like to thank Dena Dincauze, who read and edited numerous versions of this article while acting as the chair of my thesis committee at the University of Massachusetts, Amherst. I must also acknowledge the other members of my committee. Julie Brigham-Grette (Geosciences Department) and Laurie Godfrey (Department of Anthropology) offered many helpful suggestions about the text, reflected throughout.

The interpretation presented here would not have been possible without the use of information gathered in Riverside by Peter A. Thomas and Mary Lou Curran between 1977 and 1980. I am further grateful to Peter A. Thomas for providing me with ^{14}C dates for several sites.

I must also acknowledge David Cremeens who reviewed this article and offered many useful editorial suggestions. The graphics presented here were produced by the author and were adapted from other sources as cited. Any mistakes noted in the text are solely the responsibility of the author.

REFERENCES CITED

Ashley, G.M., and Stone, J.R. (1992). Ice wedge casts, pingo scars and the drainage of glacial Lake Hitchcock. In *Guidebook for Field Trips in the Connecticut Valley Region of Massachusetts and Adjacent States, New England Intercollegiate Geological Conference, Contribution No. 66, Vol. 2*, edited by P. Robinson and J.B. Brady, pp. 305–331. Department of Geology and Geography, University of Massachusetts, Amherst, Mass.

Bain, G.W., and Meyerhoff, H.A. (1963). *The Flow of Time in the Connecticut Valley, Geological Imprints*. Connecticut Valley Historical Museum, Springfield, Mass.

Brigham-Grette, J., and Wise, D.U. (1988). Glacial and deglacial landforms of the Amherst area, north central Massachusetts. In *Field Trip Guidebook, AMQUA 1988, Contribution No. 63*, edited by J. Brigham-Grette, pp. 209–244. Department of Geology and Geography, University of Massachusetts, Amherst, Mass.

Commonwealth of Massachusetts. (1988). *County Map of Massachusetts*. Form 151-1M-4/88-P802011.

Curran, M.L., and Thomas, P.A. (1979). Phase III—Data recovery: Wastewater treatment in the Riverside Archaeological District of Gill, Massachusetts, University of Vermont, Report No. 19. Manuscript on file, Department of Anthropology, University of Massachusetts, Amherst.

Dincauze, D.F. (1989). Geoarchaeology in New England: an early Holocene heat spell? *Review of Archaeology* 10(2):1–4.

Jahns, R.H. (1947). *Geologic Features of the Connecticut River Valley, Massachusetts as Related to Recent Floods*. Water Supply Paper No. 996. U.S. Department of the Interior, Washington, D.C.

Jahns, R.H. (1966). *Surficial Geologic Map of the Greenfield Quadrangle, Franklin County, Massachusetts*. U.S. Geological Survey Geol. Quadrangle. Map GQ-474.

Jefferson, M.S.W. (1898). The postglacial Connecticut at Turner's Falls, Mass. *Journal of Geology*, 1898: 463–472.

Koteff, C., Stone, J.R., Ashley, G.M., Larsen, F.D., Boothroy, J.C., and Dincauze, D.F. (1988). Glacial Lake Hitchcock, postglacial uplift and post-lake archaeology. In *Field Trip Guidebook, AMQUA 1988, Contribution No. 63*, edited by J. Brigham-Grette, pp. 169–208. Department of Geology and Geography, University of Massachusetts, Amherst, Mass.

Mulholland, M.T. (1984). Patterns of change in prehistoric southern New England: A regional approach. Unpublished doctoral dissertation, Department of Anthropology, University of Massachusetts, Amherst.

Ridge, J.C., and Larsen, F.D. (1990). Reevaluation of Antevs' New England varve chronology and new radiocarbon dates of sediments from glacial Lake Hitchcock. *Geological Society of America Bulletin* 102:889–899.

Rittenour, T.M. (1999). Drainage history of glacial Lake Hitchcock, northeastern United States. Unpublished master's thesis, Department of Anthropology, University of Massachusetts, Amherst.

Sammartino, C.F. (1981). *The Northfield Mountain Interpreter: Facts About the Mountain, the River, and the People*. Northeast Utilities, Berlin, Conn.

Stone, B.D., and Borns, H.W., Jr. (1986). Pleistocene glacial and interglacial stratigraphy of New England, Long Island, and adjacent Georges Bank and Gulf of Maine. *Quaternary Science Review* 5:39–52.

Stoughton, R.M. (1978). *History of the Town of Gill, Franklin County Massachusetts 1793–1943*. E.A. Hall & Company, Greenfield, Mass.

Stuiver, M., Reimer, P.J., Bard, E., Beck, J.W., Burr, G.S., Hughen, K.A., Kromer, B., McCormac, F.G., van der Plicht, J., and Spark, M. (1998). INTCAL98 radiocarbon age calibration 24,000–0 cal BP. *Radiocarbon* 40:1041–1083.

Thomas, P.A. (1977). A Phase I assessment of cultural resources for a proposed wastewater treatment system in the Riverside District of Gill, Massachusetts. Report submitted to Tighe and Bond Consulting Engineers.

Thomas, P.A. (1980). The Riverside District, the WMECO site, and suggestions for archaeological modeling. In *Early and Middle Archaic Cultures in the Northeast*, edited by D. Starbuck and C. Bolian, pp. 73–95. Occasional Publications in Northeast Anthropology No 7. Bethlehem, Conn.

Thorson, R.M., and Schile, C.A. (1995). Deglacial eolian regimes in New England. *Geological Society of America Bulletin* 107:751–761.

Turners Falls Reporter. (1875). *Peske-ompsk-ut; or The Falls Fight. A Series of Random Sketches Showing a Glimpse of the Early History of Turners Falls, Which Appeared in "The Turners Falls Reporter" During the Months of January and February 1875.* Turners Falls Reporter Job Office, Turners Falls, Mass.

U.S. Geological Survey. (1890). *Atlas of Massachusetts: From Topographical Surveys Made in Cooperation by the USGS and the Commissioners of the Commonwealth, 1884–1888, Preliminary Edition.* Published by the Commission.

U.S. Geological Survey. (1979). *Greenfield, Quadrangle, Franklin County, Massachusetts 7.5-Minute Series (Topographic), N4230-W7230/7.5.* AMS 6469 II SE-SERIES V814. U.S. Department of the Interior, Geological Survey.

CHAPTER 12

ARCHAEOLOGICAL SITE FORMATION IN GLACIATED SETTINGS, NEW JERSEY AND SOUTHERN NEW YORK

Donald M. Thieme

A dense urban metropolis now sits in the area on the eastern seaboard immediately north of the last glacial margin. Traces of prehistoric human activity are sparse, although formerly more abundant, having been compromised by the works of modern industry. Archaeological contexts are typically sandwiched between Pleistocene glacial deposits and very late Holocene marsh peat or disturbed land. The area is nonetheless crucial to understanding the retreat of the Laurentide ice sheet, the initial peopling of eastern North America, and the archaeological antecedents of historic tribes.

This chapter will summarize current research in both Quaternary geology and geoarchaeology, with a particular emphasis on the lower reaches of the Passaic and Hackensack Rivers in New Jersey (Figure 12.1). The objective is to show some implications of recent geological findings for archaeology and vice versa. Both disciplines currently stress fundamental understanding of the processes that generate stratified deposits. Indeed, many of the formation processes we use to explain the archaeological record are environmental processes that affect non-cultural as well as cultural deposits (Schiffer 1987:199–262).

Chronology of the Quaternary Period is another topic of common interest, and radiocarbon dating is still the most widely applied dating method. The recent calibration of the radiocarbon timescale (Stuiver et al. 1998) stretches out the close of the Pleistocene Epoch, when the Laurentide ice sheet retreated through the area addressed in this chapter. The changing climate itself influenced the abundance of ^{14}C in the atmosphere, and climatic forcing of the carbon cycle continued into the period of known archaeological cultures (Stuiver et al. 1991; van Andel 1998; van Geel et al. 1999). Episodes of landscape change identified in the stratigraphy of the multicomponent Dundee Canal site (28PA143) in the Passaic River valley possibly represent responses to such climatic forcing. Correlates identified in other archaeological and geological contexts further support the case for climatic forcing.

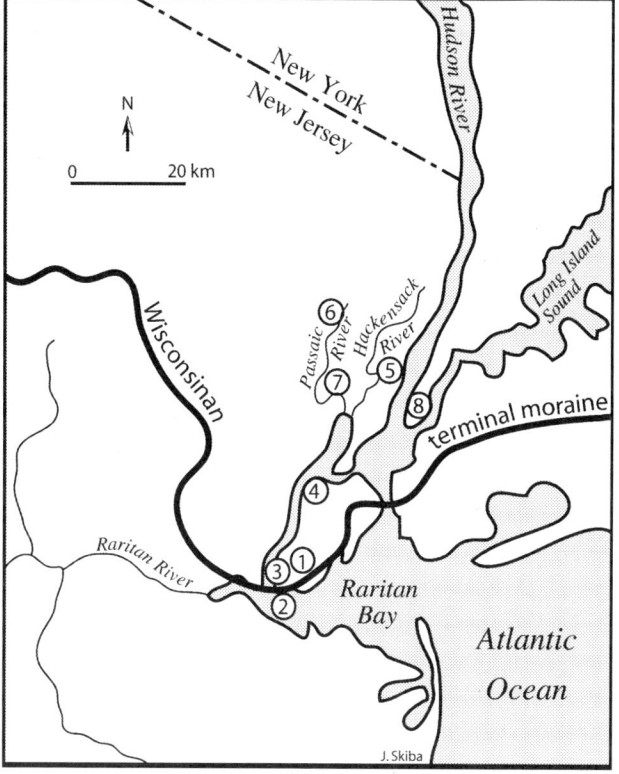

Figure 12.1. The glaciated portions of New Jersey and southern New York showing sites discussed in text.

① Richmond Hill Site (Ritchie and Funk 1971:53–54)
② Ward's Point (Ritchie and Funk 1971:50–53)
③ Hollowell (Ritchie and Fund 1971:46–49)
④ Old Place (Ritchie and Funk 1971:49–50)
⑤ North Bergen Sewer Outfalls (Thieme and Schuldenrein 1998a)
⑥ Route 21 Corridor Sites 28PA39, 28PA40, 28PA143, and 28PA145 (Thieme 1997; Tull 1997)
⑦ North Arlington Sewer Main (Thieme and Schuldenrein 1996; Thieme et al 1996)
⑧ Collect Pond (Schuldenrein 2001)

Geoarchaeology of Landscapes in the Glaciated Northeast edited by David L. Cremeens and John P. Hart. New York State Museum Bulletin 497. © 2003 by the University of the State of New York, The State Education Department, Albany, New York. All rights reserved.

GEOARCHAEOLOGY AND SITE FORMATION PROCESSES

Geoarchaeology, as defined by Butzer (1982:35), is the application of earth science methods to archaeological research problems. Earth science methods are employed to provide a regional context for archaeological finds and to determine the lateral and vertical extent of deposits from which significant finds may yet be unearthed. Recent analyses of site formation processes (Nash and Petraglia 1987; Schiffer 1987; Stein 2001) demonstrate the methodological importance of the geoarchaeological approach to the stratigraphy and physical setting of archaeological sites.

Archaeological sites are locations where past human activities took place (Deetz 1967:11–12; Knudson 1978:49). The preservation of material traces of past human activities where they were originally deposited by people (in situ) is unfortunately rare. Earth science methods must consequently be used to analyze the formation of a site, beginning with its topography and geology prior to occupation.

Glacial geology contributes less frequently to site formation analyses than other earth science subfields. Indeed, Pleistocene glacial deposits in North America are not yet known to contain prehistoric artifacts in situ. On the other hand, materials derived from glacial deposits do commonly occur in archaeological contexts. Derivation of cultural sediment by colluvial and alluvial reworking of glacial till and outwash is demonstrated below in an analysis of the multicomponent Dundee Canal site (28PA143).

Although glaciation may not have been responsible for site formation processes operating within any period of prehistoric occupation, archaeological contexts did frequently intrude glacial deposits. The formation of these deposits must be accounted for within both regional and site-specific stratigraphic frameworks. Processes observed in the vicinity of active glaciers now play a key role in the analysis of glacial landscapes and depositional environments (Boothroyd and Ashley 1975; Gustavson and Boothroyd 1987; Lundqvist 1989). In the glaciated Northeast, the concept of the "recessional morphosequence" (Koteff 1974; Koteff and Pessl 1981) has proven an effective tool for relating the Pleistocene glacial deposits to one another and to modern analogues.

REGIONAL GEOLOGY AND QUATERNARY STRATIGRAPHY

The topography of the glaciated portions of New Jersey and southern New York is dominated by north-south trending rifts in the continental crust that first developed during the breakup of Pangea (Isachsen et al. 1991:50–51). The Laurentide ice sheet advanced over these rifts at least twice during the Pleistocene Epoch (Cotter et al. 1986; Stanford 1997, 2000; Stanford and Harper 1991; Sirkin 1986), forming the valleys now occupied by the Hudson, Hackensack, and Passaic Rivers. Most of Long Island and much of Staten Island were deposited as glacial detritus. Sirkin (1982, 1986) assigned the ice-contact deposits on Long Island to early- and late-Wisconsinan advances of at least two lobes of the Laurentide ice sheet. In New Jersey, the older till has been assigned to the Illinoian or pre-Illinoian (Stanford 1997).

Sirkin obtained 29 radiocarbon dates for a section of glacial deposits at Port Washington on the north shore of Long Island (Figure 12.1). Mid-Wisconsinan ages from 36,000–25,000 years ago for shell and 44,000–35,000 years ago for wood predominate, with pollen spectra further indicating an interstadial warm period (Sirkin and Stuckenrath 1980). Inverted stratigraphy in several of the columns sampled was attributed to folding and thrusting of the sediments as they were overridden by the Late-Wisconsinan ice. A date of 21,750 ± 750 B.P. (SI-1590) was obtained for silty sediment capping outwash near the base of the section (~19 m above mean sea level). Sirkin considered this to be a maximum age for the arrival of the Hudson-Champlain lobe in western Long Island.

The terminal moraine for the Late-Wisconsinan advance is called Harbor Hill after the ice-contact deposits on this prominent feature in Brooklyn overlooking the Verrazano Narrows. As the ice sheet retreated from this terminal position, it left behind evidence in New Jersey for at least five recessional margins (see Figure 12.2). The deposits laid down in close association with each ice margin were generated by the local behavior of the ice sheet. Recent models developed for glacial deposits in the northeastern United States (Koteff 1974; Koteff and Pessl 1981; Stone et al. 1998) refer to these genetically related deposits as "recessional morphosequences."

In each morphosequence, the coarsest sediment textures are found at the up-ice "head of outwash." Bouldery till or stratified drift are most typical of true "moraines," but each head of outwash should contain material let down in place as the ice was melting. Grain size decreases and landforms are less collapsed in distal parts of a morphosequence, where the stratigraphy typically shows evidence of transport by meltwater (Boothroyd and Ashley 1975; Gustavson and Boothroyd 1987). The ice itself was always moving forward at each recessional margin, although the net balance within the sheet reflected more ice lost than gained. The impounded meltwater rose to a level defined by either a glacial lake plane or a valley knickpoint (Stone et al. 1998).

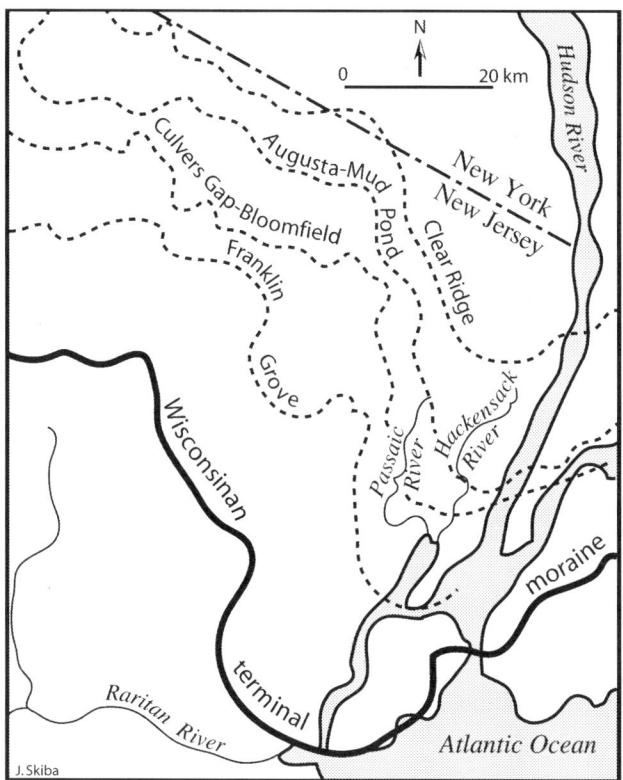

Figure 12.2. Late Pleistocene recessional margins (dashed lines) in New Jersey (after Stanford 1997).

Stepwise northward retreat of the ice sheet through the five recessional margins in Figure 12.2 resulted in a complex history of lake basins that drained up-ice as well as down-ice or laterally as spillways opened between the bedrock valleys (Stanford 1997; Stanford and Harper 1991). Due to common parent material in Newark Group sedimentary lithologies, the glaciolacustrine sediments from all of these interconnected basins are very similar. Thick winter varves of reddish brown muds alternate with more heterolithic sandy varves deposited as the ice melted during the summer. The apparently uniform lithostratigraphy led to earlier proposals of very large and long-enduring proglacial lakes (Antevs 1925; Lovegreen 1974; Reeds 1925, 1926; Salisbury 1902; Salisbury and Kummel 1893; Widmer 1964).

Lake-bed silt and clay accumulations up to 80 m thick are the result of successive meltwater impoundments in the lower reaches of the Hudson, Hackensack, and Passaic Rivers (Lovegreen 1974; Stanford 1997; Stanford and Harper 1991). Independent age constraints from radiocarbon dating and pollen stratigraphy suggest that the last pulse of meltwater occurred sometime prior to 14,000 radiocarbon years before present, or ca. cal. 17,000 B.P. (Stanford 1993; Thieme and Schuldenrein 1996; Thieme et al. 1996; Weiss 1974). This predates any known archaeological contexts from either New Jersey or southern New York by several thousand years. Nonetheless, deglaciation did have profound effects upon the physical setting and stratigraphy of archaeological sites located in both the coastal plain and the forested interior.

Within the rift basin valleys, modern river channels are relatively shallow with low gradients due to the subdued topography and clayey glaciolacustrine sediments. Floodplains evolved relatively late compared to the larger and steeper bedrock valleys of the Delaware River (Schuldenrein 1994; Stewart 1991) and the Susquehanna River (Cremeens et al. 1998; Thieme and Schuldenrein 1998b; Vento et al. 1999). The Hudson and Hackensack River valleys were scoured the deepest by glacial ice, and relict floodplain deposits appear to have been submerged during postglacial sea-level rise (Geoarcheology Research Associates 2000; LaPorta et al. 1999; Newman et al. 1969; Thieme 2000; Thieme and Schuldenrein 1998a; Weiss 1974). In addition to differences between drainages, sediment texture and bedding characteristics vary within each valley as a result of the depositional history at each recessional margin.

PLEISTOCENE-HOLOCENE BOUNDARY AND EARLY ARCHAEOLOGICAL CONTEXTS

The chronological boundary between the Pleistocene and Holocene Epochs of the Quaternary Period of the geological time scale is represented by a stratigraphic contact at the top of the glacial deposits in New Jersey and southern New York. Overlying sediment packages are typically much younger than the arbitrary date of 10,000 B.P. which has been assigned to this boundary (Hageman 1972; Harland et al. 1990). The apparent time gap at the Pleistocene-Holocene boundary can be ascribed both to localized erosion by streams choked with glacial meltwater and to a regional hiatus in the delivery of sediment. The "missing" sediment is very much missed by archaeologists working in the glaciated Northeast, because this is where the earliest prehistoric remains typically occur. The most significant beds for archaeologists are sandwiched between Pleistocene till, outwash, or lacustrine sediment and very late Holocene marsh peat or disturbed land.

This very general stratigraphic framework for identifying the deposits of interest has been employed in several recent projects of geoarchaeological prospection. One such project involved the supervision of geotechnical borings performed for sewer outfalls near North Bergen, New Jersey (Thieme and Schuldenrein 1998a). Organic sediment from a diamicton 6 m below surface was dated to 19,400 ± 60 B.P., or ca. cal. 23,000 B.P. (Table 12.1). The

Table 12.1. Radiocarbon Dates from Geoarchaeologic Investigations, New Jersey and Southern New York

Location	Elevation (mbls)	Generic Lithofacies	Material	^{14}C yr B.P.	Calibrated yr B.P. (1σ range)	Lab Number
Richmond Hill	0.50	Cultural sediment	Wood charcoal	9,360 ± 120	10,570 (10,731–10,295)	I-4929
Wards Point	1.40	Cultural sediment	Wood charcoal	8,250 ± 140	9,167 (9,468–9026)	I-5331
Wards Point	1.25	Cultural sediment	Wood charcoal	7,260 ± 125	8,092 (8,180–7944)	I-4512
Old Place	1.40	Cultural hearth	Wood charcoal	7,260 ± 140	8,092 (8,183–7884)	I-4070
Hollowell	0.90	Cultural sediment	Wood charcoal	3,110 ± 90	3,350 (3,440–3212)	I-3965
North Bergen Sewer B-10	6.00	Ice-contact diamicton	Bulk sediment	19,450 ± 60	23,061 (23,450–22,699)	Beta-112240
North Bergen Sewer B-11	3.00	Floodplain paleosol	Bulk sediment	3,650 ± 50	3,941 (4,079–3891)	Beta-112241
North Bergen Sewer B-10	2.40	Tidal marsh	Peat	1,130 ± 60	1,028 (1,166–966)	Beta-112239
28PA40 TT3 Stratum 4	1.30	Glaciolacustrine	Bulk sediment	15,390 ± 60	18,389 (18,684–18,114)	Beta-94221
28PA40 TT2, 2BC	1.00	Floodplain paleosol	Bulk sediment	2,140 ± 20	2,144 (2,302–2004)	Beta 94220
28PA40 TT4, 2C/3C	1.00	Floodplain paleosol	Bulk sediment	2,230 ± 60	2,207 (2,337–2150)	Beta-94222
28PA143, Unit B	6.50	Low-stage slackwater	Bulk sediment	13,860 ± 90	16,628 (16,686–16,382)	Beta-105557
28PA143, Block 1, 2C2	1.50	Floodplain lateral accretion	Bulk sediment	6,020 ± 50	6,821 (6,901–6756)	Beta-106377
28PA143, Block 1, 2AB	0.70	Floodplain paleosol	Bulk sediment	1,690 ± 60	1,584 (1,692–1526)	Beta-104333
28PA143, Block 1, Fea. Fill	0.50	Cultural sediment	Wood charcoal	980 ± 50	925 (947–795)	Beta-102306
28PA143, Block 1, Fea. Fill	0.50	Cultural sediment	Wood charcoal	1,010 ± 60	930 (966–802)	Beta-102307
28PA143, Block 2, 2Ab	0.50	Floodplain paleosol	Bulk sediment	2,250 ± 50	2,218 (2,339–2155)	Beta-106377
28PA143, Block 2, 2Ab	0.30	Floodplain paleosol	Bulk sediment	370 ± 60	464 (506–315)	Beta-106378
Collect Pond	9.40	Interior paleosol	Bulk sediment	4,590 ± 40	5,309 (5,435–5297)	Beta-130396
Collect Pond	8.80	Freshwater marsh	Peat	3,500 ± 50	3,761 (3,833–3691)	Beta-130395
Collect Pond	8.20	Freshwater marsh	Peat	2,490 ± 60	2,585 (2,737–2363)	Beta-130394
Collect Pond	7.60	Freshwater marsh	Peat	1,220 ± 60	1,171 (1,258–1060)	Beta-130393
North Arlington B-1	3.00	Freshwater marsh	Peat	5,030 ± 160	5,832 (5,929–5600)	Beta-80726

overlying Holocene floodplain facies fines upsection and is capped by a buried soil dated to cal. 4,000 B.P. (Figure 12.3). These relict floodplain deposits of Penhorn Creek probably continue downstream to its juncture with the Hackensack River, where local collectors have reported prehistoric artifact findspots to the New Jersey State Museum (Artemel 1979; Rutsch et al. 1978). The present land surface rests either on historic landfill or on peat emplaced once this portion of the Hackensack River valley became subject to tidal inundation. Peat from 2.4 m below surface at North Bergen was dated to ca. cal. 1000 B.P.

Calibration of radiocarbon dates using Stuiver et al. (1998) increases the time span for the depositional hiatus at the Pleistocene-Holocene boundary. Calibrated ages older than 10,000 B.P. also result for dates on some deposits that are generally considered to belong to the Holocene Epoch. This includes at least one archaeological context from southern New York that produced Late Paleoindian and Early Archaic artifacts. The time span of 8000-10,000 B.P. assigned to the Early Archaic Period based on uncalibrated radiocarbon dates (Ellis et al. 1998; Funk 1977; Ritchie and Funk 1973) now appears to have occurred at ca. cal. 9000–11,500 B.P. Although Paleoindian sites typically provide evidence for human exploitation of animals and plants which went extinct at the close of the Pleistocene Epoch (Dincauze 1993; Ellis et al. 1998; Gramly and Funk 1990; McNett 1985), this is not true for sites of the Archaic Period.

There is as yet no consensus concerning the implications of using the most recent calibration among Quaternary geologists, but it appears that the Holocene epoch should probably be considered to begin at 11,500 B.P. in calibrated calendar years. The change from Paleoindian to Archaic cultural lifeways would then still coincide with the transition from the late Pleistocene ice-age climate to the warm, temperate climate of the Holocene interglacial. Because of "plateaus" in the relationship of radiocarbon to calendar years, it has actually been suggested that changes in regional artifact styles could provide a more precise chronology spanning the transition (Ellis et al. 1998). Age controls can also be obtained from pollen, plant macrofossils, and other biostratigraphic methods (McWeeney 1994, 1999; McWeeney and Kellogg 2001; Peteet et al. 1993; Weiss 1974).

Isolated finds of fluted points and other Paleoindian artifacts are ubiquitous, particularly in southern New York. Excavations in terrestrial settings have unfortunately yet to recover a Paleoindian assemblage from a context that was not disturbed by subsequent prehistoric occupations or geological processes. Several sites on western Staten Island did clearly have substantial Paleoindian components (Kraft 1977a, 1977b; Ritchie and Funk 1971). At Port Mobil, for example, fluted points, end and side scrapers, and unifacial tools were among more than 51 lithic artifacts recovered from a sandy slope between 5 and 15 m above present mean sea level. Fluted points are also among the artifacts found on Charlestown Beach south of Port Mobil.

Four radiocarbon dates ranging from cal. 8000–11,000 B.P. were obtained from contexts at three of the archaeological sites on Staten Island (Ritchie and Funk 1971). Two of the dated sites, Richmond Hill and Wards Point, are at the southern end of the island. Well-preserved Early Archaic components are clearly present at both Wards Point and Hollowell based on the recovery of Kirk, Kanawha, LeCroy, and Stanly projectile points. Hollowell also clearly has a Late Archaic or Early Woodland component based on both the radiocarbon date and stratigraphy reported by Ritchie and Funk (1971:47).

At the Old Place site just northeast of the Goethals Bridge, charcoal from a hearth was dated to 7260 ± 140 B.P. in radiocarbon years. This calibrates to ca. cal. 8000 B.P., agreeing with the Early Archaic cultural affiliation of the Stanly, Kirk, and LeCroy projectile points found nearby. Recent geoarchaeological investigation of property on the margins of the site (Geoarcheology Research Associates 1996, 1997) indicates that the best preserved deposits date to the Middle Archaic or Late Archaic cultural periods. The excavators whose findings were reported by Ritchie and Funk (1971:49) noted several components at the site, with the Early Archaic materials coming from a relatively discrete package that has yet to be relocated.

The Old Place archaeological site is one of many locations in the study area where Holocene terrestrial deposits have been submerged beneath late Holocene tidal marsh. Rapid rates of sea-level rise were also characteristic of the late Pleistocene and early Holocene (Bloom 1983; Newman et al. 1969), and many of the locations formerly inhabited by prehistoric hunter-gatherers must consequently have been drowned within New York Harbor and Newark Bay. Such settings could theoretically contain particularly discrete and early archaeological contexts (Emery and Edwards 1966; Stewart 1999; Stright 1986; Thieme 2000). Fieldwork targeting submerged early sites is ongoing, although the studies to date have found extensive reworking during the Holocene transgression (Geoarcheology Research Associates 2000; LaPorta et al. 1999; Stright 1990).

In addition to prospection in underwater and nearshore settings seaward of the extant early contexts, identification of relict freshwater marshes and floodplain settings within the proglacial lake basins is crucial to finding further stratified prehistoric sites. Archaeologically sensitive deposits such as those sealed beneath tidal marsh peat in the Hackensack Meadowlands

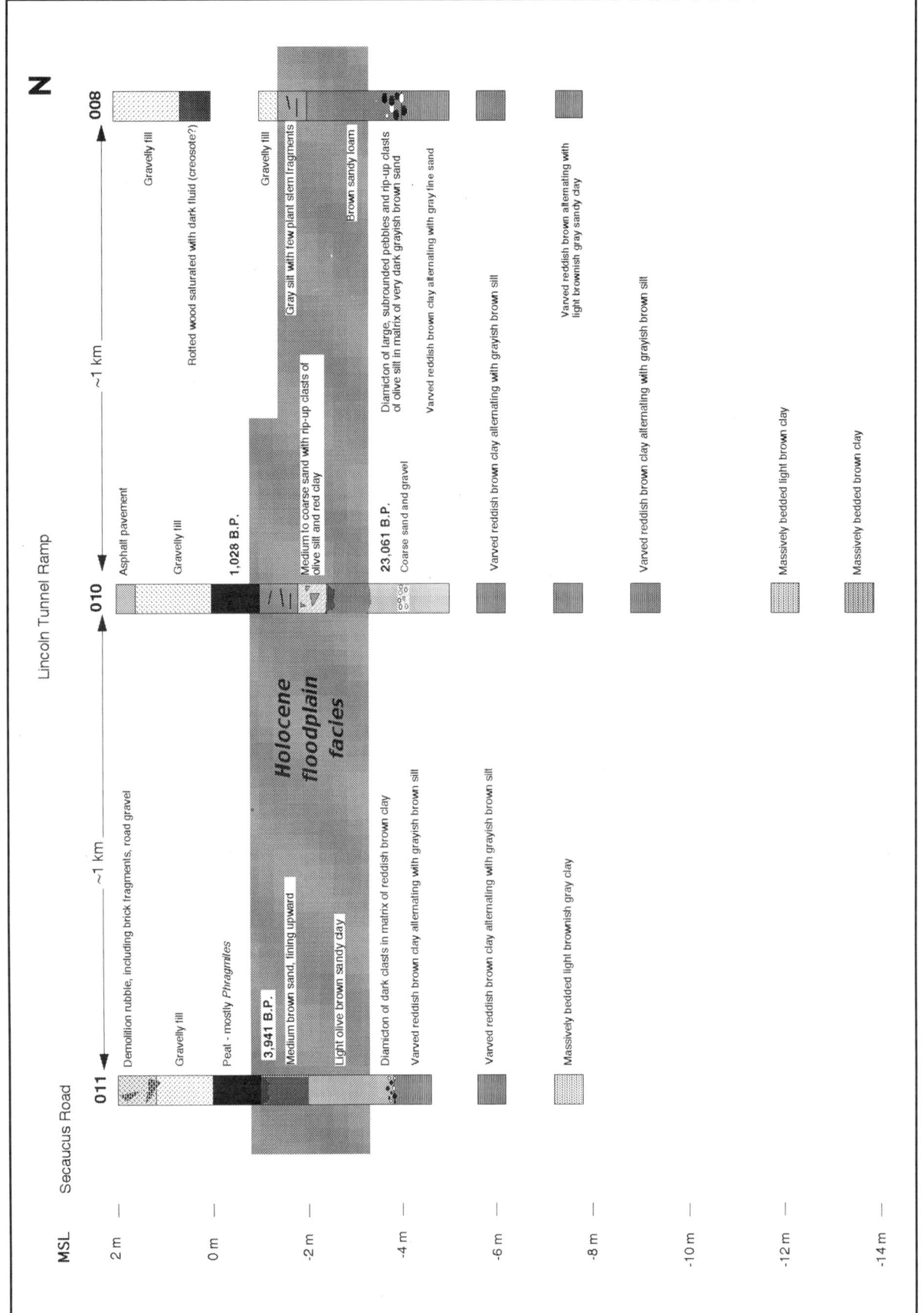

Figure 12.3. Stratigraphic cross-section east of Penhorn Creek in the Hackensack Meadowlands, North Bergen, New Jersey (after Thieme and Schuldenrein 1998a).

(Thieme and Schuldenrein 1996, 1998a) are difficult to identify using traditional archaeological techniques. Geological field and laboratory studies help by refining the stratigraphic framework and identifying sedimentological and geochemical signatures for the contexts in which prehistoric cultural materials have been found.

SITE FORMATION PROCESSES IN THE ROUTE 21 CORRIDOR, PASSAIC RIVER VALLEY

Large construction projects such as for highways, bridges, or dams sometimes provide archaeologists with the opportunity to study a number of prehistoric sites at the same time. In recent years it has become common practice to include a geomorphologist or geoarchaeologist on the project team. It is obviously important to develop a generic stratigraphy applicable to the project area in order to decide where and how deep to dig. Above and beyond such practical tasks, a geoarchaeological study can relate the materials recovered by the archaeologists to the surrounding physical landscape at the time of occupation. Processes in the physical environment often affect the location and integrity of the archaeological contexts, and such site formation processes are increasingly of interest to archaeologists working in North America (Nash and Petraglia 1987; Schiffer 1987).

The recent realignment of Route 21 near Passaic, New Jersey (Figure 12.4), is one such large construction project in which geoarchaeology played a crucial role in the excavation and interpretation of several stratified prehistoric sites. Initial (Phase I and Phase II) investigations were performed by Historic Conservation and Interpretation, Inc., in 1987 and 1988 (Mueller 1987; Rutsch et al. 1988a, 1988b). In 1996 and 1997, URS Greiner archaeologists conducted additional Phase II and Phase III excavations at four prehistoric sites (Slaughter 1997; Tull 1997; Tull et al. 1999, 2002). The geoarchaeological studies at sites 28PA39, 28PA40, 28PA145, and the Dundee Canal site (28PA143) were performed by the author and Joseph Schuldenrein of Geoarcheology Research Associates.

The Route 21 Corridor trends north-south within the Passaic River valley. The large loops in the river at Passaic (Figure 12.4) and farther downstream at the entrance to Newark Bay (Figures 12.1 and 12.2) are derangements produced by the ice sheet when it stood at the Culvers Gap-Bloomfield recessional margin (Stanford 1997). Most of the glacial deposits within the Route 21 Corridor were emplaced as the ice sheet retreated to the Augusta–Mud Pond margin and then the Cherry Ridge margin. The glaciolacustrine sediments were deposited in the proglacial Lake Paramus basin (Stanford 1993), which extended for several kilometers to the northeast. Proglacial Lake Passaic was considerably larger, but it spilled to the south into the Raritan River valley (Salisbury and Kummel 1893; Meyerson 1970). The radiocarbon date of 15,500 ± 60 B.P. (ca. cal. 18,000 B.P.) for glaciolacustrine sediment from site 28PA40 corroborates Stanford's model of a relatively small and relatively late impoundment in this part of New Jersey. Proglacial Lake Paramus submerged both this portion of the Passaic River valley and most of the Saddle River valley, which now drains east to join the Hackensack River.

Laminated reddish brown (5YR4/4) medium-to-coarse glaciolacustrine sand was also encountered in backhoe trenches at site 28PA145, capped by a coarse bouldery diamicton that was deposited in direct contact with the ice (Shaw 1985:29–46). This appears to be the Augusta–Mud Pond till moraine of Stanford (1997), with cobbles and boulders of diabase, gneiss, granite, and metagraywacke up to 50 cm long in a matrix of

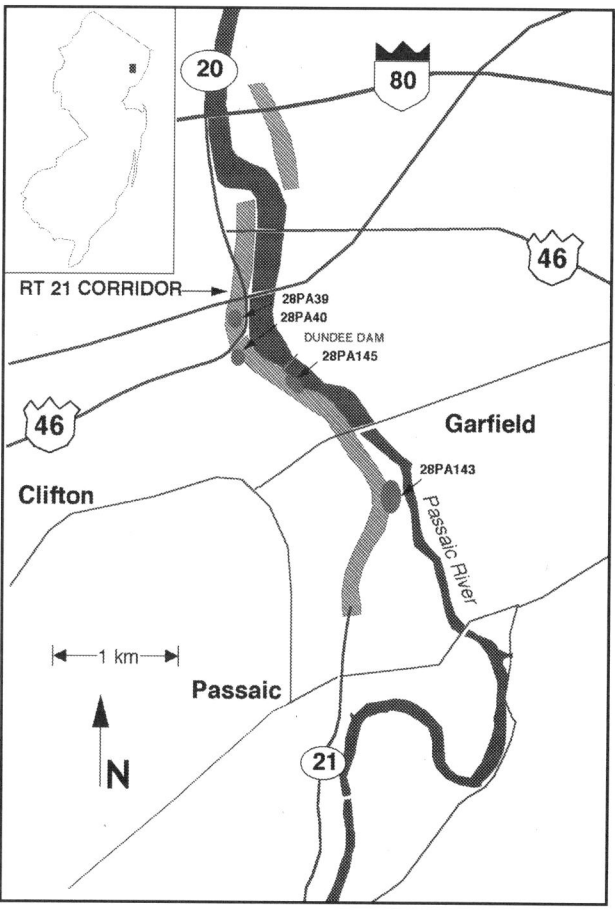

Figure 12.4. Prehistoric archaeological sites along Route 21 Corridor, Passaic County, New Jersey.

brown (7.5YR5/4) sand or sandy clay. The sequence also resembles deposits attributed to the Tappan Readvance by Averill et al. (1980), although a local readvance is fully compatible with the "recessional morphosequence" model of Koteff (1974).

Outwash deposited by melting ice during the retreat of the ice sheet through these recessional margins underlies the bed of the Dundee Canal bordering site 28PA145 and site 28PA143. Riverward of the canal bed, the late Pleistocene sediments were evidently reworked considerably as a meltwater sluice incised the channel occupied by the modern Passaic River. Basal deposits at the Dundee Canal site (28PA143) are channel sands and gravels with occasional muddy interbeds representing low-stage slackwater facies sediments of a braided stream (Boothroyd and Ashley 1975; Brakenridge 1988:141). Bulk sediment from one of these interbeds was radiocarbon dated to 13,860 ± 90 B.P., ca. cal. 16,600 B.P. This is coherent with the internal site stratigraphy, although regional deglaciation history suggests that the Passaic River was no longer receiving meltwater from the Laurentide ice sheet by this time.

Holocene sedimentation varied considerably moving both laterally away from the river and up- and downstream between the sites investigated. Tributary stream junctures functioned as traps for finer sediments, and one particularly significant trap was located in the immediate vicinity of the Dundee Canal site (28PA143). Colluvial reworking of till and outwash resulted in new landforms composed of much older materials. Sedimentation rates generally increased in the Mid through Late Holocene then slowed after 2,000 years when a soil developed on the Passaic River terrace at sites 28PA40, 28PA145, and 28PA143. Subsequent alluviation and historic landfilling have buried this soil and associated archaeological contexts from 50 cm to several meters below the present land surface.

DUNDEE CANAL SITE (28PA143) STRATIGRAPHY AND SEDIMENTOLOGY

The most important archaeological discovery within the Route 21 Corridor was the Dundee Canal site (28PA143), the first stratified prehistoric site to be professionally excavated in the Passaic River drainage. Late Archaic through Woodland components were housed in a complex sequence of alluvial and colluvial deposits dating back to the Late Pleistocene. The Woodland components were particularly well preserved, due to the presence of pit features. The pit features had been dug into a preexisting buried soil. Limited traces of earlier prehistoric activities were found beneath this buried soil.

There were two large blocks of excavation units at Dundee Canal. A deep depression between the two blocks represents the former channel of Weasel Brook. This tributary to the Passaic River was captured during the 19th century to feed a tail race in the Dundee Canal. Weasel Brook would certainly have been an active tributary to the Passaic River during the prehistoric occupations at Dundee Canal. Subsequent to its use for a canal race, the depression was used for the basements of several buildings and then filled with construction rubble.

There were four key packages of sediment common to the stratigraphy of the two excavation blocks (see Figure 12.5). The basal package consists of cross-bedded channel sand and gravel dated to ca. cal. 16,600 B.P. At the upper boundary of this package, the laterally accreted gravelly sands give way to fine sand or silt overbank flood deposits. A radiocarbon date of 6020 ± 50 B.P., or cal. 6800 B.P., was obtained for this boundary from a bulk sample of the 2C2 horizon in Block 1. Overlying sediments fine upward to a buried soil, the 2AB horizon in the Block 1 composite stratigraphy (Figure 12.6).

The third package of sediment is more variable across the site and results from cultural inputs and colluviation as well as continued episodic flooding of the terrace surface. All of the pit features and most of the artifacts at Dundee Canal were found at the base of a relict plow zone at the top of this package. Capping the Holocene alluvium, colluvium, and prehistoric midden, there was an upper package of municipal refuse and construction debris. The site was used for the Botany Mills lanolin retrieval plant in the 19th century and most recently for the Passaic town dump.

The footprint of the lanolin retrieval plant actually encompassed most of Block 2, where prehistoric pit features and other archaeological contexts were found by excavating within the plant's concrete foundations. Recent pavement and over a meter of fill were stripped from an area totaling more than 80 m^2 at which point the relict plow zone was recognized as a brown (7.5YR4/4-5/4) fine sandy sediment with slightly darker stains indicating the pit features. Datable charcoal was not found in good context in Block 2, but ceramics and other temporally diagnostic artifacts were similar for the Woodland components in the two blocks (Slaughter 1997; Tull et al. 2002).

Artifacts of the Woodland archaeological culture (1000 B.C.–A.D. 1500) were directly associated with materials that were radiocarbon dated. Radiocarbon dates of cal. A.D. 925 and A.D. 930 (Stuiver et al. 1998) were obtained for charcoal from pit features in Block 1 (Table 12.1). The Dundee Canal site also appears to have been used during the time of the earlier, Late Archaic culture (3000–1000 B.C.). Lamoka and Orient fishtail projectile points (Justice 1987; Ritchie 1969, 1971) as well

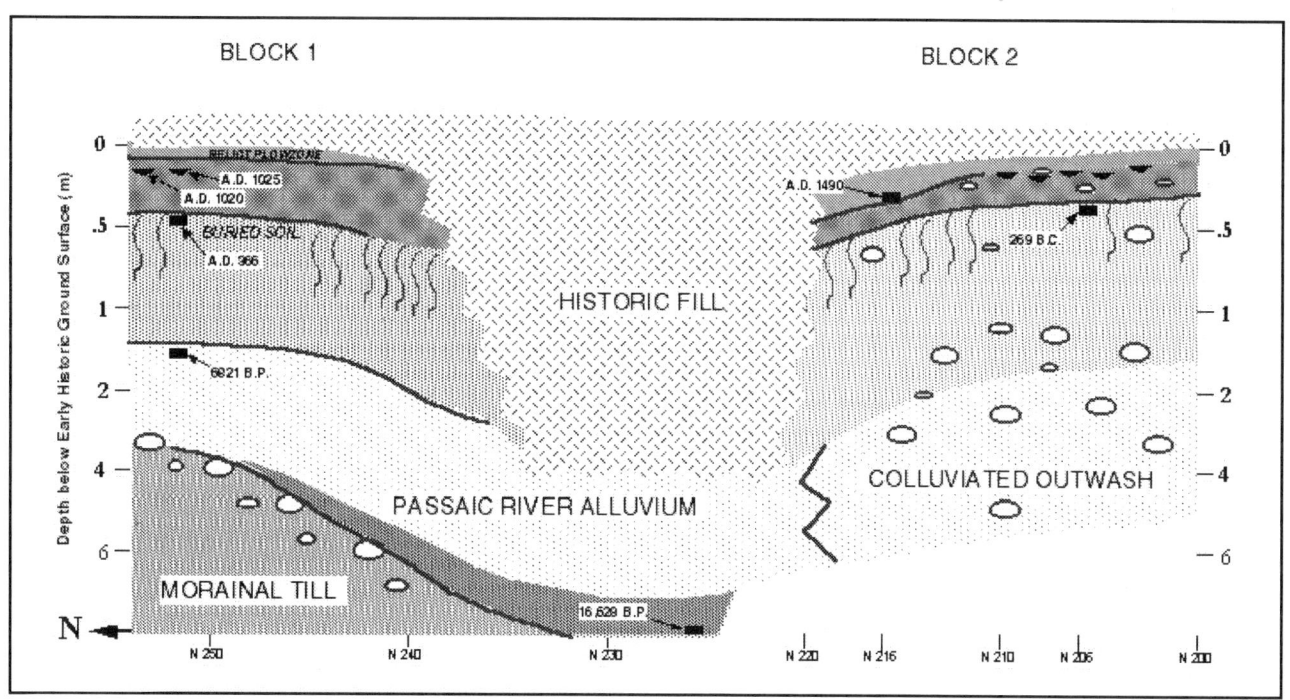

Figure 12.5. Stratigraphy of the Dundee Canal site (28PA143), Passaic County, New Jersey.

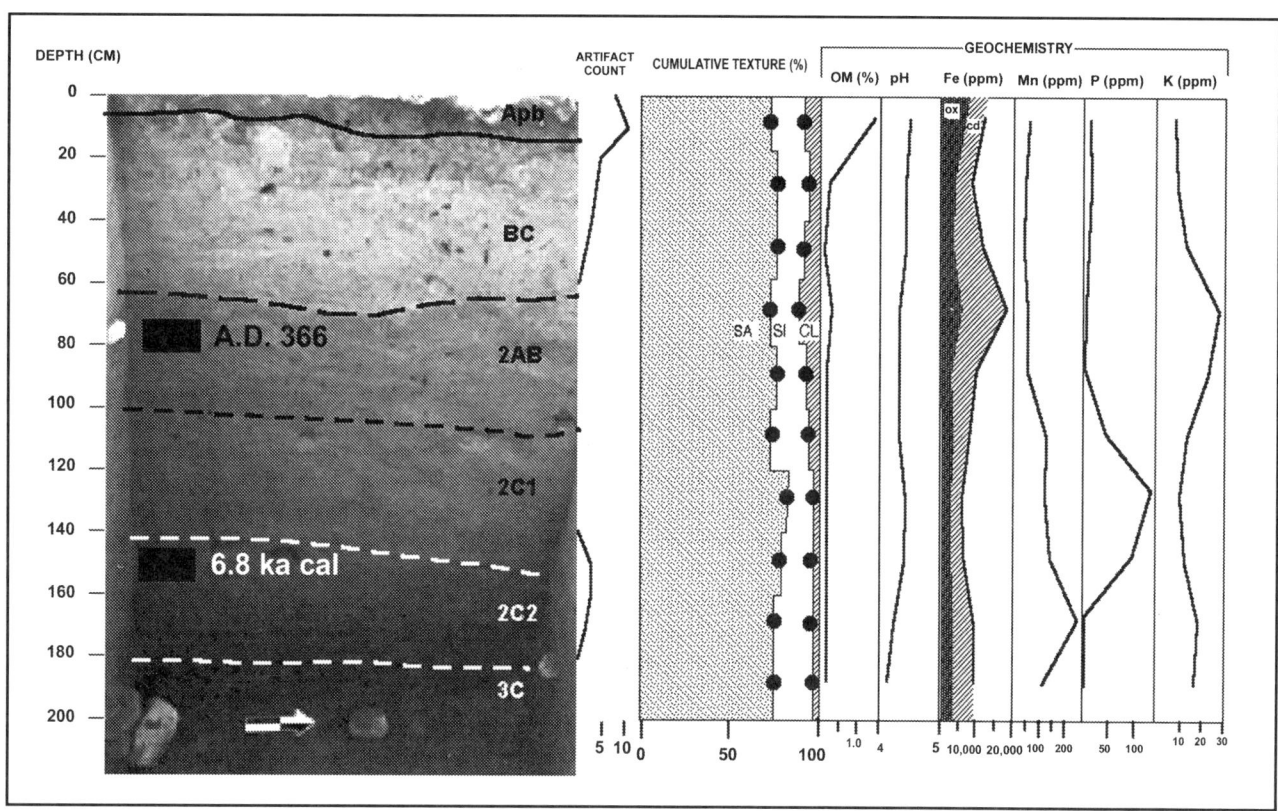

Figure 12.6. Composite stratigraphy for Block 1 at the Dundee Canal site (28PA143).

as a steatite bowl fragment were recovered from midden and colluvial deposits in Block 2.

Artifacts from the earliest components at Dundee Canal tended to have had the most complex pathways prior to entering the contexts in which they were found. Above and beyond the successive occupations at the site, this points to greater stream power and erosional activity. The sediment textures support this hypothesis in that coarser and more poorly sorted sands underlie the buried soil dated to ca. cal. A.D. 366 (1690 ± 60 B.P.) in Block 1 (2AB horizon) and to ca. cal. 269 B.C (2250 ± 50 B.P.) in Block 2 (2Ab horizon).

Most of the Holocene sediments of the Passaic River terrace were probably deposited by colluvial reworking of late Pleistocene till or outwash, and this is clearly true of the Block 2 deposits at Dundee Canal. Block 1 was in a unique setting upstream of the mouth of Weasel Brook, and this may partly account for the abundance of fine sand and silt deposited by overbank floods. All 10 samples analyzed from the Block 1 stratigraphic column were at least 70 percent sand, but there are two distinct fining-upward trends (Figure 12.7). The mean grain size decreases (increasing phi) from the 2C1 to the 2AB horizon and from the BC to the Apb horizon. The coarsest sediments are loamy sand in the 2C1 and 2C2, and their mean grain size is also greater on a clay-free basis (Gale and Hoare 1991:68; Catt 1987:493). Sorting, as measured by inclusive graphic standard deviation (Friedman and Sanders 1978:75), follows opposite trends on a composite as compared to a clay-free basis. This suggests that clay has been translocated into both the Apb and 2AB horizons.

The soil pH is extremely acid (i.e., < 4.5) (Soil Survey Staff 1951:235) for the entire Block 1 stratigraphic column. The pH generally decreases down profile but is slightly elevated in the coarser-textured 2C1 and 2C2 sediments. This may result from lateral groundwater movement along bedding planes in this sediment. The pH could also be buffered by calcium carbonate dissolved off of limestone clasts.

Total "free" iron was analyzed for all 10 samples by citrate-dithionite extraction (Birkeland 1999:90–92; McKeague et al. 1971). Acid-oxalate extraction was also performed to obtain the "active" Fe fraction complexed

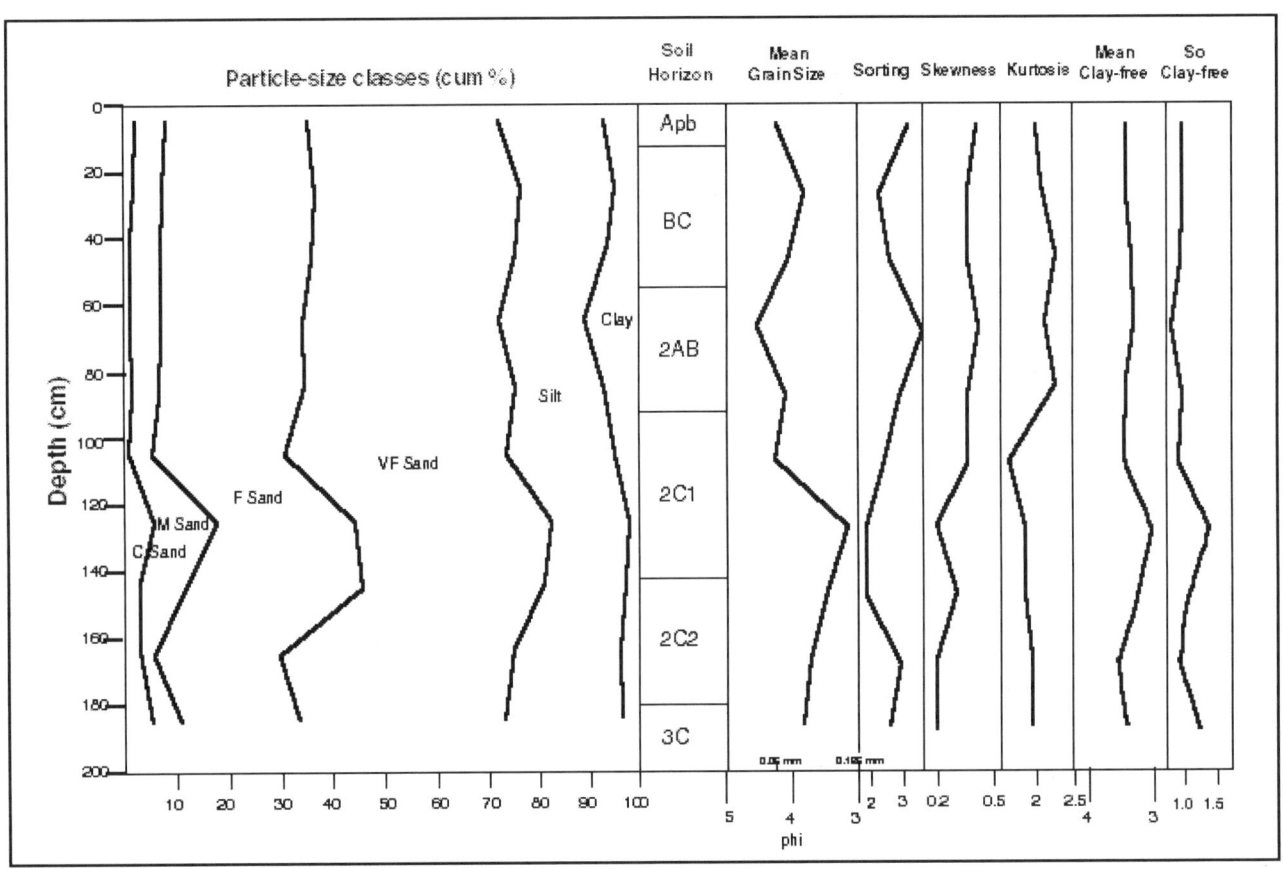

Figure 12.7. Grain-size trends in the Block 1 stratigraphic column at the Dundee Canal site (28PA143).

in amorphous hydroxides and organic molecules, a significant portion of the total free Fe not included in silicate minerals. The active Fe in ppm is the "ox" region of the graph in Figure 12.6, peaking in the Apb horizon. A disproportionate increase in the "cd" curve in the upper part of the buried soil (2AB horizon) suggests the presence of more oxidized forms such as hematite or goethite.

Results for manganese, phosphorous, and potassium show relatively little correlation with cultural activity at the Dundee Canal site but do further elucidate depositional history and pedogenic trends. Manganese increases down the column to a peak in the 2C2 horizon, probably precipitating as a function of water table fluctuation. Phosphorous peaks in the coarse-textured 2C1 horizon, probably occurring as a phosphatic coating on grain surfaces (Birkeland 1999:137). Potassium tends to follow the clay curve, in this case peaking in the 2AB horizon.

In spite of the continuous use of the property by industry and for disposal of municipal refuse, Dundee Canal provided significant new data concerning prehistoric cultural activity in the Passaic River valley. Both the massive overburden of 19th- and 20th-century debris and the complex basal contact with Pleistocene glacial deposits are in fact very typical of project areas currently being investigated by archaeologists working in the glaciated Northeast. Previous studies of soil chemistry and particle size in the region (Bilzi and Ciolkosz 1977; Ciolkosz et al. 1993; Foss 1977) suggest that the iron and potassium increase in buried soils is due to the formation of specific soil mineral phases. The effects of parent material and groundwater on the other laboratory results summarized above have also been observed in previous studies.

CONCLUSIONS: SITE FORMATION AND REGIONAL ENVIRONMENTAL PROCESSES

Since the initial proposal of geoarchaeology by Butzer (1971, 1982), and conceivably since the first scientific excavations of prehistoric sites, archaeologists have been aware of the effects of the regional environment on human culture and its material remains. The potential of archaeological excavations to provide significant information about Quaternary environmental change has also been recognized by a few pioneering researchers (Antevs 1948; Haynes 1991; Laville 1975). There have been only tentative efforts at interdisciplinary collaboration, however, between archaeologists and glacial geologists investigating the retreat of the Laurentide ice sheet in eastern North America. The findings presented above will, I hope, stimulate further collaboration along these lines and prompt some consideration of the role that glaciation may play in both regional and site-specific phenomena.

The Quaternary Period is characterized above all by glaciation, the Holocene Epoch being but one of many interglacials in the period (Holliday 2001; Williams et al. 1998:18). The ice caps that remain at the north and south pole affect atmospheric circulation (Bryson 1966; COHMAP Members 1988; Kutzbach and Guetter 1986), ocean circulation (Broecker 1991, 1994), and thereby the present, past, and future climate of the earth. It is quite reasonable to hypothesize that regionally synchronous events in the Quaternary stratigraphic record are somehow related to global forcing mechanisms recorded by ice accumulation rates (Dansgaard et al. 1993; Meese et al. 1994; O'Brien et al. 1995; Petit et al. 1990). Many high-resolution records of global climate change have been obtained, for example, from varved lacustrine sediments containing pollen and other environmentally sensitive fossils (Bradley 1999:324–326; Eicher and Siegenthaler 1976; Maenza-Gmelch 1997; McWeeney 1994; Peteet et al. 1993).

The stratification in glacial deposits, alluvium, and colluvium is much coarser than that in lake sediments. Nonetheless, correlations with external forcing mechanisms can be proposed and rates for some processes can be estimated. One glacial geologist has proposed, for example, that the sediments at a given recessional margin were deposited in a period measured in decades (Koteff 1974:124). If this were true, it would have taken no more than 500 years for the Laurentide ice sheet to retreat through all five of the margins mapped in New Jersey. This is excessively rapid given what is known about deglaciation elsewhere in the Northeast (Muller and Calkin 1993; Ridge et al. 1999), but it is a starting point to compare with the estimates from radiocarbon dating. Dates presented above and by Stanford (1997) suggest that the ice had retreated from New Jersey by cal. 17,000 B.P.

The Dundee Canal stratigraphy demonstrates that there are in fact some alluvial deposits in the study area that belong to the interval between deglaciation and the Pleistocene-Holocene boundary. As argued above on both biostratigraphic and archaeological grounds, this boundary should now be placed at cal. 11,500 B.P. Radiocarbon dates from pre-Clovis contexts at Cactus Hill in Virginia (Goodyear 1999:435–436; McAvoy and McAvoy 1997) and Meadowcroft Rockshelter in Pennsylvania (Adovasio et al. 1977, 1978, 1999) suggest that humans may have reached eastern North America several millennia before cal. 11,500 B.P. Laurentide ice was still present in some portions of eastern North America at the time these more controversial sites are inferred to have been occupied. Some Pleistocene glacial deposits may therefore conceivably contain artifacts and other remains of prehistoric cultures, although there have been no in situ finds to date.

The geometry of the basins that held the meltwater from the Laurentide ice sheet resulted in several particularly large "catastrophic" discharges into both the Gulf of Mexico (Leventer et al. 1982) and the Atlantic Ocean (Broecker et al. 1989; Teller 1990). Recent studies of marine sediment cores (Bond et al. 1996, 1997; DeMenocal and Bond 1997; Marchitto et al. 1998) indicate that meltwater discharges and ice-rafting events caused abrupt changes in the earth's climate. In particular, the Younger Dryas abrupt global cooling at ca. cal. 12,500 B.P. can be attributed to a disruption of the production of North Atlantic Deep Water (Broecker et al. 1988). Pollen and plant macrofossil evidence for the Younger Dryas has been found in lake sediment cores from the glaciated Northeast (Maenza-Gmelch 1997; McWeeney 1994; Peteet et al. 1993).

Deposits of Younger Dryas age are of considerable interest to archaeologists because of their potential to contain the remains of Clovis and other Paleoindian cultures. Unfortunately, sites containing Paleoindian materials in good stratigraphic context have yet to be excavated in the study area. Cold intervals in the glaciated Northeast tend also to have been dry in terms of effective precipitation (Webb et al. 1998). This might mean that water tables would have been depressed in the interior, and we know that regional sea levels were still at least 30 m lower than at present (Bloom 1983; Newman et al. 1969). As suggested by McWeeney and Kellogg (2001), the deposits of archaeological interest in the interior river valleys may very well be a meter or more below the modern water table. The submerged terrestrial deposits in New York Harbor and Newark Bay continue to attract archaeological attention because of their potential to contain discrete, early contexts (Geoarchaeology Research Associates 2000; LaPorta et al. 1999; Stright 1986, 1990).

Recent models of the geometry of glacial deposits (Koteff 1974; Koteff and Pessl 1981; Stone et al. 1998) can be used to predict the variation in sediment textures along the downstream axis of river valleys in the glaciated Northeast. In the Passaic River valley, the morainal deposits at the recessional margins stand out because the channel gradient is somewhat steeper and flanked by particularly resistant terrace outcrops (Stanford 1993). This variation along the downstream axis affected the distribution of prehistoric features and cultural materials among the four sites investigated within the Route 21 Corridor. Although the relationships were determined after completion of the field investigations, it clearly is possible to use the results of the Quaternary surficial mapping in New Jersey to explain and predict certain processes of archaeological site formation.

The greater stream power and erosional activity evident in the Early through Mid-Holocene deposits at Dundee Canal is consistent with paleo-environmental proxy data from the study area and the glaciated Northeast as a whole. The warmest interval within the Holocene epoch, the Hypsithermal, was locally characterized by abundant precipitation that favored both deciduous forests and freshwater marshes (Rue and Traverse 1997; Webb et al. 1998). Peat dated to 5800 cal. B.P. at North Arlington, for example, records the development of an early freshwater marsh in the Hackensack Meadowlands (Thieme et al. 1996). At the Collect Pond on Manhattan Island, freshwater peat accumulated on top of a buried soil that dates to cal. 5300 B.P. (Schuldenrein 2002). Farther afield, freshwater marsh replaced closed-canopy forest at Robbins Swamp in northwestern Connecticut at ca. 8000 B.P. (Nicholas 1998). This is approximately the same time as the highstand at Lake Owasco in north-central New York reported by Dwyer et al. (1996).

Whereas overbank deposition occurred at Dundee Canal under conditions of high stream power and effective moisture, the buried soil formed during what appears to have been a cold, dry interval in the regional climate. As recently observed by Anderson (2001), the first several centuries of the Woodland Period were relatively cold, with two fairly dramatic short-term cold events. The radiocarbon dates for the Dundee Canal buried soil are actually nearly a thousand years younger than these proposed abrupt global cooling events at 3100 years ago and 2800 years ago (Baillie 1988; Bond et al. 1997; van Geel et al. 1998, 1999). Perhaps it was really the sequence of rapid oscillations from cool-dry to warm-wet climate modes in the late Holocene that produced the buried soil.

By knowing more about these sorts of regional environmental processes that affect the formation of archaeological sites, we are able to make better decisions about where and how deep to dig in order to save on the costly and tedious labor of hand excavation. For urbanized settings in the glaciated Northeast, this means being able to identify intact Holocene deposits beneath the historic overburden. It also means being able to identify glacial deposits beneath the sediments of archaeological interest. Geoarchaeological knowledge further makes it possible to place artifact assemblages in stratigraphic sequence and relate them to changes in the regional environment. To the extent that we are able to determine the rates at which materials were eroded from or deposited on the land surface, as attempted above for Dundee Canal, we can relate the relative integrity of the archaeological contexts to regional environmental change.

REFERENCES CITED

Adovasio, J.M., Gunn, J.D., Donahue, J., and Stuckenrath, R. (1977). Meadowcroft rockshelter: A 16,000 year chronicle. In *Amerinds and Their Paleoenvironments in Northeastern North America*, edited by W.S. Newman and B. Salwen, pp. 137–159. Annals of the New York Academy of Sciences **288**, Albany.

Adovasio, J.M., Gunn, J.D., Donahue, J., and Stuckenrath, R. (1978). Meadowcroft Rockshelter, 1977, an overview. *American Antiquity* **43**:632–651.

Adovasio, J.M., Pedler, D., Donahue, J., and Stuckenrath, R. (1999). No vestige of a beginning nor prospect for an end: Two decades of debate on Meadowcroft Rockshelter. In *Ice Age People of North America: Environments, Origins, and Adaptations*, edited by R. Bonnichsen and K.L. Turnmire, pp. 416–431. Center for the Study of the First Americans, Corvallis, Oreg.

Anderson, D.G. (2001). Climate and culture change in prehistoric and early historic eastern North America. *Archaeology of Eastern North America* **29**:143–186.

Antevs, E.V. (1925). Condition of formation of the varved glacial clay. *Geological Society of America Bulletin* **36**.

Antevs, E.V. (1948). Climatic changes and pre-white man. *University of Utah Bulletin* **38**:168–191.

Artemel, J. (1979). *Historic and Archaeological Resources of the Northeast Corridor, New Jersey*. Report prepared for DeLeuw, Cather/Parsons, Newton, N.J.

Averill, S.P., Pardi, R.R., Newman, W.S., and Dineen, R.J. (1980). Late Wisconsin-Holocene history of the lower Hudson region: New evidence from the Hackensack and Hudson River valleys. In *Field Studies of New Jersey Geology and Guide to Field Trips, 52nd Annual Meeting of New York State Geological Survey*, edited by W. Manspeizer, pp. 160–186. Rutgers Press, New Brunswick, N.J.

Baillie, M.G.L. (1988). Irish oaks record volcanic dust veils drama. *Archaeology Ireland* **2**:71–74.

Bilzi, A.F., and Ciolkosz, E.J. (1977). Time as a factor in the genesis of four soils developed in recent alluvium in Pennsylvania. *Soil Science Society of America Journal* **41**:122–127.

Birkeland, P.W. (1999). *Soils and Geomorphology*. Oxford University Press, New York.

Bloom, A.L. (1983). Sea level and coastal changes. In *Late-Quaternary Environments of the United States, the Holocene*, edited by H.E. Wright, Jr., pp. 42–51. University of Minnesota Press, Minneapolis.

Bond, G., deMenocal, P., and Showers, W. (1996). Abrupt climate shifts on sub-Milankovitch timescales in the North Atlantic during the Holocene and the last Glacial-Interglacial cycle. *EOS* **77**(22):F-15.

Bond, G., Showers, W., Cheseby, M., Lotti, R., Almasi, P., deMenocal, P., Priore, P., Cullen, H., Hajdas, I., and Bonani, G. (1997). A pervasive millennial-scale cycle in north Atlantic Holocene and Glacial climates. *Science* **278**:1257–1266.

Boothroyd, J.C., and Ashley, G.M. (1975). Processes, bar morphology, and sedimentary structures on braided outwash fans, northeastern Gulf of Alaska. In *Glaciofluvial and Glaciolacustrine Sedimentation*, edited by A.V. Jopling and B.C. McDonald, pp. 193–222. Society for Economic Paleontologists and Mineralogists, Tulsa, Okla.

Bradley, R.S. (1999). *Paleoclimatology: Reconstructing Climates of the Quaternary*. Academic Press, San Diego.

Brakenridge, G.R. (1988). River flood regime and floodplain stratigraphy. In *Flood Geomorphology*, edited by V.R. Baker, R.C. Kochel, and P.C. Patton, pp. 139–156. Wiley-Interscience, New York.

Broecker, W.S. (1991). The great ocean conveyor. *Oceanography* **4**:79–89.

Broecker, W.S. (1994). Massive iceberg discharges as triggers for global climate change. *Nature* **372**:421–424.

Broecker, W.S., Andree, M., Wolfli, W., Oeschger, H., Bonani, G., Kennett, J., and Peteet, D. (1988). The chronology of the last deglaciation: implications to the cause of the Younger Dryas event. *Paleoceanography* **3**:1–20.

Broecker, W.S., Kennett, J.P., Flower, B.P., Teller, J.T., Trumbore, S., Bonani, G., and Wolfli, W. (1989). Routing of meltwater from the Laurentide ice sheet during the Younger Dryas cold episode. *Nature* **341**:318–321.

Bryson, R.A. (1966). Air masses, stream lines, and the boreal forest. *Geographical Bulletin* **8**:228–269.

Butzer, K.W. (1971). *Environment and Archaeology*. Aldine, Chicago.

Butzer, K.W. (1982). *Archaeology as Human Ecology: Method and Theory for a Contextual Approach*. Cambridge University Press, Cambridge.

Catt, J.A. (1987). Palaeosols. *Progress in Physical Geography* **11**:487–510.

Ciolkosz, E.J., Waltman, W.J., and Thurman, N. C. (1993). *Iron and Aluminium in Pennsylvania Soils*. Department of Agronomy, Pennsylvania State University, University Park.

COHMAP Members. (1988). Climate changes of the last 18,000 years: Observations and model simulations. *Science* **241**:1043–1052.

Cotter, J.F.P., Ridge, J.C., Evenson, E.B., Sevon, W.D., Sirkin, L.A., and Stuckenrath, R. (1986). The Wisconsinan history of the Great Valley, Pennsylvania and New Jersey and the age of the "terminal moraine." In *The Wisconsinan Stage of the First Geological District, Eastern New York*, edited by D.H. Cadwell, pp. 22–49. New York State Museum Bulletin 455. The University of the State of New York, Albany.

Cremeens, D.L., Hart, J.P., and Darmody, R.G. (1998). Complex pedostratigraphy of a terrace fragipan at the Memorial Park site, central Pennsylvania. *Geoarchaeology* **13**:339–359.

Dansgaard, W., Johnsen, S.J., Clausen, H.B., Dahl-Jensen, D., Gundestrup, N.S., Hammer, C.U., Hvidberg, C.S., Steffensen, J.P., Sveinbjörnsdottir, A.E., Jouzel, J., and Bond, G. (1993). Evidence for general instability of past climate from a 250-kyr ice-core record. *Nature* **364**:218–220.

Deetz, J.A. (1967). *Invitation to Archaeology*. Natural History Press, Garden City, New York.

DeMenocal, P., and Bond, G. (1997). Holocene climate less stable than previously thought. *EOS* **78**:447–454.

Dincauze, D.F. (1993). Pioneering in the Pleistocene: Large Paleoindian sites in the Northeast. In *Archaeology of Eastern North America: Papers in Honor of Stephen Williams*, edited by J.B. Stoltman, pp. 42–59. Mississippi Department of Archives and History, Jackson.

Dwyer, T.R., Mullins, H.T., and Good, S.C. (1996). Paleoclimatic implications of Holocene lake-level fluctuations, Owasco Lake, New York. *Geology* **24**:519–522.

Eicher, U., and Siegenthaler, U. (1976). Palynological and oxygen isotopic investigations on late-Glacial sediment cores from Swiss lakes. *Boreas* **5**:109–117.

Ellis, C., Goodyear, A.C., Morse, D.F., and Tankersley, K.B. (1998). Archaeology of the Pleistocene-Holocene transition in eastern North America. *Quaternary International* **49/50**:151–166.

Emery, K.O., and Edwards, R.L. (1966). Archaeological potential of the Atlantic continental shelf. *American Antiquity* **31**:733–737.

Foss, J.E. (1977). The pedological record at several Paleoindian sites in the Northeast. In *Amerinds and Their Paleoenvironments in Northeastern North America*, edited by W.S. Newman and B. Salwen. *Annals of the New York Academy of Sciences* **228**:234–244.

Friedman, G.M., and Sanders, J.E. (1978). *Principles of Sedimentology*. John Wiley & Sons, New York.

Funk, R.E. (1977). Early cultures in the Hudson drainage basin. In *Amerinds and Their Paleoenvironments in Northeastern North America*, edited by W.S. Newman and B. Salwen. *Annals of the New York Academy of Sciences* **228**:316–332.

Gale, S.J., and Hoare, P.G. (1991). *Quaternary Sediments*. Belhaven Press, London.

Geoarchaeology Research Associates (GRA). (1996). *Staten Island Bridges Program: Modernization and Capacity Enhancement Project, Phase IB Geomorphological Analysis, Final Report of Field Investigations*. Report prepared for Parsons, Brinckerhoff, Quade, and Douglas, Inc., New York.

Geoarchaeology Research Associates (GRA). (1997). *Enhancement Project, Phase IB/3 Geomorphological Analysis, Final Report of Field Investigations: Report on Coring and Additional Radiocarbon Dating*. Report prepared for Parsons, Brinckerhoff, Quade, and Douglas, Inc., New York.

Geoarchaeology Research Associates (GRA). (2000). *Geomorphological and Archeological Study of New York and New Jersey Harbor Navigation Channels*. Report prepared for the U.S. Army Corps of Engineers, New York District.

Goodyear, A.C. (1999). The early Holocene occupation of the southeastern United States: A geoarchaeological summary. In *Ice Age People of North America: Environments, Origins, and Adaptations*, edited by R. Bonnichsen and K.L. Turnmire, pp. 432–481. Center for the Study of the First Americans, Corvallis, Oreg.

Gramly, R.M., and Funk, R.E. (1990). What is known and not known about the human occupation of the northeastern United States until 10,000 B.P. *Archaeology of Eastern North America* **18**:5–32.

Gustavson, T.C., and Boothroyd, J.C. (1987). A depositional model for outwash, sediment sources, and hydrologic characteristics, Malaspina Glacier, Alaska: A modern analog of the southeastern margin of the Laurentide ice sheet. *Geological Society of America Bulletin* **99**:187–200.

Hageman, B.P. (1972). *Reports of the International Quaternary Association Subcommission on the Study of the Holocene* No. 6.

Harland, W.B., Cox, A.V., Llewellyn, P.G., Pickton, C.A.G., Smith, A.G., and Smith, D.G. (1990). *A Geologic Time Scale, 1989*. Cambridge University Press, Cambridge.

Haynes, C.V., Jr. (1991). Geoarchaeological and paleohydrological evidence for a Clovis-age drought in North America and its bearing on extinction. *Quaternary Research* **35**:438–450.

Holliday, V.T. (2001). Quaternary geoscience in archaeology. In *Earth Sciences and Archaeology*, edited by P. Goldberg, V.T. Holliday, and C.R. Ferring, pp. 3–35. Kluwer/Plenum, New York.

Isachsen, Y.W., Landing, E., Lauber, J.M., Rickard, L.V., and Rogers, W.B. (1991). *Geology of New York, A Simplified Account*. New York State Museum Education Leaflet No. 28. New York State Museum, Albany.

Justice, N.D. (1987). *Stone Age Spear and Arrow Points of the Midcontinental and Eastern United States: A Modern Survey and Reference*. Indiana University Press, Bloomington.

Knudson, S.J. (1978). *Culture in Retrospect: An Introduction to Archaeology*. Rand McNally, Chicago.

Koteff, C. (1974). The morphologic sequence concept and deglaciation of southern New England. In *Glacial Geomorphology*, edited by D.R. Coates, pp. 121–146. State University of New York, Binghamton.

Koteff, C., and Pessl, F., Jr. (1981). *Systematic Ice Retreat in New England*. U.S. Geological Survey Professional Paper No. 1179. Washington, D.C.

Kraft, H.C. (1977a). Paleoindians in New Jersey. In *Amerinds and Their Paleoenvironments in northeastern North America*, edited by W.S. Newman and B. Salwen. *Annals of the New York Academy of Sciences* **288**:264–281.

Kraft, H.C. (1977b). The Paleo-Indian sites at Port Mobil, Staten Island. In *Current Perspectives in Northeastern Archaeology: Essays in Honor of William A. Ritchie*, edited by R.A. Funk and C.F. Hayes III, pp. 1–19. Researches and Transactions of the New York State Archaeological Association No. 17(1).

Kutzbach, J.E., and Guetter, P.J. (1986). The influence of changing orbital parameters and surface boundary conditions on climate simulations for the past 18,000 years. *Journal of Atmospheric Sciences* **43**:1726–1759.

LaPorta, P.C., Sohl, L.E., Brewer, M.C., Elder, K.L., Franks, C.E., Bryant, V.M., Jr., Jones, J., Marshall, D., and Glees, M. (1999). *Cultural Resource Assessment of Proposed Dredge Material Management Alternative Sites in the New York Harbor-Apex Region*. Report prepared for the U.S. Army Corps of Engineers, New York District.

Laville, H. (1975). Climatologie et chronologie du Paleolithique en Perigord: Etude sedimentologique de sepots en grottes et Sous Abris. *Etudes Quaternaires, Memoire* **4**, Universite de Provence, France.

Leventer, A., Williams, D.F., and Kennett, J.P. (1982). Dynamics of the Laurentide ice sheet during the last deglaciation: evidence from the Gulf of Mexico. *Earth and Planetary Science Letters* **59**:11–17.

Lovegren, J.R. (1974). Paleodrainage history of the Hudson Estuary. Unpublished master's thesis, Columbia University, New York.

Lundqvist, J. (1989). Glacigenic processes, deposits, and landforms. In *Genetic Classification of Glacigenic Deposits*, edited by R.P. Goldthwait and C.L. Matsch, pp. 3–16. A.A. Balkema, Rotterdam, Netherlands.

Maenza-Gmelch, T. (1997). Vegetation, climate, and fire during the late-glacial-Holocene transition at Spruce Pond, southeastern New York, U.S.A. *Journal of Quaternary Science* **12**:14–24.

Marchitto, T.M., Jr., Curry, W.B., and Oppo, D.W. (1998). Millennial-scale changes in North Atlantic circulation since the last glaciation. *Nature* **383**:557–561.

McAvoy, J.M., and McAvoy, L.D. (1997). *Archaeological Investigations of 44SX2202, Cactus Hill, Sussex County, Virginia.* Research Report Series No. 8. Virginia Department of Historic Resources, Richmond.

McKeague, J.A., Brydon, J.E., and Miles, N.M. (1971). Differentiation of forms of extractable iron and aluminum in soils. *Soil Science Society of America Proceedings* **35**:33–38.

McNett, C.W. (1985). *Shawnee Minisink: A Stratified Paleoindian-Archaic Site in the Upper Delaware Valley of Pennsylvania.* Academic Press, New York.

McWeeney, L.J. (1994). *Archaeological Settlement Patterns and Vegetation Dynamics in Southern New England in the Late Quaternary.* Doctoral dissertation, Yale University. University Microfilms, Ann Arbor, Mich.

McWeeney, L.J. (1999). A review of late Pleistocene and Holocene climate changes in southern New England. *Bulletin of the Archaeological Society of Connecticut* **62**:3–18.

McWeeney, L.J., and Kellogg, D.C. (2001). Early and middle Holocene climate changes and settlement patterns along the eastern coast of North America. *Archaeology of Eastern North America* **29**:187–212.

Meese, D.A., Gow, A.J., Grootes, P., Mayewski, P.A., Ram, M., Stuiver, M., Taylor, K.C., Waddington, E.D., and Zielinski, G.A. (1994). The accumulation record from the GISP2 core as an indicator of climate change throughout the Holocene. *Science* **266**:1680–1682.

Meyerson, A.L. (1970). Glacial Lake Passaic: Palynological evidence for draining of the Great Swamp Stage. *New Jersey Academy of Sciences Bulletin* **15**:10–12.

Mueller, R.G. (1987). *Environmental Reconstruction, Soils, and Geomorphic History at Sites of the Route 21 Project, Passaic County, New Jersey.* Report prepared for Historic Conservation and Interpretation, Inc. Newton, N.J.

Muller, E.E., and Calkin, P. (1993). Timing of Pleistocene glacial events in NY State. *Canadian Journal of Earth Sciences* **30**:1829–1849.

Nash, D.T., and Petraglia, M.D. (1987). *Natural Formation Processes and the Archaeological Record.* BAR International Series 352. London.

Newman, W.S., Thurber, D.S., Zeiss, H.S., Rokach, A., and Musich, A.L. (1969). Late Quaternary geology of the Hudson River estuary: A preliminary report. *Transactions of the New York Academy of Sciences* **31**:548–570.

Nicholas, G.P. (1998). Assessing climatic influences on human affairs: Wetlands and the maximum Holocene warming in the Northeast. *Journal of Middle Atlantic Archaeology* **14**:147–160.

O'Brien, S.R., Mayewski, A., Meeker, L.D., Meese, D.A., Twickler, M.S., and Whitlow, S.I. (1995). Complexity of Holocene climate as reconstructed from a Greenland ice core. *Science* **270**:1962–1964.

Peteet, D.M., Daniels, R.A., Heusser, L.E., Vogel, J.S.,. Southon, J.R., and Nelson, D. (1993). Late-glacial pollen macrofossils and fish remains in northeastern U.S.A.: The Younger Dryas oscillation. *Quaternary Science Reviews* **12**:597–612.

Petit, J.R., Mounier, I., Jouzel, J., Kortkevich, Y.S., Kotlyakov, V.I., and Lorius, C. (1990). Paleoclimatological and chronological implications of the Vostok core dust record. *Nature* **343**:56–58.

Reeds, C.A. (1925). Glacial Lake Hackensack and adjacent lakes. *Geological Society of America Bulletin* **36**:155.

Reeds, C.A. (1926). The varved clays at Little Ferry, New Jersey. *American Museum Novitates* **209**:1–16.

Ridge, J.C., Besonnen, M.R., Brochu, M., Brown, S.L., Callahan, J.W., Cook, G.J., Nicholson, R.S., and Toll, N.J. (1999). Varve, paleomagnetic and ^{14}C chronologies for late Pleistocene events in New Hampshire and Vermont (U.S.A.). *Geographie Physique et Quaternaire* **53**:79–107.

Ritchie, W.A. (1969). *The Archaeology of New York State.* Natural History Press, New York.

Ritchie, W.A. (1971). *New York Projectile Points: A Typology and Nomenclature.* New York State Museum Bulletin 384. The University of the State of New York, Albany.

Ritchie, W.A., and Funk, R.E. (1971). Evidence for Early Archaic occupations on Staten Island. *Pennsylvania Archaeologist* **41**(3):45–59.

Ritchie, W.A., and Funk, R.E. (1973). *Aboriginal Settlement Patterns in the Northeast.* New York State Museum and Science Service Memoir 20. The University of the State of New York, Albany.

Rue, D.J., and Traverse, A. (1997). Pollen analysis of the Hackensack, New Jersey Meadowlands tidal marsh. *Northeastern Geology and Environmental Science* **19**:211–215.

Rutsch, E., Sandy, W., Bianchi, L., and Condell, P. (1978). *Stage 1A Cultural Resource Survey for the Hudson County Sewerage Authority 201 Wastewater Facility Plan—District I, Jersey City, North Bergen, Secaucus, and Kearny.* Report prepared for Havens and Emerson, Inc., Newton, N.J.

Rutsch, E., Sandy, W., Bianchi, L., and Condell, P. (1988a). *Archeological Cultural Resources Survey of the Proposed Route 21 Alignments in Passaic County, New Jersey. Phase I: Historical Research and Preliminary Identification of Potentially Significant Cultural Resources.* Report prepared for Federal Highway Administration, Region 1 and New Jersey Department of Transportation, Trenton.

Rutsch, E., Sandy, W., Bianchi, L., and Condell, P. (1988b). *Infield Testing Portion of the Archeological Cultural Resources Survey of the Proposed Route 21 Alignments in Passaic County, New Jersey.* Report prepared for Federal Highway Administration, Region 1 and New Jersey Department of Transportation, Trenton.

Salisbury, R.D. (1902). *The Glacial Geology of New Jersey.* Final Report. Trenton, N.J.

Salisbury, R.D., and Kummel, H.B. (1893). *Lake Passaic: An Extinct Glacial Lake.* Annual Report of the State Geologist of New Jersey, Section VI:225–328.

Schiffer, M.B. (1987). *Formation Processes of the Archaeological Record.* University of New Mexico Press, Albuquerque.

Schuldenrein, J. (1994). Alluvial site geoarcheology of the middle Delaware valley: A fluvial systems paradigm. *Journal of Middle Atlantic Archaeology* **10**:1–21.

Schuldenrein, J. (2001). Stratigraphy of the Collect Pond based on borings at Foley Square, New York. Report prepared for J. Geismar on file at Geoarchaeology Research Associates, Riverdale, N.Y.

Schuldenrein, J. (2002). *Stratigraphy of the Collect Pond based on borings at Foley Square, New York*. Report prepared for J. Geismar.

Shaw, J. (1985). Subglacial and ice marginal environments. In *Glacial Sedimentary Environments*, edited by G.M. Ashley, J. Shaw, and N.D. Smith, pp. 7–84. Society for Economic Paleontologists and Mineralogists, Tulsa, Okla.

Sirkin, L.A. (1982). Wisconsinan glaciation of Long Island, New York, to Block Island, Rhode Island. In *Late Wisconsinan Glaciation of New England*, edited by G.L. Larson and B.S. Stone, pp. 35–59. Kendall/Hunt, Dubuque, Iowa.

Sirkin, L.A. (1986). Pleistocene stratigraphy of Long Island, New York. In *The Wisconsinan Stage of the First Geological District, Eastern New York*, edited by D.H. Cadwell, pp. 6–21. New York State Museum Bulletin 455. The University of the State of New York, Albany.

Sirkin, L.A., and Stuckenrath, R. (1980). The Portwashingtonian warm interval in the northern Atlantic coastal plain. *Geological Society of America Bulletin* 91:332–336.

Slaughter, B. (1997). Archaeological investigations at the Dundee Canal site (28PA143), Passaic, New Jersey. *Abstracts, 63rd Eastern States Archaeological Federation Conference*, p. 19.

Soil Survey Staff. (1951). *Soil Survey Manual*. U.S. Department of Agriculture Handbook No. 18, U.S. Government Printing Office, Washington, D.C.

Stanford, S.D. (1993). *Surficial Geology of the Weehawken and Central Park Quadrangles, Bergen, Hudson, and Passaic Counties, New Jersey*. New Jersey Geological Survey, Trenton.

Stanford, S.D. (1997). Pliocene–Quaternary geology of northern New Jersey: An overview. In *Pliocene–Quaternary Geology of Northern New Jersey, Guidebook for the 60th Annual Reunion of the Northeastern Friends of the Pleistocene*, pp. 1-1–1-26. New jersey Geological Survey, Trenton.

Stanford, S.D. (2000). Pliocene-Pleistocene discharge of the Hudson to the New York Bight: The view from land. *GSA Abstracts* 22:53.

Stanford, S.D., and Harper, D.P. (1991). Glacial lakes of the lower Passaic, Hackensack, and lower Hudson valleys, New Jersey and New York. *Northeastern Geology* 13:277–286.

Stein, J.K. (2001). A review of site formation processes and their relevance to geoarchaeology. In *Earth Sciences and Archaeology*, edited by P. Goldberg, V.T. Holliday, and C.R. Ferring, pp. 3–35. Kluwer/Plenum, New York.

Stewart, D.J. (1999). Formation processes affecting submerged archaeological sites: An overview. *Geoarchaeology* 14:569–587.

Stewart, R.M. (1991). Archaeology and environment in the upper Delaware River valley. In *The People of Minisink*, edited by D.G. Orr and D. Campana, pp. 79–116: National Park Service, Mid-Atlantic Region, Philadelphia.

Stone, J.R., DiGiacomo-Cohen, M., Lewis, R.S., and Goldsmith, R. (1998). Recessional moraines and the associated deglacial record of southeastern Connecticut and Long Island Sound. In *Guidebook to Field Trips in Rhode Island and Adjacent Regions of Connecticut and Massachusetts*, edited by D.P. Murray, pp. B-5:1-20. 90th Annual Meeting, New England Intercollegiate Geological Conference, Kingston, R.I.

Stright, M.J. (1986). Human occupation of the continental shelf during the late Pleistocene/early Holocene, methods for site location. *Geoarchaeology* 1:347–363.

Stright, M.J. (1990). Archaeological sites on the North American continental shelf. In *Archaeological Geology of North America*, edited by N. Lasca and J. Donahue, pp. 439–465. Geological Society of America, Boulder, Colo.

Stuiver, M., Reimer, P.J., Bard, E., Beck, J.W., Burr, G.S., Hughen, K.A., Kromer, B., McCormac, F.G., van der Plicht, J., and Spark, M. (1998). INTCAL98 radiocarbon age calibration 24,000–0 cal. BP. *Radiocarbon* 40:1041–1083.

Stuiver, M., Reimer, P.J., Braziunas, T.F., Becker, B., and Kromer, B. (1991). Climatic, solar, oceanic, and geomagnetic influences on late-glacial and Holocene atmospheric $^{14}C/^{12}C$ change: *Quaternary Research* 35:1–24.

Teller, J.T. (1990). Volume and routing of late-glacial runoff from the southern Laurentide ice sheet. *Quaternary Research* 34:12–23.

Thieme, D.M. (1997). Archaeological stratigraphy of the Dundee Canal site (28PA1433) and other sites in the Route 21 corridor. Abstracts, 63rd Eastern States Archaeological Federation Conference, p. 20. Mount Laurel, N.J.

Thieme, D.M. (2000). Paleoenvironmental and archaeological contexts in the New York and New Jersey harbor region. *SAA Abstracts, 65th Annual Meeting*, p. 329.

Thieme, D. M., and Schuldenrein, J. (1996). *Quaternary Paleoenvironments in the Hackensack Meadowlands: A Geological and Palynological Study of Borings for the Proposed North Arlington Force Main and Pumping Station, North Arlington, N.J.* Report prepared for Neglia Engineering Associates, Lyndhurst, N.J.

Thieme, D.M., Schuldenrein, J., and Maenza-Gmelch, T. (1996). Mid-Holocene warming and development of the Hackensack tidal marsh on the bed of Glacial Lake Hackensack. *Abstracts, American Quaternary Association, 14th Biennial Meeting*, p. 134.

Thieme, D.M., and Schuldenrein, J. (1998a). *Paleoenvironmental analysis of the combined sewer overflow planning study, planning area IA, North Bergen (West), Hudson County, New Jersey*. Report prepared for Richard Grubb and Associates, Inc., Cranbury, N.J.

Thieme, D.M., and Schuldenrein, J. (1998b). Wyoming Valley landscape evolution and the emergence of the Wyoming Valley culture. *Pennsylvania Archaeologist* 68(2):1–17.

Tull, S. (1997). *Cultural Change in a Floodplain Setting During the Woodland Period: A View from the Dundee Canal Site, Passaic County, New Jersey*. Abstracts, 63rd Eastern States Archaeological Federation Conference, p. 21.

Tull, S.W., Slaughter, B., Scholl, M., Picadio, D., Brown, M., and Sterling, B. (1999). *Archaeological Investigations at Sites 28PA39, 28PA40, 28PA145, the Early Dundee Canal Terminus Area, and the Eagle Foundry Site, Route 21 Extension Cultural Resources Mitigation, Passaic County, New Jersey*. Report prepared for New Jersey Department of Transportation, Trenton.

Tull, S.W., Slaughter, B. et al. (2002). *Final Report of Phase II and Phase III Investigations at the Dundee Canal Site (28PA143), Route 21(9) Cultural Resources Mitigation, Passaic County, New Jersey*. Report prepared for New Jersey Department of Transportation, Trenton.

van Andel, T. (1998). Middle and upper Paleolithic environments and the calibration of ^{14}C dates beyond 10,000 BP. *Antiquity* **72**:26–33.

van Geel, B., van der Plicht, J., Kilian, M.R., Klaver, E.R., Kouwenberg, J.H.M., Renssen, H., Reynaud-Farrera, I., and Waterbolk, H.T. (1998). The sharp rise of delta-^{14}C ca. 800 cal. BC: Possible causes, related climatic teleconnections, and the impact on human environments. *Radiocarbon* **40**(2):535–550.

van Geel, B., Raspopov, O.M., Renssen, H., van der Plicht, J., Dergachev, V.A., and Meijer H.A.J. (1999). The role of solar forcing upon climate change. *Quaternary Science Reviews* **18**:331–338.

Vento, F.J., Donahue, J., and Adovasio, J.M. (1999). Geoarchaeology. In *The Geology of Pennsylvania*, edited by C.H. Schultz, pp. 770–777, Pennsylvania Geological Survey and Pittsburgh Geological Society, Harrisburg.

Webb, T. III, Anderson, K.H., Bartlein, P.J., and Webb, R.S. (1998). Late Quaternary climate change in eastern North America: a comparison of pollen-derived estimates with climate model results. *Quaternary Science Reviews* **7**:587–606.

Weiss, D. (1974). Late Pleistocene stratigraphy and paleoecology of the lower Hudson River estuary. *Geological Society of America Bulletin* **85**:1561–1580.

Widmer, K. (1964). *The Geology and Geography of New Jersey*. Van Nostrand, Princeton, N.J.

Williams, M., Dunkerly, D., De Deckker, P., Kershaw, P., and Chappell, J. (1998). *Quaternary Environments*. Arnold, London.

CHAPTER 13

LANDSCAPE CHANGE, HUMAN OCCUPATION, AND ARCHAEOLOGICAL SITE PRESERVATION AT THE GLACIAL MARGIN: GEOARCHAEOLOGICAL PERSPECTIVES FROM THE SANDTS EDDY SITE (36Nm12), MIDDLE DELAWARE VALLEY, PENNSYLVANIA

Joseph Schuldenrein

The archaeology of glaciated landscapes is difficult to interpret because of a paucity of stratified sites and the magnitude and extent of geomorphic activity at the Pleistocene-Holocene transition. Human and landscape reconstructions are challenging across the North American continent and especially problematic in the northeastern United States. Here the broad reach of Euroamerican impacts across the landscape has resulted in whole-scale recontouring of prehistoric terrain. Because the greatest potential for integrating depositional and occupational records lies in ancient floodplains and these are also the land segments most widely damaged by development, the long-term archaeological record in the Northeast is not as well documented as in other parts of the Eastern Woodlands, especially the midcontinent (i.e., Mississippi and Ohio drainages) and the Southeast. Equally critical are late glacial and postglacial stream dynamics where high discharges and outwash completely overhauled valley floor and margin settings, thus inhibiting the potential for discovery of key Paleoindian and Archaic sites.

Against this backdrop, the geoarchaeology of the Upper and Middle Delaware Valley opens up a critical window for understanding Holocene alluvial stratigraphy. Archaeological and geological exploration of the floodplain and terrace sequences has been ongoing for well over 50 years (Epstein 1969; Kinsey 1972; Leverett 1934) (Figure 13.1). Access to the buried Late Quaternary record of the Delaware River system is facilitated by a relatively narrow valley that contains a two-terrace system within the glaciated portion and a one-terrace succession south of the glacial margin (Schuldenrein 1994; Witte 1997). Accordingly, lateral migrations of the pre-Holocene stream were limited by the width of the bedrock valley, and a "stacked vertical succession" of deposits is typical of the alluvial sequence. Prehistoric deposits can be correlated by thickness and type of depositional units (and less by lateral position) along the terrestrial landscapes spanning the 130-km valley length between the mountainous Appalachian Highlands and the Coastal Plain. The oldest, most sustained valley sequences—Paleoindian through Archaic, Woodland, and Euroamerican contact—are at Shawnee-Minisink (McNett 1985) and Shawnee Island (Stewart et al. 1991), immediately north of the glacial margin.

Sandts Eddy is a multicomponent, stratified site containing limited amounts of later prehistoric material (Woodland and Euroamerican contact) but significant, intact, and deeply buried early prehistoric components (Early and Middle Archaic Periods). It is located just south of the late Woodfordian glacial margin (Figure 13.1). The early Holocene cultural horizons are preserved in coarse and weathered sands indicative of a once-meandering, high-energy stream antecedent to the current stream course. The site's potential for landscape reconstruction became apparent as progressively deeper horizons isolated discrete alluvial and anthropogenic sediment packages that provided contexts to the artifact assemblages. Cultural materials were preserved at depths in excess of 4 m (Bergman and Doershuk 1994; Bergman et al. 1998). The objectives of the geoarchaeological research at this site were threefold: (1) reconstructions of the Holocene floodplain corresponding to more than 10,000 years of human activity; (2) unraveling site formation processes affecting the individual occupation strata, with emphasis on the Early and Middle Archaic Periods (10,000–6000 B.P.); and (3) modeling the regional geoarchaeology of the Middle Delaware Valley based on correlation and projection with existing cultural and alluvial stratigraphies.

Geoarchaeology of Landscapes in the Glaciated Northeast edited by David L. Cremeens and John P. Hart. New York State Museum Bulletin 497. © 2003 by the University of the State of New York, The State Education Department, Albany, New York. All rights reserved.

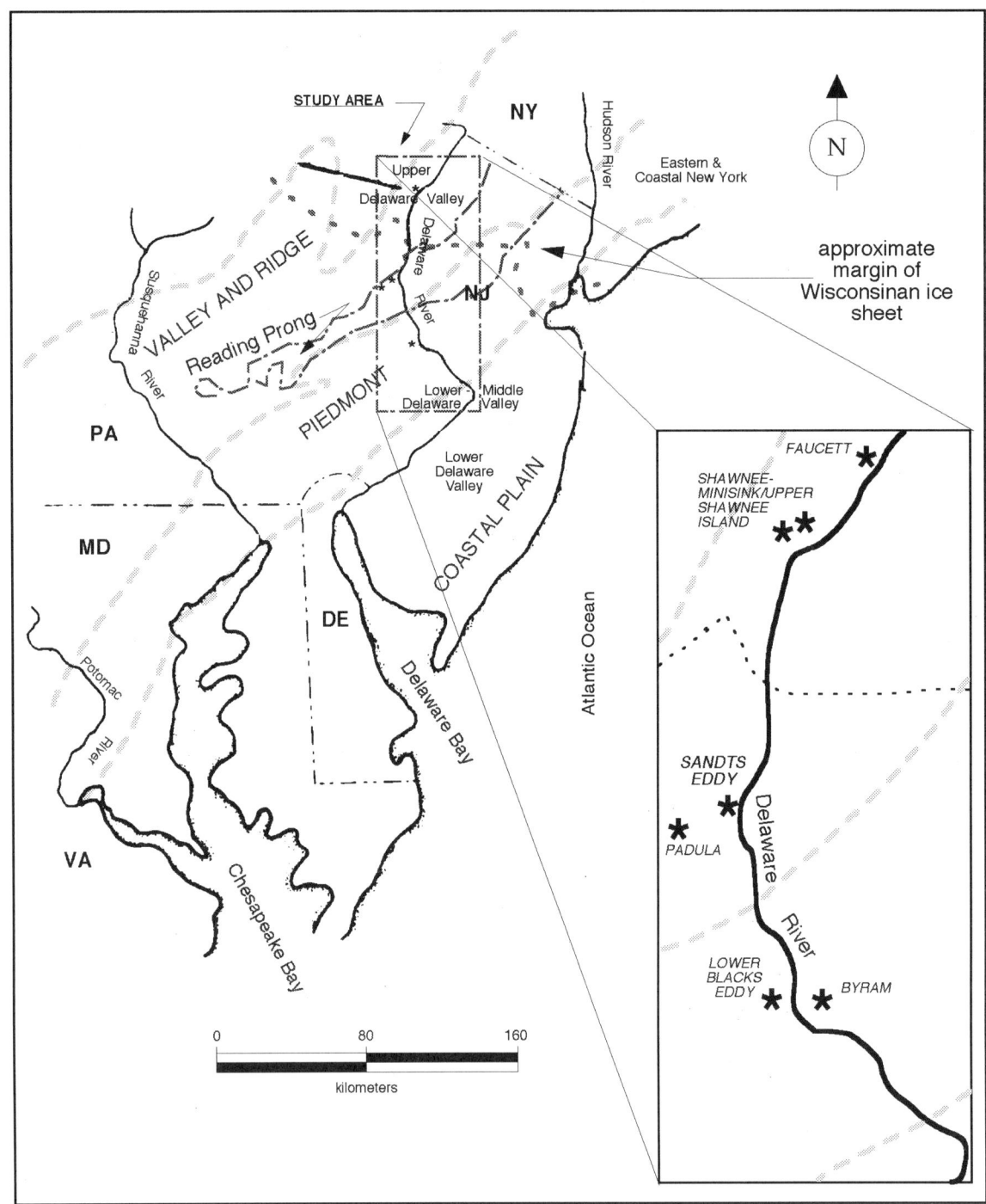

Figure 13.1. Physiographic setting of Sandts Eddy (36Nm12), Valley and Ridge province. Inset illustrates locations of stratified floodplain sites discussed in text.

GEOLOGICAL AND GEOMORPHIC SETTING

The upper and middle reaches of the Delaware River drain a portion of the Great Valley section of the Ridge and Valley province in eastern Pennsylvania (Fenneman 1938; Berg et al. 1989). This physiographic region straddles the northern boundary of the Reading Prong, the highland terrain of northeast-trending ridges of gneiss and hard crystalline rocks that separates the Ridge and Valley from the Piedmont (Parker et al. 1964).

Local topography is dominated by broad, moderately dissected valleys with undulating surfaces. Tributary flow is south-southeast to the trunk stream (Delaware River) (Figures 13.2 and 13.3). The most prominent local drainages are Mud Run, which debouches into the Delaware, just south of 36Nm12, and Martins Creek, which empties into the Delaware several kilometers upstream. The site occupies a first terrace (T-1) ca. 5 m (16.5 ft) above mean stream level. A lower, 1-2 m floodplain bench (T-0) is irregularly banked against the T-1 (Figure 13.4).

Figure 13.3 shows that the archaeological site was originally mapped as two landform and morpho-stratigraphic units, the contemporary alluvium (Qal) and the outwash (Qwo) (Davis et al. 1967; Miller et al. 1939). The alluvium and outwash, although not formally differentiated by Davis et al. (1967), conform, respectively, to the site's Holocene first terrace (T-1) and an older (Late Pleistocene) alluvial fan. The outwash is part of the T-2, recognized farther north. The Mud Creek alluvial fan consists of a poorly sorted outwash that incorporates slope residuum and exfoliated bedrock slabs derived from the northwestern valley slopes; bedrock is chiefly (Lower Ordovician) Epler Formation gray dolomite and cryptogranular gray limestone (Davis et al. 1967) (Oe on Figure 13.3). Its nodular and cryptocrystalline cherts are widely represented in the site's lithic raw material assemblages (Bergman and Doershuk 1994; Bergman et al. 1998).

The most recent regional mapping of Late Quaternary deposits (Crowl and Sevon 1980) traced the Late Wisconsinan glacial border within several kilometers (north) of Sandts Eddy. Extensive pockets of Olean ground moraine (Crowl and Sevon 1980, Plate 1) consist of a reddish brown to yellowish brown poorly sorted till of clays, silts, sands, cobbles, and boulders. Localized accumulations of Olean outwash were identified along Martins Creek as well. These deposits date to between 15,000 and 10,000 B.P. (Crowl and Sevon 1980). Significantly, sedimentological and descriptive comparisons between Crowl and Sevon's (1980) outwash and the present fan and T-1 terrace deposits did not suggest chrono-stratigraphic correlations.

This study refines the previous mapping. It is proposed that the Sandts Eddy fan is of uncertain Wisconsinan age and that the earliest aggradational phase for the postglacial Delaware is that preserved at the base of the T-1, more than 4 m below the present T-1 surface.

At Sandts Eddy the T-1 has been extensively modified by 19th- and 20th-century landscaping, which drastically altered the original surface contours. The proximal portion of the T-1 has been truncated by a boat ramp.

Figure 13.2. Aerial photo of Sandts Eddy site and surrounding terrain.

The natural landform is approximately 30 m wide (north-south) and grades down to isolated spurs of the contemporary floodplain (T-0). The archaeological excavations (Figure 13.5) were oriented across a distance of 60 m (east-west), following near-level contours on the T-1, but the landform is continuous along the upstream and downstream ends.

PREVIOUS INVESTIGATIONS

Three separate phases of archaeological investigation were performed at Sandts Eddy, each of which identified a relationship between particular cultural horizons and sediment and/or soil types. Earliest excavations were in 1969 and concentrated on the east end of the site (Figure 13.5, "SPA excavations") (Fehr et al. 1971). Three stratigraphic zones were recognized to a maximum depth of 1 m. The uppermost is a plow zone (Ap horizon; 0–0.3 m below surface), recognized by excavators as extensively disturbed. Highest artifact densities were preserved in underlying Zone 2, a yellow sand (0.3–0.5 m) that produced the largest concentrations of artifacts as well as discrete features from the Archaic, Transitional, and Woodland Periods. Unfortunately horizon separation within Zone 2 was not reported. Zone 3 was

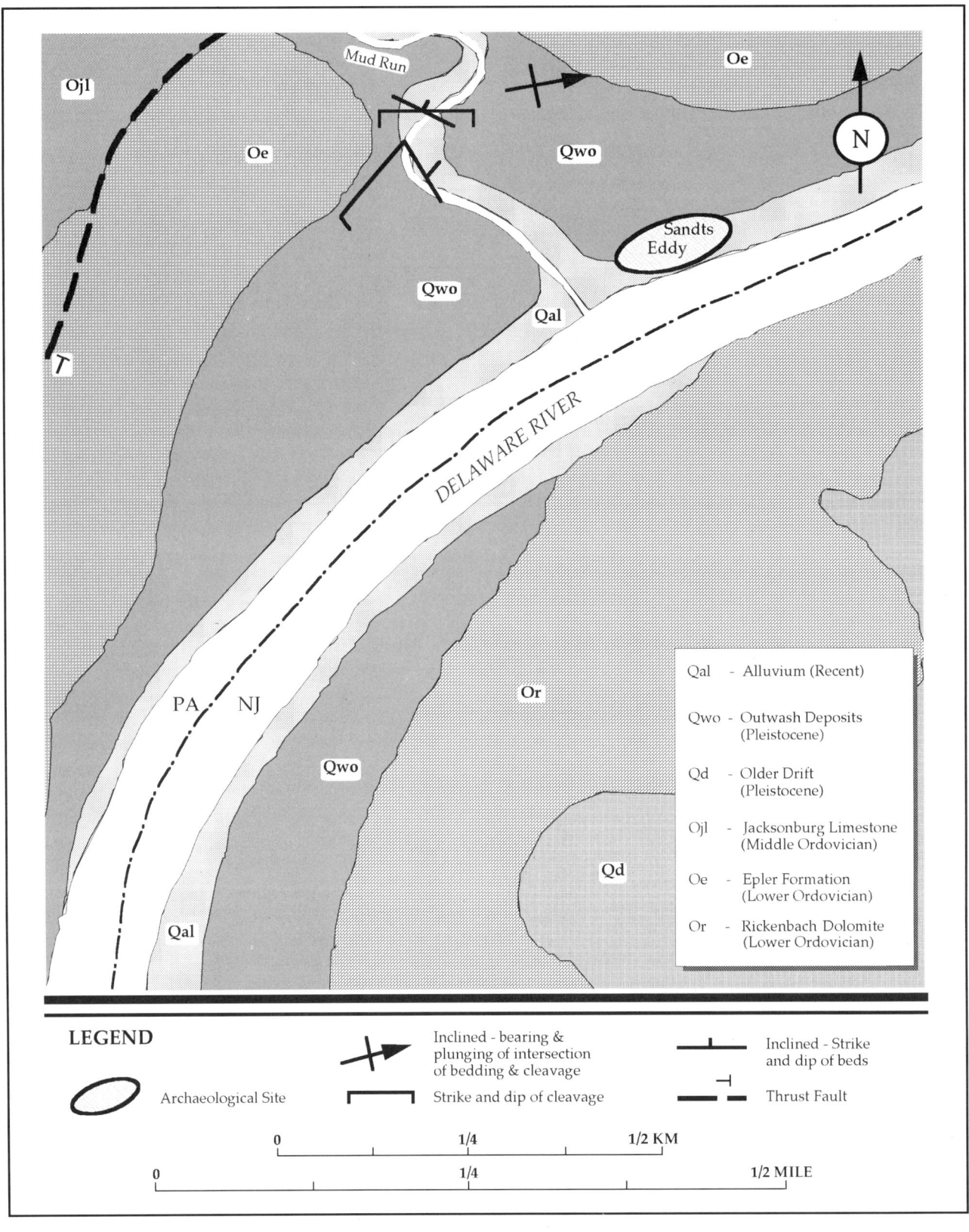

Figure 13.3. Geomorphic map of Sandts Eddy site (36Nm12), Middle Delaware Valley, Pennsylvania and New Jersey (modified after Davis et al. 1967). Note: T-1 is not formally mapped but is most accurately placed on the distal margins of the area mapped "Qal" (see text).

Figure 13.4. Morphology of the T-1 terrace. Note effects of historic landscaping and irregular surface contours. T-0 (partially visible) runs along left side of photo (see treeline). Photo by C. Bergman.

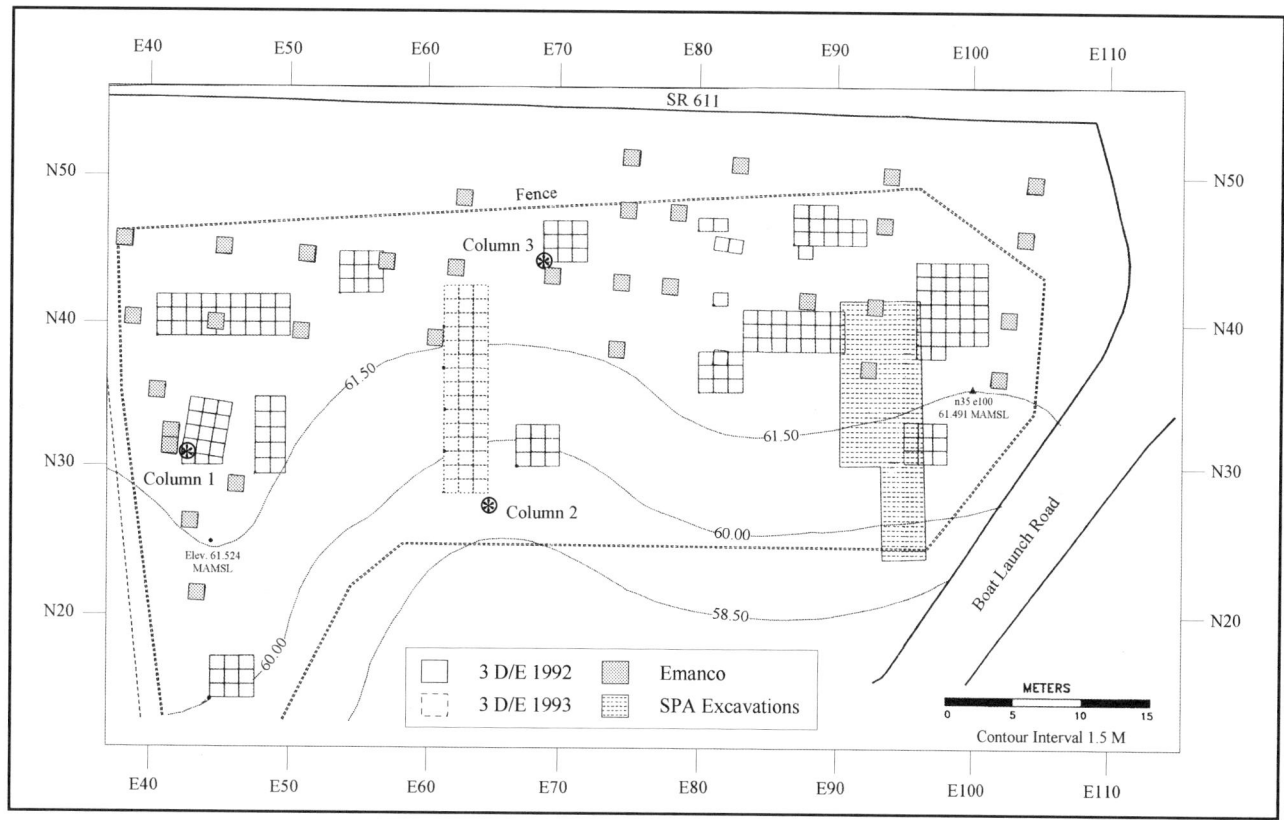

Figure 13.5. Plan of Sandts Eddy site excavations with contours of site. Note locations of soil-sediment columns 1, 2, and 3.

"... a deposit of alluvial sand containing red soil veins ... to an undetermined depth" (Fehr et al. 1971:41); probes extended to 1.0 m. Artifacts within that deposit were considered to be intrusive. These red soil veins are subsoil lamellae, or subhorizontal distributions of iron, silica, and clay enriched fine sands of depositional and/or pedogenic origin (see discussion below).

Follow-up Phase II cultural resources excavations were undertaken between 1985 and 1989 (Figure 13.5, "Emanco excavations"). Excavation units covered the northern and western portion of the site with a series of aligned 1-by-1-m tests (Weed et al. 1990:55-106). The study developed a more formal stratigraphy, identifying five master (I–V) and five subordinate lithological units. Systematic subsurface excavation did not extend beneath 1.0–1.5 m, although isolated borings reached in excess of 3 m. Strata I and II were historic, and Stratum III was interdigitated with the uppermost prehistoric (i.e., Woodland) deposits. This general sequence (Units I–III) was contained in the uppermost 50 cm across the site (Weed et al. 1990:Figure 20). Stratum IV (0.7–1.2 m below surface) offsets the historic and aboriginal occupations and was considered an A/B2 soil horizon, presumably maintaining organic and humic properties of the original ground cover; it contains the subsoil lamellae. Stratum V (> 1.15 m below surface) is the basal olive sand and contained no artifacts.

Weed et al. (1990:Figure 21) report peak prehistoric activity levels in Strata III and IV. Observations purportedly confirmed a stable surface in Stratum IV and disruption to the uppermost prehistoric deposits (Stratum III) early in the Historic Period. The interface between Strata III and IV marked the Archaic to Woodland transition; dated features in Stratum IV are of Terminal Archaic and Late Archaic age (3,970 ± 80 yr B.P. and 3,200 ± 90 yr B.P.; Weed et al. 1990, pp. 91 and 95).

Table 13.1 correlates the stratigraphic observations presented in the Phase I and Phase II reports (Fehr et al.

Table 13.1. Stratigraphic Correlations for Three Phases of Excavation at Sandts Eddy

Alluvial Unit (Phase III)[a]	Phase I Sequence[b]		Phase II Sequence[c]		Cultural Component/ Time Range
	Depth (cm)	Stratigraphy	Depth (cm)	Stratigraphy	
1	0-20	Zone 1 – "plow zone"	0-5	I – "Mottled sod/recently disturbed"	Historic
	20-30	Zone 2 – "yellow sand"	5-25 (discontinuous)	II – Graded historic fill	
			25-50 (discontinuous)	IIIa,b – Historic/aboriginal midden filles (interdigitated)	
2/2a	30->100	Zone 3 – "alluvial sand with red soil veins"	50-70	IIIc,d – Truncated aboriginal midden fill	Mixed Woodland
			70-90	IV – "A/B2; silty sands" (Eroded Bw soil)	Transitional/ Late Archaic
			90-115	IVa – "B2a silty sands with reddish bands" (Fining upward alluvium with lamellae)	Late Archaic
3	N/A	N/A	>115	V – "Basal olive sands" (massively bedded alluvium)	Early-Middle Archaic
4	N/A	N/A	N/A	N/A	Early Archaic? Paleoindian?

[a]This chapter.
[b]Fehr et al. (1971).
[c]Weed et al. (1990).

1971; Weed et al. 1990). Column 1 indexes the observations with the alluvial units designated in the present study. Descriptive terms and stratum nomenclature used by authors of the previous studies are cited in quotations. The earlier studies did not identify the potential for preservation for deeply stratified Early Holocene archaeological deposits, because excavation did not exceed 1.2 m. Thus no cultural assignments predating the Late Archaic period were established.

The present excavations extended approximately 2–3 m deeper than the previous efforts. Geoarchaeological studies of composite sections extending to more than 4 m were performed between 1991 and 1993 in conjunction with Phase III excavations led by 3D Environmental Corporation (Bergman and Doershuk 1994). Figure 13.5 illustrates the extent of the excavations. Typically, Middle and Early Archaic assemblages were recovered in the central and lower portions of most excavation blocks (in Alluvial Unit 3). The present investigations extended and revised previous stratigraphic interpretations using the most widely accepted geological and archaeological stratigraphic taxonomies (ISG 1997; NASCN 1983).

FIELD AND RESEARCH METHODS

Field Work

Four principal alluvial units were recognized across the site. Stratigraphic nomenclature was standardized by field spits originally designated by Roman numerals (I–XX; based on excavation levels) (Bergman and Doershuk 1994; Bergman et al. 1998). These were eventually correlated with alluvial units (Figure 13.6). Soil horizons were recorded and described as well. Several cycles of soil formation were recognized, especially in the rich Early to Middle Archaic horizons within Alluvial Unit 3.

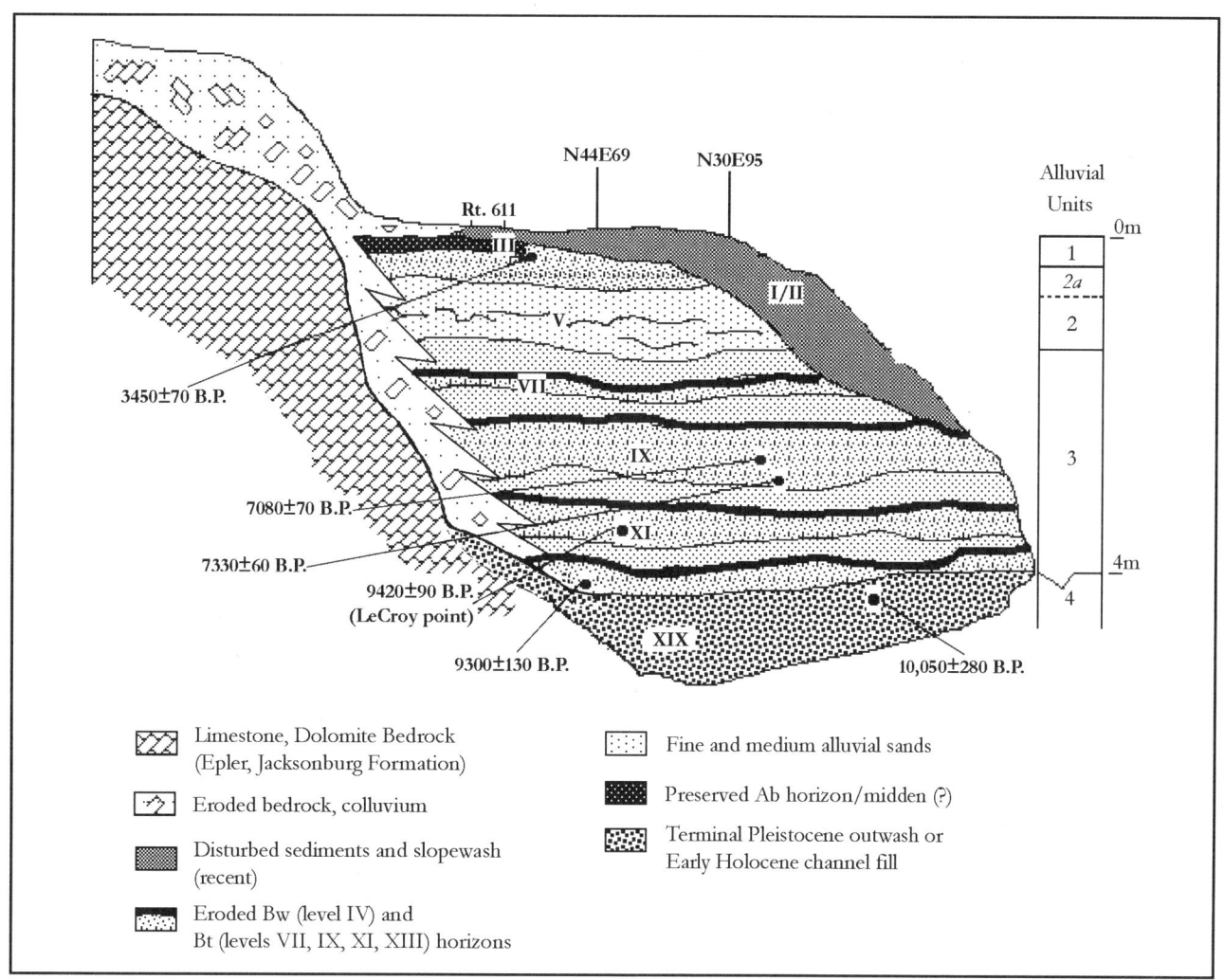

Figure 13.6. Sandts Eddy generalized stratigraphic profile.

Subparallel horizonation of multiple but discrete lateral accretion facies within this unit were segregated by degree of weathering.

Sampling and mapping involved procurement of grain-size samples to isolate depositional variability; selection of Bw and Bt horizons for geochemical and micromorphological specimens; and sampling of soil humates in Ab, AB horizons and charcoal and/or soil humates from cultural features for radiometric dating. Features and bulk organic sediments were also taken from Bw or Bt horizons.

Laboratory Analysis

Granulometry and geochemical testing were performed on three complete stratigraphic columns. These correspond to the most complete sequence at the highest site elevations (N44 E69); a column on the lowest portion of the landform (N27 E65); and a profile along the western site margin (N31 E43) (Figure 13.5).

Composite granulometry was run on sediment samples from these sequences using a modification of the hydrometer method for the silt and clay fractions (peptizing in a 5 percent hexametaphosphate solution), followed by wet-sieve calibration of the sand fraction at 1/2 Ø intervals (Gee and Bauder 1986; Head 1980). Parameters of sorting (So), skewness (Sk), and kurtosis (Kg) were calculated using the method of moments (after Friedman and Sanders 1978; also Folk 1974).

A battery of quantitative geochemical tests were applied to soil horizons to obtain signatures of both weathering on the T-1 and evidence for human occupation in the form of disaggregated cultural residues. This was most apparent for the lower Archaic components that were preserved in subtle AB or even Bw and Bt horizons. The elements tested to identify weathering and anthropogenic additions to the profile included calcium (Ca), magnesium (Mg), potassium (K), and phosphorous (P). The most common cultural residues isolated by these ion tests are bone, wood ash, excreta, and animal meat and tubers (Cook and Heizer 1965; Anderson and Schuldenrein 1985; Kolb et al. 1990; Schuldenrein 1995). To examine the degree of weathering and oxidation/reduction in the various different sola (here Bw and Bt), relative concentrations of mobile iron (Fe) and manganese (Mn) were measured along with organic matter (OM) and pH. Covarying trends can help to determine if vertical changes in the profile are attributable to soil-forming processes, human inputs into the sediments, or combinations of pedogenic and anthropogenic transformations to the matrix.

Micromorphology was performed on subsoil lamellae or red bands preserved within Stratum V at N39 E41. This is possibly the most precise technique for charting weathering patterns in a vertical sequence (Courty et al. 1989). The objective was to determine whether or not lamellar formation is a depositional or pedogenic phenomenon. The method also helped determine stream energy thresholds and the depositional mechanisms most responsible for the propagation of lamellae in the changing stream environment (overbanking vs. lateral accretion).

Composite granulometry and geochemistry was performed at the Soils and Physical Geography Laboratory at the University of Wisconsin, Milwaukee. Micromorphological analyses were undertaken by Dr. Paul Goldberg, Boston University.

GENERAL STRATIGRAPHY

Four separate alluvial units were recognized in the composite stratigraphy, differentiated largely by structural and textural changes in unit lithology. Soils also helped to define separation between alluvial units. In Alluvial Units 2 and 3, where successive soils (or sequaa) formed on successively accumulating stream deposits, the soil chronology was relative, informing on changes specific to the pedon or locus specific soil body. The boundaries of the alluvial units were ultimately defined on the basis of larger scale changes in sedimentary and hydrographic environments that were confirmed by sedimentological analysis.

The most prominent stratigraphic interface across the site was that separating the Early and Middle Archaic horizons from the Late Archaic through Woodland deposits. The former were linked to more active, higher-energy stream environments while the latter represented channel overbank sediments accumulated on an elevated floodplain above the channel floor. The Middle and Early Archaic sediments are associated with Alluvial Unit 3, whereas the later prehistoric to historic occupations were preserved in Alluvial Units 2a, 2, and 1. Key radiocarbon dates are depicted alongside stratigraphic units in Figure 13.6. A complete list of dates with proveniences is presented in Table 13.2.

Pedogenic development was also more pronounced in the coarser, near-channel sediments of Alluvial Unit 3. Thus the densest lower Archaic deposits were housed in rubefied (reddened), mineralized, and clay enriched soil horizons that formed over more massively bedded, poorly structured, and unweathered parent alluvium. Recurrent cycles of sand deposition overlain by stabilized, artifact-rich soil horizons characterized the basal 2–3 m (i.e., Early Holocene) component of the T-1 stratigraphic record. Detailed descriptions of the lithostratigraphic units follow below (Figure 13.6).

Table 13.2. Radiocarbon Dates from 36Nm12.

Stratum	Provienience	Sample Number	Conventional Radiocarbon Age	Calibrated Age Range cal. B.C.[a]
III / IV interface	bulk soil sample, N34 E42	Beta-50730	3450 ± 70 B.P	1939 (1743) 1532 B.C.
VII	charcoal, N35 E63, 58.799–58.699 m[b]	Beta-61773	7500 ± 170 B.P.	6650 (6392) 6013 B.C.
IX	Charcoal, N31.75 E97.45, 59.662 m	Beta-51500	7080 ± 70 B.P.	6155 (5984, 5944, 5924) 5797 B.C.
IX	carbonized hazel nuts, N36 E64, 58.188–58.148 m	Beta-61582 (CAMS-5834)	7330 ± 60	B.P. 6375 (6216, 6167, 6164) 6029 B.C.
IX	charcoal, N27.14 E62.97, 58.374 m	Beta-61332	8450 ± 130	B.P. 7735 (7537) 7142 B.C.
XI	charcoal, N34 E45, 57.764 m	Beta-51501	9420 ± 90 B.P.	9137 (8721, 8708, 8691, 8659, 8650) 8345 B.C.
XI	bulk soil sample, N32 E44	Beta-53142	9300 ± 130 B.P.	9112 (8549, 8487, 8483) 8264 B.C.
XVIII	charcoal, N28.64 E64.41, 56.074 m	Beta-61413	10,150 ± 180 B.P.	10,837 (9944, 9914, 9797, 9764, 9751) 9247 B.C.
XIX	charcoal, N27 E63, 55.775–55.675 m	Beta-61744	10,050 ± 280 B.P.	10,873 (9604, 9547, 9518, 9491) 8747 B.C.

[a] Calibrated age range based on 2 sigma using CALIB 4.3 program (Stuiver et al. 1998). Value in parentheses is intercept of radiocarbon age with calibration curve.
[b] Elevation in meters above mean sea level.

Alluvial Unit 1

The uppermost 50 cm consist of admixtures of historic slopewash, mudflows, alluvium, and disaggregated historic, clearance, and construction debris. Deposits are thin on level surfaces but have accumulated as a minor debris fan along a north-south axis on the riverward slope toward the bankline (Figure 13.5). In places they have overridden the terrace edge and are eroding—by backwearing and headward retreat—into the river. They are unconformably underlain by the uppermost intact stream deposition, Alluvial Unit 2. Most landscaping occurred since the mid-19th century but was most extensive over the past 40 years.

Alluvial Unit 2 (2a)

Sediments consist of between 1.0 and 1.5 m of silt loam alluvium. Disposition of the unit is generally level, but it thins riverward within 12–16 m of the landform crest (north-south); erosion has truncated the uppermost levels. The sediment matrices seal in the stabilized terrace landform abutting the toeslopes of the bedrock valley (Figure 13.3).

Laminar bedding structures confirm that stream sands accumulated as classic "top stratum" suspended load fines (Brakenridge 1988; Friedman and Sanders 1978), laid down by sustained overbanking coincident with the more stabilized channel postdating the Late Archaic. A uniform cambic (Bw) profile is pervasive but variably capped by a complex of organic, humic horizons. This paleosol has been offset as Subunit 2a in the composite section (Figure 13.6), because it has not been truncated as have most exposures. Optimal preservation is due to location well above the former and present floodlines. A single determination of 3450 ± 70 B.P. is consistent with the Late Archaic, confirmed by discrete vertical and laterally separated components for the Late Archaic through Woodland Periods. Thicknesses of individual cultural components from these time frames are on the order of tens of centimeters.

Underlying Subunit 2a is Unit 2, a series of moderately well sorted silty sands. Although they generally fine upward, the lithologic body is typically one- to two-size grades coarser than that of 2a. These parent sands were laid down in a somewhat higher-energy stream environment. Analysis of Alluvial Unit 2 sediments suggests recurrent short-term soil and sedimentation hemicycles punctuated by lamellae or laterally thin red bands (see subsequent section). Formation of the lamellae may be related to stabilization of alluvial surfaces after a single or series of episodic floods ceased laying down sediment. Because fining upward is the active depositional process, fine sands and silts settled out in response to flood recession. Sands are displaced by silts and clays in the upper levels of a given deposition. When the deposition stopped and surfaces stabilized, soil formation commenced, initially by disaggregation of A horizons and subsequently by translocation of iron and aluminum to form reddish Bw horizons of the characteristic cambic profiles. After a period of nondeposition and rooting of the humic horizon, a new hemicycle of alluviation was initiated.

Cultural materials within Alluvial Unit 2 were contained exclusively within the 2a and the uppermost 10 cm of Unit 2, indicative of thin Transitional and Late Archaic surfaces near the top of the unit. A higher-energy floodplain below that may have sustained marginal vegetation but could also have been removed from the axis of human activity.

Alluvial Unit 3

Considerably coarser sediments are characteristic of Alluvial Unit 3. Unit 3 sediments beneath the interface average 20 percent higher sand content and proportionately lower silt and clay concentrations. Within Unit 3, four to seven discrete weathering horizons were recognized, depending on location on the T-1. Horizons were considered soils following the criteria of the U.S. Department of Agriculture (1975:25) in which "lamellae 1 cm or more thick and enough of them to make a total thickness of something like 15 cm or more, should be present to constitute an argillic horizon in a sand or loamy sand." Birkeland (1999) proposes that the lamellae tend to merge into Bt horizons with time. Based on thicknesses of lamellae, soil horizons were separated by moderately sorted bodies of medium alluvial sands. The sola may be considered either strong Bw (cambic) or weak Bt (argillic) horizons with thicknesses varying between 10 and 60 cm. The considerably deeper red colors of the soils as well as their firm, prismatic peds are evidence of sustained soil formation, rather than the shorter-term stabilization inferred by thinner lamellae. Examination of these thicker B horizons disclosed abundant tubular infillings and clustered pore networks signifying that vegetation and root matting stabilized during longer term pedogenesis.

Alluvial Unit 3 accounts for between 1.2 and 2.0 m of depth in the composite section (Figure 13.6). Along the north-south axis of the T-1, basal field levels were tightly clustered in the northern (higher) elevations and leveled off and thickened in a riverward direction. This appears to be a function of progressive southern and eastern migration of the channel, accompanied by more tractive load sedimentation near the center of the active channel. The morphology of these basal buried surfaces mimicked that of the eroded underlying Pleistocene-Holocene gravels (Alluvial Unit 4) that rose up to 0.5 m higher in the north than in the south. These also sloped 5-10° to the southeast and may have formed a more logistically accessible and better-drained activity locus for the earliest (i.e., Early Archaic) inhabitants of the site. However, it remains difficult to trace some of the lowermost southern field levels to the north and to determine whether or not they are truncated or otherwise incorporated within discontinuous and localized alluvial facies.

Alluvial Unit 4

Alluvial Unit 4 incorporates the uppermost Pleistocene-Holocene contact gravels. Only 0.2 m of exposure was observed. Invariably these consist of pebble- to boulder-sized clasts of exogenous origin. They include a variety of rounded and abraded erratics. Matrices typically feature interdigitations of coarse sandy pockets and abundant ferro-manganese staining. It is unclear whether or not the top of Unit 4 is defined by the base of gravel entrained sands of Unit 3 or if the transition to Unit 3 is more gradational, depending on lateral position on the T-1 landform.

Summary of Alluvial Sequence

Alluvial Units 1–4 were recorded in most sections at 36Nm12 but varied in facies, thickness, and degree of soil development depending on position on the landform. The facies chronology of the alluvial sequence is summarized as follows (top to bottom): (1) interdigitated slopewash, alluvium, and historic fill (Recent); (2/2a) overbank vertical accretion deposits with lamellae and buried A horizons (late Holocene); (3) cyclic sequences of upward fining coarse sands capped by Bt horizons (early Holocene); (4) high-energy stream sands and gravels (terminal Pleistocene).

Changes in unit lithology signaled, first and foremost, threshold shifts in stream energy, competence, discharge, and hydrography of the trunk stream (the Delaware). By extension, causative mechanisms in the overhaul of the drainage may be attributable to climatic mechanisms (i.e., forcing) and attendant paleo-ecological developments including vegetation succession.

Stream environments emerged as torrential late glacial discharges (Alluvial Unit 4). Subsequent evolution was characterized by successive migrations of a sinuous channel (Alluvial Unit 3), which eventually stabilized in the present Delaware stream bed where vertical accretion of fines was responsible for incremental alluviation (Alluvial Unit 2). The complex historic and recent sediment redistributions comprise the only body of nonalluvial sediment presently sealing in the site (Unit 1).

SEDIMENTOLOGY AND GEOCHEMISTRY

Preservation conditions at the site were such that it was often difficult to differentiate archaeological sediments from soil horizons. As discussed, the latter were either Bw or Bt horizons that imparted a rubefied chroma to the matrix, often obscuring features or archaeosedimentary boundaries. Textural and geochemical testing often helped to distinguish culturally imparted sediment signatures.

Results of the soil and sediment analysis for three representative stratigraphic columns are illustrated in Figures 13.7, 13.8, and 13.9 (see Figure 13.5 for grid location). Columns 1 (Figure 13.7; N31 E43) and 3 (Figure 13.9; N27 E65) are at the western and northernmost portions of the site and preserve the most complete and intact sediment packages for the T-1; Alluvial Units 1–3 are represented. Column 2 is at the southern margins of the excavation and preserves the thickest studied Early Holocene stratigraphy (10,000–6000 B.P.) in the Delaware Valley to date.

The columns depict changes in the stratigraphy for four discrete data sets. The first presents the depositional sequence and soil horizons grouped by alluvial unit, field levels, and radiometric dates. Immediately to the right, archaeological components are identified. The second set of data shows the particle size distribution (granulometry; by weight) for four fractions: coarse sands (< 1 Ø), medium-fine sands (< 4 Ø), silts (< 8 Ø), and clays (> 8 Ø). Grain size parameters of mean size (Mz), sorting (So), skewness (Sk), and kurtosis (Kg) are also presented. The third data set includes geochemical plots for phosphorous (P), magnesium (Mg), potassium (K), iron (Fe), manganese (Mn), pH, organic matter (OM), and calcium carbonate ($CaCO_3$).

Beginning with the upper profiles–Columns 1 and 3 (Figures 13.7 and 13.9)–the upper 40 cm incorporate a series of historic and subrecent lanscaping layers. At the base (Alluvial Unit 1/Level III) isolated features and an irregular Woodland Period horizon were identified. Underlying the Woodland is the uppermost buried soil, a cambic (Bw) horizon that preserves Transitional and Late Archaic cultural materials. The soil formed on a series of overbank deposits in which lamellae formed. These overlie the thickest floodplain deposits—Alluvial Unit 3—that extend to the base of the sequence. Three soils—with Bt (i.e., Level IX) and Bw sola—were recognized in the field; degree of pedogenesis is variable by elevation and slope position. Columns 1 and 3 preserve the best-developed soils.

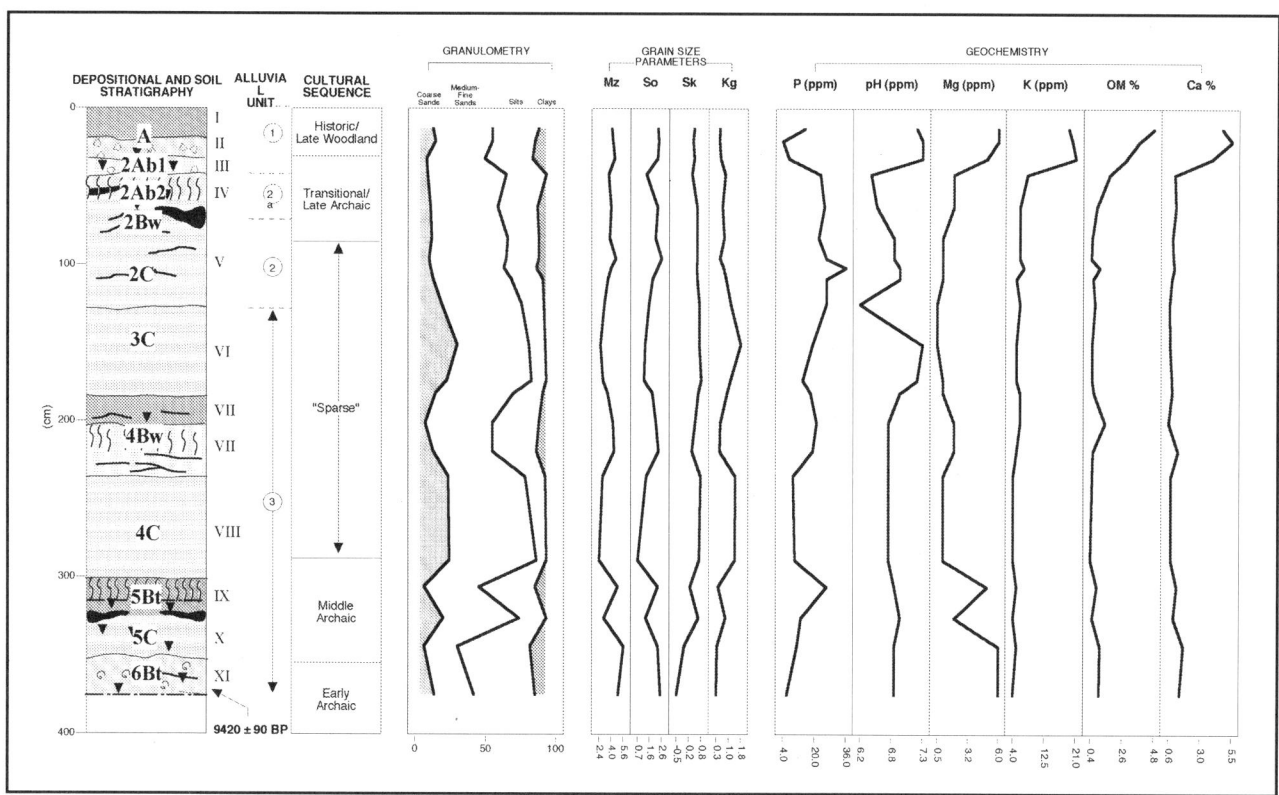

Figure 13.7. Stratigraphy and soil-sediment analysis: N31 E43, Column 1.

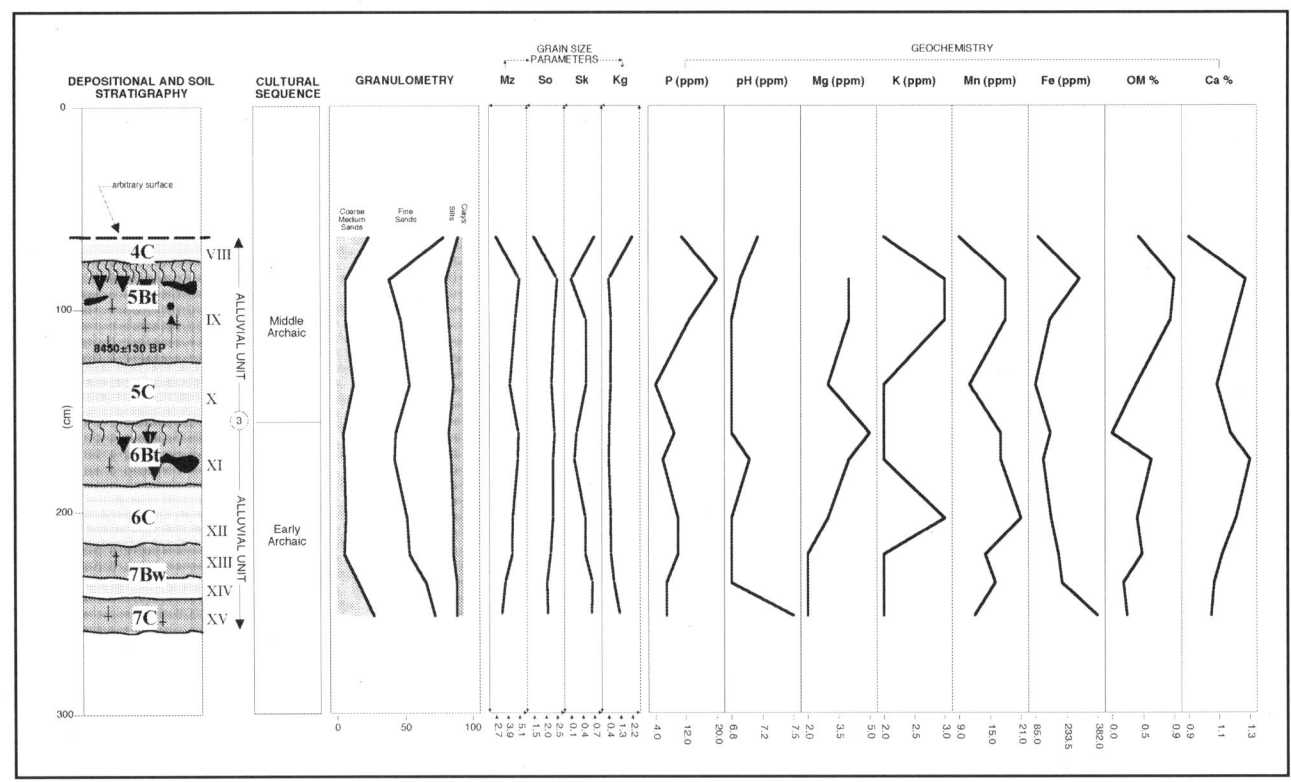

Figure 13.8. Stratigraphy and soil-sediment analysis: N27 E65, Column 2.

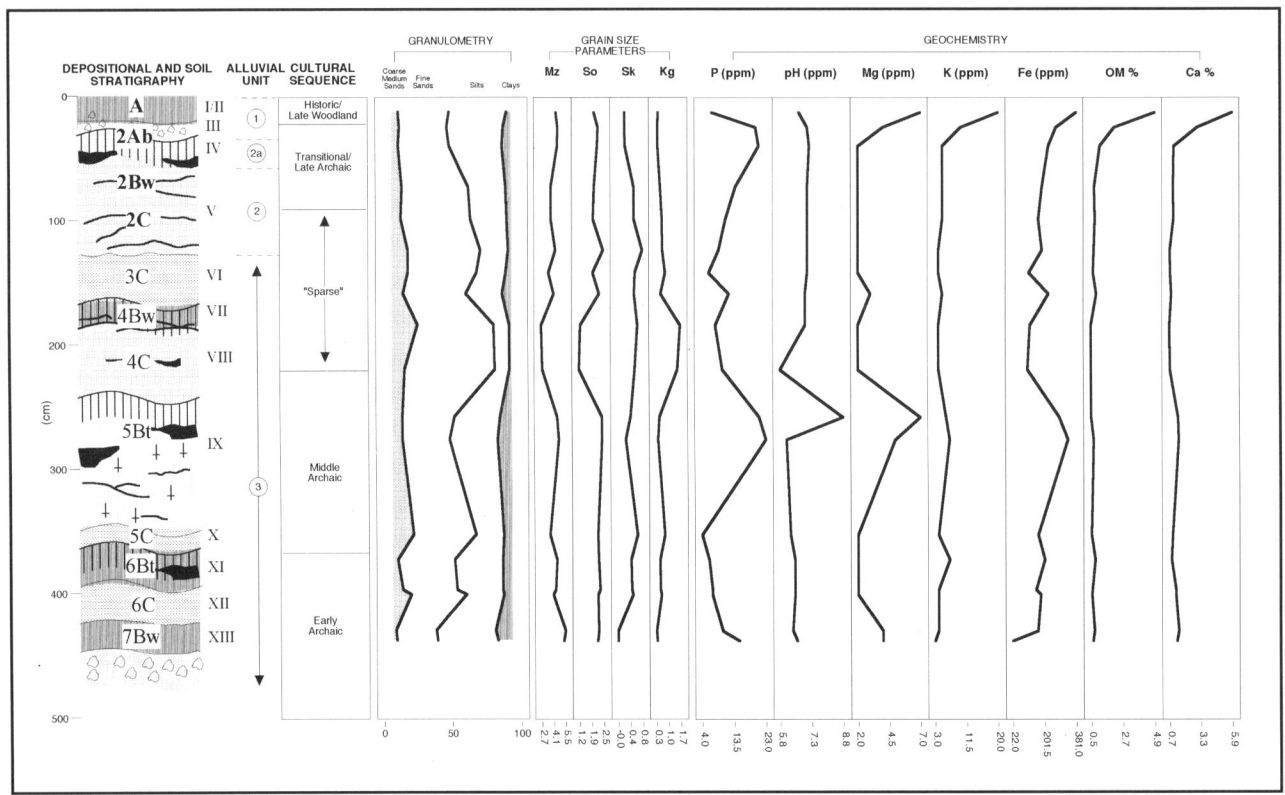

Figure 13.9. Stratigraphy and soil-sediment analysis: N44 E69, Column 3.

The most striking geochemical trends are the peak frequencies for most elements in the upper strata. They reflect contemporary exposure of surface horizons to fertilizer and soil enhancing chemicals as well as enrichment of reworked surface debris. Weathering and prehistoric chemical inputs can be tracked only beneath field Stratum III. The degree of modification to older sediments was striking. One to 2 m of culturally sterile sediment (designated "sparse" in the cultural sequence columns of Figures 13.7 and 13.9) spanned upper Alluvial Unit 2 to near the base of Alluvial Unit 3 and separated the Middle from Late Archaic depositions. It was therefore possible to filter out anthropogenic inputs from the sediments by monitoring changing concentrations of the individual chemical indicators for the culturally sterile sediments and soils. Pedogenic inputs are also minimal, because only a single, thin soil horizon is bracketed within the accumulation. Within this culturally sterile alluvium, concentrations of most elements are relatively uniform. pH increases and then diminishes and stabilizes in the central portion. There are minor Mg, K, OM, and Ca bulges.

More typically, however, there is minimal geochemical variability within the sterile alluvium. For the overlying and underlying cultural horizons, vertical trends are more instructive. Phosphorous (P) is the most diagnostic anthropogenic element. In Column 3 (Figure 13.9) and to a lesser degree in Column 1 (Figure 13.7), peak P concentrations characterize overlying Transitional Archaic (Level IV) and underlying Middle Archaic (Level IX) sediments. Increasing P values are also noted for the Early Archaic horizon (Level XI). For Level IX, the surge in P concentration is paralleled by high values for Mg and pH as well as mobile Fe. Increased Fe is due to mobilization of iron oxides during weathering, a measure of soil formation. Although similar trends in P and Mg in Columns 2 and 3 (Figures 13.8 and 13.9) may be interpreted as additional evidence, P is not typically a measure of soil formation, and Mg is linked to more humid weathering environments. Thus anthropogenic origins are the most likely explanations for these trends. Magnesium increases in Level XI in Column 1 are noteworthy, because these mimic values for the anthropogenically enhanced near-surface sediments (cf. Levels I/II and XI). Both Mg and P are common by-products of animal bone disaggregation, wood ash, and food preparation by burning (Anderson and Schuldenrein 1985; Butzer 1982; Eidt 1977; Schuldenrein 1995), activities that could have been practiced by the Archaic occupants on the site.

A departure from the relatively weak levels of human activity in the archaeological sediments is noted in the deepest and oldest sequence, Column 2. Both Middle Archaic Level IX and Early Archaic Level XI (Figure 13.8) peak for all parameters including P, K, OM, Ca, and Mn and Fe. The latter two are diagnostic of illuviation and oxidation-reduction patterns in temperate and moist environments. The converging trends of both weathering and anthropogenic indicators may signify occupation on well-developed soils at Sandts Eddy, at a time when surfaces were stable.

Although the geochemical results underscored probable human impacts to the substrate, the variability in grain sizes reflect changing sedimentation patterns (cf. Figures 13.7, 13.8, and 13.9). As noted earlier, the principal break in the T-1 sequence is the passage from a lateral accretion to overbanking environment. This is most evident at the transition from Alluvial Unit 2 to Alluvial Unit 3. Mean grain sizes in Unit 2 are in the very fine sand grade. For Unit 3 they are considerably more irregular, because of the sorting characteristics of a lateral accretion regime, termination of the alluvial cycle, and subsequent surface stabilization (and pedogenesis).

Fining upward sequences are recorded in the cyclic depositions of Alluvial Unit 3 capped by the soils of Levels XIII, IX, and VII. Although alluvial fining can be masked by soil-forming clays, inspection of the grain-size distributions show that for all of the affected levels, increases in illuvial fines correspond with more dramatic additions of silts and fine sands; the latter were apparently laid down by floodwaters. In Alluvial Unit 2, there is a unimodal increase in fine sands, a function of a vertical accretion regime that stabilized after the T-1 was built up to 4 to 6 m levels.

There is considerable variability in mean grain sizes, with the coarsest sediments linked to the base of nonpedogenically altered lateral accretion sets. These trends are, not surprisingly, paralleled by the optimal size sorting and weakly skewed distributions. The paleosols in the basal horizons (Levels IX and XI) are characterized by coarse skewing, because illuvial clays were translocated onto medium-to-fine sandy parent materials.

For the earlier Archaic levels, occupation was not sufficiently dense to impart a signature on the grain-size distributions. This is in contrast with loci of intense human activity (i.e., middens or archaeological mounds) where extensive trampling tends to produce grain-size curves that diverge broadly from the normal distribution. At Sandts Eddy departures from fluvial and overbank sedimentation are largely attributable to soil formation. Column 1 (Figure 13.7), for example, illustrates that in the 2-m accumulation of "sparse" cultural activity, the only break in a relatively uniform build-up of medium-fine sands was the interval of soil formation correlative with Level VII.

Figure 13.10 highlights patterned differences in grain-size distributions for lateral accretion (Alluvial Unit 3) and overbanking deposits (Alluvial Unit 2). The triaxial plot shows that overbank sediments cluster in a narrow band featuring 60–80 percent sands and 15–30 percent silts. On average, Alluvial Unit 2 contains 20 percent higher sands than Alluvial Unit 3, principally in the fine-sand grade. Alluvial Unit 3 is a more diverse sediment population, with sand concentrations varying between 40 and 90 percent. Alluvial Unit 3 sediments do not even overlap with those of the vertical accretion sediments. They extend to coarser and finer grades. Alluvial Unit 1 features an unexpectedly heterogeneous population because of its origin as a historic fill. The broad distributions within Alluvial Unit 3 reflect the variable energy environments associated with a meandering stream and the duration of this cycle of sedimentation.

MICROMORPHOLOGY AND THE GENESIS OF THE LAMELLAE

The origins of the lamellae were a key question, because interpretations of their origins were needed to identify stable surfaces (if origins were pedogenic) or flooding patterns (if sedimentary). At Sandts Eddy, the thicker, more extensive mineralized horizons (within Unit 3) were considered to be pedogenic, but the variability in the morphology, density, and horizonation of the thinner rubefied bands in Units 2 and 1 suggested that their origins were more equivocal. Extensive sedimentological studies have been undertaken to establish the depositional versus pedogenic origins of lamellae (Courty et al. 1989; Dijkerman et al. 1967; Foss and Segovia 1984; Larsen and Schuldenrein 1990; Schuldenrein 1988; Torrent et al. 1980). Although many argue for pedogenic mechanisms, evidence is also strong that conditions promoting lamellar development are sedimentological and derive from sorting related to bedding. In the southeastern Piedmont, Larsen and Schuldenrein (1990) have linked lamellar thicknesses and clustering patterns to fining upward alluvial sequences. They also chronicled "peak climatic conditions" favorable to lamellar formation during the Middle Archaic to Late Archaic transition, ca. 6000 B.P. Most recently Rawling (2000) has proposed that lamellar genesis should be considered on a site-specific basis because of the diverse depositional and soil-forming environments in which they evolve; either or both processes may be at work in a given geological setting.

Locally lamellae have been observed in nearly all T-1 substrate of the Middle and Lower Delaware Valley, ever since their presence was noted the length of the drainage by Kinsey (1972). Increased thicknesses and deepest accumulations of lamellae have been correlated with downstream fining of alluvial fill along the Middle and Lower Delaware (Schuldenrein 1994; Schuldenrein et al. 1991).

At Sandts Eddy, laterally continuous lamellae are typically preserved in 1- to 2.5-m-thick fine sands of Alluvial Unit 2. Underlying sediments (Unit 3), grading thicker by 0.5–1 Ø, have considerably more diffuse lamellae. Alluvial Unit 2 is correlated with a Terminal Archaic occupation and an overbanking depositional regime that extended to the Late Archaic horizons. Underlying sediments are coarser, of Early Holocene age, and attributed to lateral accretion. To explore the conditions of lamellar evolution, a series of specimens was collected for micromorphological study. The detailed micromorphological study is presented in Bergman and Doershuk (1994: Appendix C). In this summary key observations are presented.

Five samples were collected from location N39 E41 (Figure 13.5), within a vertical column of 1.2 m. Specimens from Alluvial Units 1, 2, and 3 and field Levels II, III, IV, V, and VI were prepared on slides and examined. Descriptions follow the terminology of Bullock et al. (1985) and interpretations were provided by Dr. Paul Goldberg and the author.

All of the samples display the same basic quartzitic composition, consistent with their common fluvial source and a relatively short interval of deposition. Moreover, all exhibit the same type and style of textural pedofeatures, which take the form of dark reddish brown dusty clay void coatings and fillings that range in thickness from ~15–30 μm. These types of coatings are generally localized around individual voids, and are

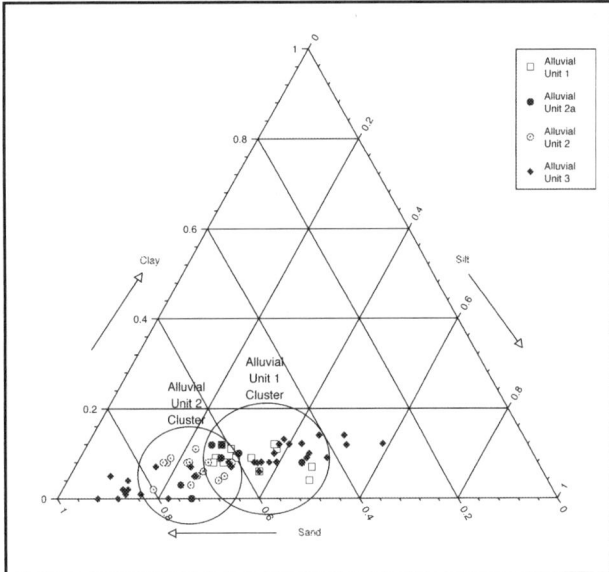

Figure 13.10. Triaxial plot (sand, silt, clay) of particle-size distributions by Alluvial Unit, Sandts Eddy site.

typical for Bt horizons. It would not be possible to distinguish these from clay coatings found in typical Bt horizons.

Whereas all samples displayed clay coatings, only the deepest specimen (Alluvial Unit 3; Level VII) exhibited fine silty textural accumulations. It had strong indications of soil illuviation, indicated by high concentrations of clays in voids of a fine sand and silt parent sediment (Figure 13.11). The same pattern of illuviation was expressed in the upper sediment (Alluvial Unit 2; Level V), although clays coatings were not as prominent (Figure 13.12). Both sediment types are commonly associated with soils evolving in environments exposed to periodic flooding (FitzPatrick 1993), as is the case here. Soil formation is indicated for both samples and within both alluvial units. The reasons for the singular occurrence of fine sandy matrix in Sample VII are not clear, particularly because this unit seems to be characterized by sustained soil formation, rather than the shorter-term stabilization more typically registered by lamellae that occur above this sample. This may be characteristic of "converging lamellae," as discussed earlier. It is also possible that the type of channel flow or position in the fluvial landscape was different during the time of accumulation of this silty material.

Perhaps the most significant observation generated by the thin section analysis is that characteristics for Bt horizonation (clay illuviation) are preserved in soils that are relatively younger (i.e., < 6000 B.P.) than those typically associated with argillic profiles (Birkeland 1999). These observations bolster interpretations for cyclic sed-

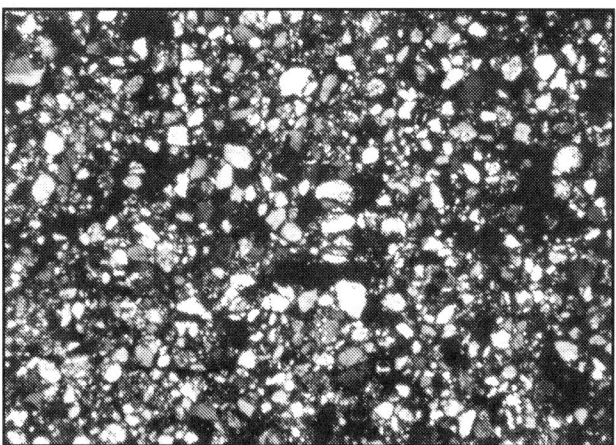

Figure 13.12. Photomicrograph of lamella sample, Alluvial Unit 2. Silty sand with broad, diffuse band of dark clay infilling (center). Note large horizontal voids (middle). XPL. Field of view ~6.4 mm. Photo by P. Goldberg.

imentation of fining upward alluvium punctuated by episodes of intense soil formation of relatively brief duration.

DEPOSITIONAL AND SOIL-FORMING ENVIRONMENTS

Table 13.3 integrates the stratigraphy and archaeological sequences at Sandts Eddy with reconstructed sedimentary and soil environments. The primary alluvial units (Column 1) are tracked transverse to the T-1 by average depths, north to south (Columns 2 and 3). The baseline exposure used for the stratigraphy spanned an exposure from N42 to N28 along E65, measuring from the highest to lowest portions of the landform respectively (refer to Figure 13.5). Because the subsurface contours did not necessarily conform to those of the present surface, field Levels V–VIII are truncated along the north-south axis.

The thickest sediment accumulation of more than 3.5 m is in Alluvial Unit 3. In addition to containing 14 field levels, 4 soil sequaa have been identified (Column 5). Column 6 indexes the absolute site chronology with radiocarbon dates. Eight of these bracket more than 3000 years of alluvial history and prehistoric occupation between 10,500 and 7000 B.P.

Column 7 identifies the cultural chronology based on the lithic inventory (Bergman and Doershuk 1994; Bergman et al. 1998). Alluvial Unit 2 is culturally sterile, perhaps reflecting the passage from the actively migrating to the entrenched, overbanking channel. Occupations in Alluvial Unit 3 are discreet but sparse, be-

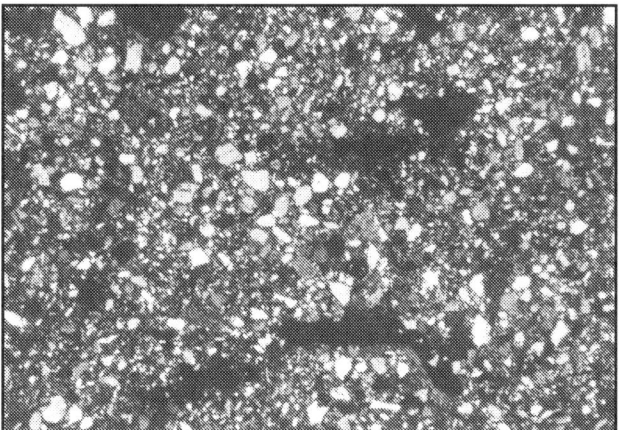

Figure 13.11. Photomicrograph of lamella sample, Alluvial Unit 3. Compact silty sands exhibit reddish-brown clay coatings up to 30 µ thick (upper section). Note elongated void coated with silt and some clay (lower section). XPL. Field of view ~6.4 mm. Photo by P. Goldberg.

Table 13.3. Integrated Cultural and Sedimentary Stratigraphy, Sandts Eddy

Unit	North Depth (cm)	South Depth (cm)	Field Levels	Soil Stratigraphy	Radiocarbon Dates (B.P.)	Cultural Chronology	Sedimentary and Soil Environments
1	0-20	0-50	I, II, III	A, Ap, 2Ab1	N/A	Historic, Mixed Woodland	Destabilization of late Holocene paleosol; slopewash interdigitated with alluvium and construction fill; debris fan formed at southern slope apex; headwater retreat at T-1/T-0 bankline
2a	20-45	n/A	IV	2Ab2-2Bw	3450 ± 70	Mixed Woodland, Transitional Archaic, Late Archaie	Lower portion of late prehistoric midden or series of surfaces; partially intact (Woodland or Transitional Archaic); cambic (Bw) soil horizon; vertical accretion (overbank) sedimentary environment; stabilized channel
2	45-135	50-55	V	2C	N/A	None	Basal sediments mark initiation of low energy, fining upward sedimentation; subsoil lamellae identify soil forming environments; latter stages of migrating channel
3	135-490	55-350	VI-XIX	3C-4Bw-4C-5C-6Bt-6C-7Bw-7C-8Bt-8C	7599±170 (VII) 7080±70 (IX) 7330±60 (IX) 8450±130 (IX) 9420±90 (XI) 9300±130 (XIII) 10,150±180 (XVII) 10,500±280 (XIX)	Middle Archaic, Early Archaic, Paleoindian(?)	Episodic high energy sedimentation; mixed load deposition of actively migrating and incising stream; depositional intervals punctuated by soil formation; oxidized, weak to moderately developed argillic horizons; "A-horizons' truncated; discrete prehistoric loci
4	>490	>350	XX	9C	N/A	None(?)	Meltwater and torrential channel sedimentation as Delaware channel geometry and hydrographic systems emerge during Pleistocene to Holocene transition

coming denser to the top. Column 8 synthesizes key elements of the depositional and soil-forming chronology.

Figure 13.13 is the composite site-wide stratigraphy, west-east transect along the center of the site (N30 axis on Figure 13.5). The depression near E67 reflects historic recontouring. The section discloses at least eight separate surfaces either capped by paleosols or truncated by unconformities. The complexities of the sequence are heightened by the "compressed" soil successions. Given that at least 12,000 years of landscape evolution and human occupation are preserved, the duration of soil-forming intervals is on the order of at least 1500 years. As discussed, some of the weathered sola are considered Bt horizons because of converging lamellae. The rate of soil development is rapid, a controversial hypothesis, because it is often held that the diagnostic properties of argillic soils—clay skins, prismatic structures, red chromas—require 10,000 years to form at a minimum (see Birkeland 1999).

A possible explanation for this discrepancy lies in the developmental history of the T-1. As detailed in Table 13.3, each successive alluvial unit represents drastic alteration of the extant landscape. The floodplain may be

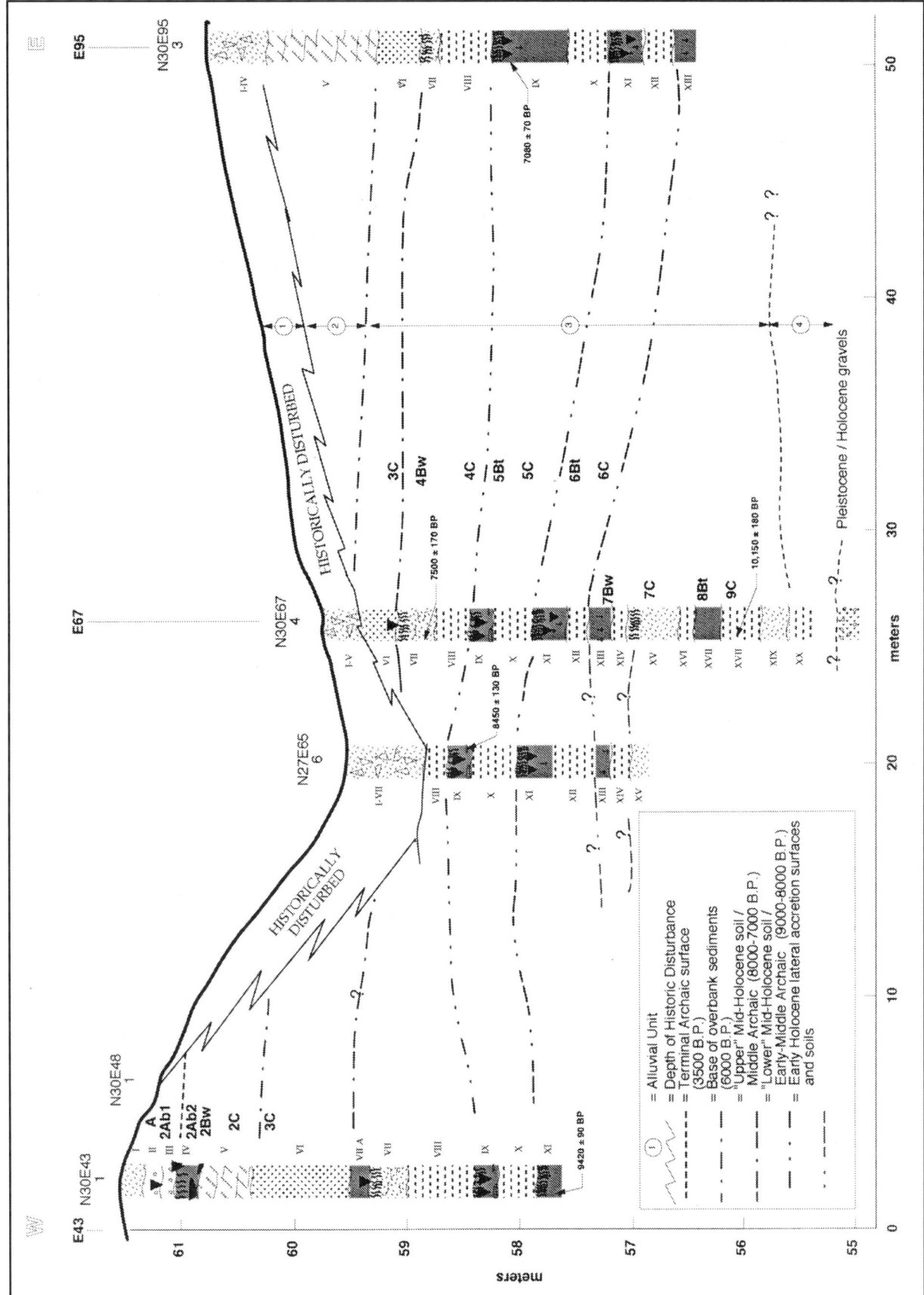

Figure 13.13. Composite Holocene stratigraphy, Sandts Eddy site: 36Nm12 (N30 axis).

visualized as a progressively stabilized setting since its postglacial emergence. The sequence is that of a "net" constructional landform over more than 10,000 years. More significant is the vector of construction that clearly proceeds from west to east, as the channel migrated across the Delaware. Incremental or even episodic migrations of the channel away from the earliest available land surfaces—those on the western flanks of the bottoms at Sandts Eddy—would have opened up those tracts to soil development, stabilization, and occupation. These conditions produced the lush landscapes favored by prehistoric populations. It has been estimated that under certain climatic and drainage conditions, argillic (Bt) horizons can form in Pennsylvania alluvium in as little as 2,000 years (Bilzi and Ciolkoscz 1977). The Sandts Eddy situation would appear to conform to such circumstances.

TAPHONOMY AND SITE FORMATION

To examine questions of archaeological sedimentation, taphonomy, and site formation, it is instructive to differentiate between Alluvial Unit 3 (Early and Middle Archaic) and Alluvial Units 2 and 2a (Late Archaic-Euroamerican contact). The transition from a dynamic fluvial discharge regime characterized by lateral accretion (Unit 3) to more quiescent overbanking (Units 2/2a) produced differential patterns of site maintenance and sediment storage. Also contributing to the general "preservation signature" is the sealing effect of soils that maintain the integrity but may blur the articulation of discrete archaeological assemblages.

The T-1 prior to 6000 B.P. was evolving laterally, as the Delaware River migrated eastward, exposing an expansive floodplain along its inner bank. Principal deposits were the coarse-to-medium sands (Alluvial Unit 3) forming the basal surfaces on which the Early and Middle Archaic settlers lived. The specific occupation loci were raised knolls on the general floodplain that overlooked the stream microenvironment. The Early Archaic assemblages of Stratum XI are typical. Artifact clusters (at N29 E64 and N36-37 E65) spanned a formerly sloping surface (ca. 9500 B.P.) (Bergman and Doershuk 1994: chapter 15; Bergman et al. 1998).

High-discharge events including periodic floods and aperiodic channel splays and cut-off chutes would have eroded the archaeological deposits. Because occupations were generally small during earlier prehistoric periods, the susceptibility of any given locus to erosion and/or site burial was high. This accounts for the diffuse distributions of artifacts in Stratum XI and would also be an accurate barometer of Early Archaic site expectation and preservation in other portions of the landform. As noted, one of the main factors responsible for sealing in the deposits is the firm structure of the argillic paleosol (Bt horizon) that evolved over several thousand years and compacted the archaeological deposits within it. Subsequent to initial occupation, the soil incorporated the artifact or sediment matrix and "overprinted" it, thereby rendering it more resistant to erosion. Overprinting is the product of long-term postoccupational weathering during which the archaeological deposits are transformed chemically within the evolving Bt (see Ferring 1992; Waters 1992). The elevation of the original site locus above the general floodline was also critical in preservation.

Similar mechanisms were in effect during the Middle Archaic (ca. 7500 B.P.), when the Stratum IX surface formed. At this time, however, the stream had migrated even farther to the east, while its dynamism became more subdued. This is demonstrated sedimentologically in average fining of mean grain sizes for the later Middle Holocene sequence (Figure 13.9). There is a more homogeneous distribution of artifacts across the T-1 (Bergman and Doershuk 1994: chapter 15; Bergman et al. 1998), suggesting that a more extensive surface was suitable for occupation while erosion slowed.

After 6000 B.P. (Alluvial Unit 3/2 transition) the former floodplain (T-0) evolved to a raised landform position—or T-1—elevated above levels of all but the highest amplitude floods. Archaeological deposits are sparse in the transitional lateral accretion to overbank horizons, but subsequently the density of occupation increases in the Late to Transitional Archaic and Woodland horizons in Alluvial Unit 2a.

Patterns of Late Holocene archaeological sedimentation are considerably different from those in the Early Holocene fills. While the density of occupation increases, the distribution of cultural materials expands across broader surfaces but in thinner vertical accumulations. This is because rates of natural sedimentation diminished exponentially through time as the constructional T-1 built up. This trend is inversely correlated with occupation that tended to expand as a result of the availability of a well-drained landform, less susceptible to flooding. It afforded logistic advantages to riverine and aquatic resource bases.

Accordingly, a compressed cultural stratigraphy spans the last 3000–4000 years of occupation in Alluvial Unit 2a. It is contained in dense anthropogenic sediment matrices, 30–80 cm thick. These consist of either discrete, activity specific features or more complex interdigitations of cultural lenses and the upper cambic soil (2Ab-2Bw pedon). Such stratification is best viewed as a series of palimpsests or superpositions of cultural features and natural surfaces.

Table 13.4 correlates the landscape history and preser-

Table 13.4. Landscape History and Archaeological Preservation Potential by Archaeological Component, Sandts Eddy

Archaeological Component	Alluvial Unit	Occupational Landscape (T-1 Landform)	Cultural Sedimentation Pattern	Preservation Matrix	Preservation Potential
Historic	1	Extensive erosion and surface grading of T-1 by industrial age and 20th-century development projects on landform; truncation of T-1 foreslope and headwearing and sedimentation; aggregation of T-0	Diffuse and occasionally dense clusters of industrial and building debris; "inverted stratigraphy" resulting from periodic sediment overturning and grading activities	Organic and inorganic fills in secondary context	High
Woodland	2a	Marginal vertical sedimentation and landform construction for 1000-2000 years; flooding very infrequent on T-1; new cycle of cutting and filling (to lower levels) associated with T-0 surface	Palimpsests of thin to deeper cultural features superimposed on one another and intrusive into cambic paleosol (Bw horizon); diffuse artifacts displaced by bioturbation and pedoturbation processes	Organic feature fills occasionally contaminated by overlying fill sediments; some displacement of upper layers; lateral stratification and truncation of subadjacent fills; entertainment of features and subassemblages in upper cambic soil (Bw horizon)	Moderate-High
Transitional and Late Archaic	2a/2(?)	Aggregated T-1 surface several meters above waterline; entrenched channel; infrequently flooded; general terrace morphology begins to approximate present contours; levee construction	Significant diagnostic artifacts and dispersed activity areas; features and perishables; occupation on a weak cambic (Bw) paleosol interdigitated with organic lensing and more extensive midden feature	Limited feature preservation and superposition of discrete feature fills with buried organic soil (2Ab horizon) and possible midden	Moderate-High
Middle Archaic	3	Laterally and vertically constructed landform with concavo-convex floodplain building above T-0 to T-1 elevations; above mean annual flood lines but exposed to limited flooding; periodic channel shifts	Discrete activity loci and assemblage clusters; concentrations of perishables; limited number of discrete features	Strongest features articulated because of dominance of preserved vegetal remains and organics; more typically "overprinting" of artifacts by strong argillic (Bt horizon)	Moderate
Early Archaic	3	Island and/or point bar surfaces aggrading laterally across floodplain; gentle relief above waterline; slackwater basins and overflow chutes traverse floodplain	Diffuse cultural debris, with occasional concentrations of lithic materials; limited perishable residues	Progressively stronger soil development coupled with more diffuse occupational evidence and acidic environmental results in dominant overprint signature	Low-Moderate
Paleoindian	4	Isolated channel bars and islands overlooking braided stream	N/A	N/A	Low

vation potential for individual archaeological components at Sandts Eddy. An estimate of preservation potential is based on reconstructed landscape events, cultural formation processes, and the soils and sediments housing the archaeological materials. The trend is for preservation potential to be enhanced up the sequence (from Low to High) in direct relation to increased landscape stability and diminished sedimentation through time. Significantly, Low-Moderate to moderate preservation potentials, for the Early and Middle Archaic respectively, are the only valid projections for sites of this period, because Sandts Eddly is one of the few sites in the Delaware Valley with Early Holocene assemblages in pristine context.

SYNTHESIS: A MODEL OF LANDSCAPE ARCHAEOLOGY

A nearly 12,000-yr record of occupation and landscape evolution is preserved in the more than 4-m-thick alluvium underlying the T-1 surface at Sandts Eddy. A diachronic site formation model registering these changes and their preserved archaeological contexts is presented in Figures 13.14a and 13.14b. The model is a time transgressive transect across the T-1 and the flanking Delaware River. It chronicles aggradation of the primary landform and subsequent depositional, erosional, and soil-forming cycles beginning with the Pleistocene gravels.

Seven reconstructions of site geography are depicted for the Paleoindian, Early Archaic, Middle Archaic, Late Archaic, Terminal Archaic, Woodland, and Historic periods. Field levels for the major occupation link particular sediment bodies and soil horizons to the archaeological record.

As noted earlier, the landscape succession at Sandts Eddy is best viewed in terms of changing thresholds of fluvial and alluvial dynamism. Over the long term the most active Delaware stream environments were displaced by a mature alluvial system. Geomorphically, this is signaled by the initial formation of the T-1 as a floodplain surface, its 4-6-m vertical construction, and subsequent entrenchment as a new base level was reached and a second floodplain (T-0) emerged.

Cycle 1 is initiated by depositions of the basal gravels (Alluvial Unit 4). The gravels are relict features of torrential channel activity, at a time when base levels were lowered dramatically and the adjusting channel laid down a broad series of gravel bars on the stream floor. At this time the Delaware freely traversed the floodplain. The outer floodbelt was contained by the steep-sided valley walls, approximately 0.25 km to the east.

Cycle 2 (Alluvial Unit 3) began ca. 11,000 B.P., when a migrating stream laid down finer-grained channel deposits. Flow levels stabilized, as channel migration produced shifting point bars, constructing a laterally discontinuous floodplain (T-0). The channel moved eastward as the bottoms became habitable segments, separated by anastamosing flow lines. The relatively limited Early Archaic occupation and its sloping surfaces mark Early Holocene sand lenses and bottom-set coarse sands and mixed bedload deposits. These indicate that stream activity was once dominant around probable point bars, the loci of Early Archaic activity. It is probable that Stratum XI represents a series of elevated surfaces that were islands at the time of occupation. Although channel movement was generally to the east and south, during high discharge periods overflow lines traversed the lower surfaces in the vicinity of the occupation.

Cycle 3 (upper Alluvial Unit 3) witnessed a slowing of net aggradation, stream flow, and progressive channel adjustment. This resulted in the coalescence of fluvial landforms (channel and point bars) to a laterally continuous floodplain. Vertical accretion dominated near the end of the period, as a concavo-convex floodplain became a T-1, reaching heights in excess of 3 m. The Early Archaic occupations were overridden by coarse sand splays. Subsequent Middle Archaic settlement emerged on a more continuous and level landform, considerably less susceptible to inundations. The dominant occupation (Stratum IX) is preserved in an argillic (Bt) horizon. Occasional flooding was confined to back-slopes on the valley margins.

By 6000–5500 B.P. (Late Archaic, Cycle 4), Alluvial Unit 2 was laid down as an overbank sediment. Soil formation was more limited as rates of sedimentation exceeded pedogenesis. Underlying sediments, and their occupations, were buried relatively intact. The Late Archaic occupations extended across newly accreted surfaces. Archaeological sedimentation dominated over natural aggradation as dense accumulations of anthropogenic deposit interdigitated with the cambic soil (Bw) horizon that began to form.

Prehistoric occupation intensified during Transitional Archaic times (Cycle 5, Alluvial Unit 2a). After 3500 B.P. surfaces were rarely overridden, except during periods of highest discharge. The Delaware was entrenched in its channel and ongoing vertical accretion resulted in levee construction. Habitation was nearly continuous. This produced a complex cultural stratigraphy in which a discontinuous midden was intruded by more discrete features related to seasonal occupations. A series of palimpsests characterize the dominant site-formation processes.

Intensive anthropogenic sedimentation persisted to 1500 B.P. (Cycle 6, upper Alluvial Unit 2a). Although the T-1 was effectively stabilized by the levee, there is evi-

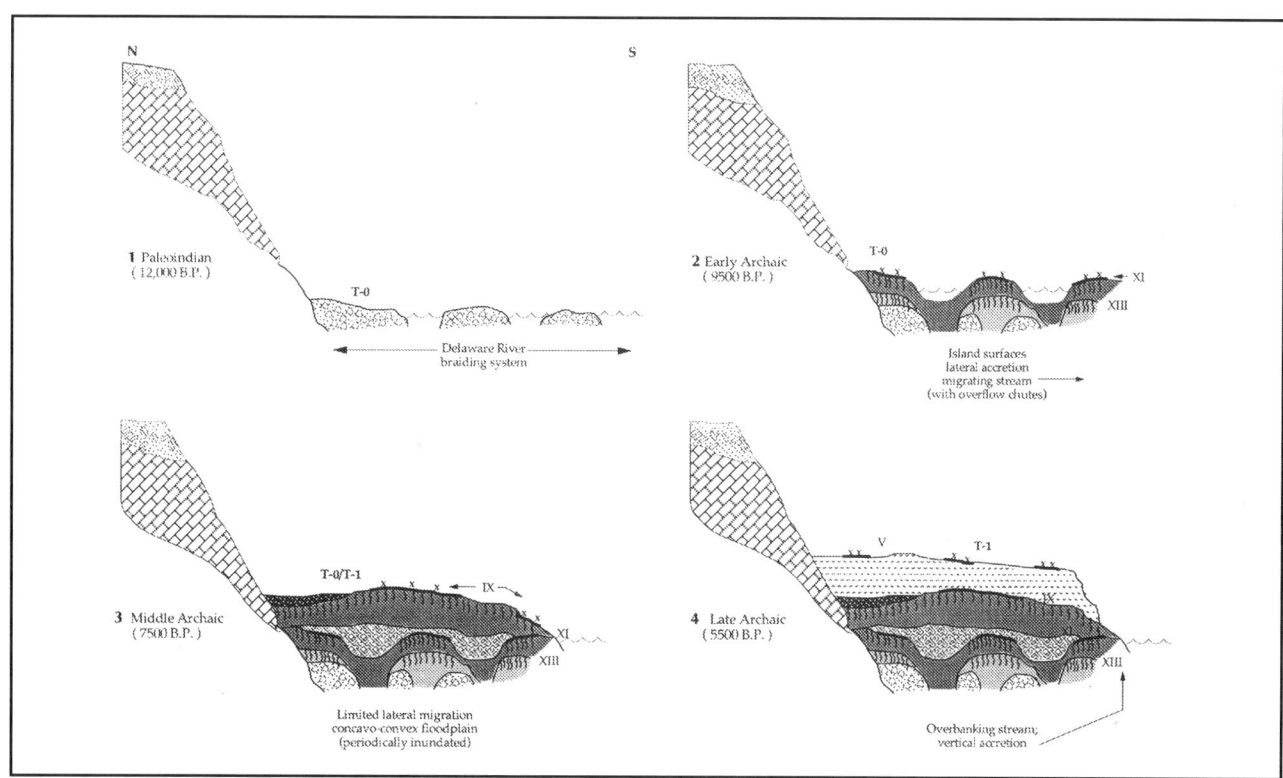

Figure 13.14a. Diachronic cyclical model of landform evolution, occupation and site preservation, Sandts Eddy (36Nm12), Pennsylvania: Cycles 1–4 (12,000–5500 B.P.).

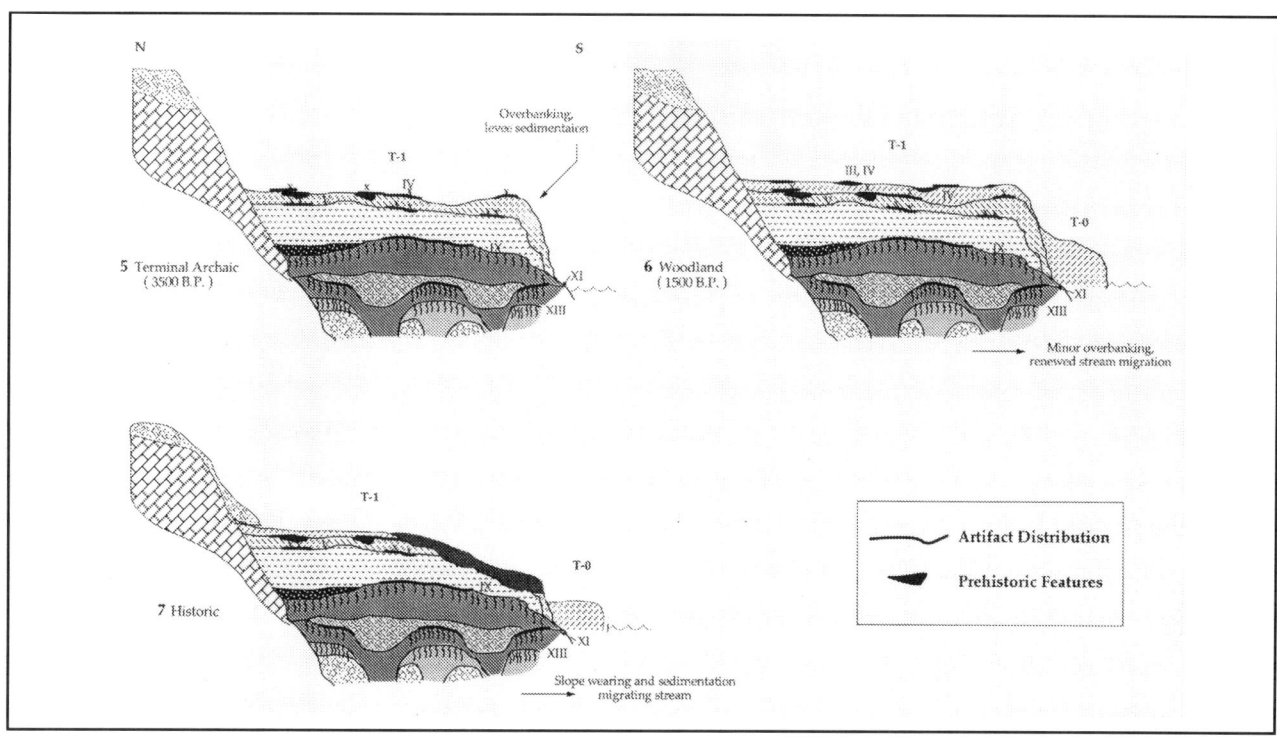

Figure 13.14b. Diachronic cyclical model of landform evolution, occupation, and site preservation, Sandts Eddy (36Nm12), Pennsylvania: Cycles 5–7 (3500 B.P.–present).

Chapter 13 *Landscape Change, Human Occupation, and Archaeological Site Preservation at the Glacial Margin* **201** dence for limited cutting and filling on a lower-lying surface, the T-0, that was beginning to build up along the terrace margins. The landform is inset against the bank, sloping riverward and indicating ongoing channel migration. There is no evidence of occupation on the T-0; upstream and downstream of the site it has been variously eroded.

Finally, in historic times (Cycle 7, Alluvial Unit 1) the upper surfaces have been periodically reworked by construction activity and landscaping. A by-product of this activity has been intensive slope erosion and sediment loss. Archaeological assemblages in underlying strata have been variously removed from primary contexts resulting in an irregular surface. Stream migration has been ongoing.

SANDTS EDDY AND THE ALLUVIAL GEOARCHAEOLOGY OF THE DELAWARE VALLEY

Sedimentation and Holocene Climates

The near continuous and well-dated Holocene alluviation record at Sandts Eddy underscores changing fluvial morphology and flooding regimes that bear on broader paleo-climatic trends. Ritter et al. (1973) generated a study of Delaware Valley sedimentation rates for the Faucett and Byram sites (Figure 13.1, inset) based on a discontinuous 6,000-yr record. His findings, which were reanalyzed by Ferring (1986), demonstrated a nearly perfect linear correlation between depth of sediment and age for the latter Holocene ($r = 0.969$). Sandts Eddy extends the range of investigation by 4,000 years and rounds out the timeline. Over the past 20 years additional stratified sites have been excavated along the Delaware and can be incorporated into analysis.

Ferring (1986) explored the usefulness of alluviation rates for interpreting patterns of archaeological site construction and for assessing within and between site variability. Only limited efforts have addressed the method's potential for addressing paleo-climatic issues, although some research has linked variable deposition and erosion rates to ancient moisture regimes and circulation patterns (see Bull 1988; Church and Slaymaker 1989; Knox 1983, 1999).

The analysis presented here begins with the assembly of raw sedimentation rate data for the most deeply stratified and accurately reported archaeological sites along the Delaware Valley. Six sites were selected that spanned a linear distance of 160 stream km. The location of the sites is shown in Figure 13.1. Along the downstream axis, the sample included Faucett (Kinsey 1972, 1975; Ritter et al. 1973), Shawnee Minisink (McNett 1985), Upper Shawnee Island (Stewart et al. 1991), Sandts Eddy (Bergman and Doershuk 1994; Bergman et al. 1998), Byram (Kinsey 1975), and Lower Blacks Eddy (Schuldenrein et al. 1991). Alluviation rates were typically not compiled by the principal researchers. In some cases, however, sufficient numbers of well-provenienced and tightly spaced radiocarbon dates provided an index of sediment accumulation. This was especially true of Faucett, Byram, and Lower Blacks Eddy where in excess of five absolute dates per site were available. Other sites (Upper Shawnee Island, Shawnee Minisink) contained few absolute dates, but temporally diagnostic features provided age control allowing for interpolation of sedimentation chronology. It is cautioned further that the plots do not depict alluviation rates *sensu stricto*, because they do not account for intervals of soil formation. Collectively, however, these diverse chrono-stratigraphic markers provide a broad index for calibrating net rates of T-1 build-up within the greater Middle Delaware Valley.

Figure 13.15 plots the set of actual and interpolated dates for the six sites by depth. Each curve begins with the earliest radiocarbon date available for the site. Relative slope steepness is the measure of the sedimentation rate. The data converge around several temporally reinforcing trends. First, dated sequences older than 4000 B.P. apply only to Upper and Middle Delaware Valley sites. As discussed in the subsequent section, this is a function of basin morphology. Second, sedimentation rates appear to be highest for the basal sequences at most sites and typically level off toward the tops of sequences, in Late Holocene levels. This is consistent with the overbanking regimes that have been confirmed for Sandts Eddy and which were recognized for the Upper Valley by Ritter et al. (1973). Third, the general slope trends are remarkably parallel between sites suggesting overall uniformity in sedimentation for extensive stretches of the drainage.

Significantly there is pronounced steepening in slope for the sequences older than 5500 B.P. Alluviation rates are considerably higher at the base of each cultural sequence. Although these results are preliminary—because of the small sample of dates ($n = 12$) and the even smaller number of absolute ages ($n = 10$)—the data are in accord with the Sandts Eddy pattern of aggradation that demonstrated highest discharge regimes and thickest accumulations in the basal Holocene horizons (lower Alluvial Unit 3).

To expand these observations and explore valley-based associations between sedimentation and Holocene climates and moisture regimes, it is instructive to develop an index for changing sedimentation rates for the sampled sites. An appropriate index is the running mean, determined for 1,000-yr intervals spanning the Holocene. The running mean is calculated by aver-

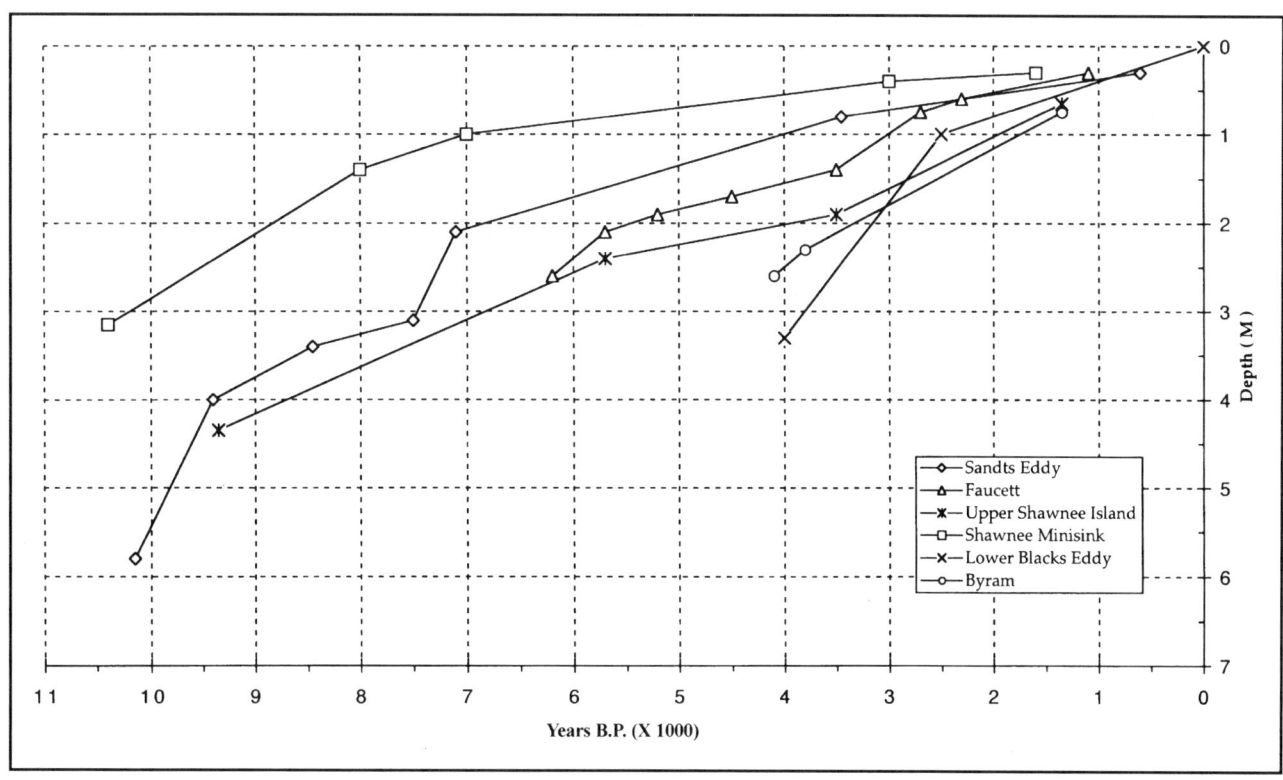

Figure 13.15. Holocene sedimentation rates: Delaware Valley alluvial sites.

aging the sedimentation rate for all plots (in Figure 13.15). Gross correlations with paleo-climatic cycles, specifically changing regimes of available moisture, can then be superimposed on a widely used paleo-climatic model linking flood frequency, sedimentation, and climatic response (after Wendland and Bryson 1974). Figure 13.16 graphs the running mean on the moist-dry cycles projected from the climatic model. The plots highlight the trend to accelerated alluviation during the Early Holocene (11,000–8000 B.P.), which subsequently diminishes (to 7500 B.P.), accelerates rapidly for the Middle Holocene (to 4500 B.P.), falls off again (at 3500 B.P.), and is succeeded by a final phase of accelerated deposition. These trends are followed closely by the alternating moist-dry cycles that demonstrate a close fit between diminished sedimentation rates and drying trends. Conversely high levels of deposition appear to be closely linked to moist Holocene climatic cycles. Although it is premature to generalize the significance of the climate-sedimentation correspondence, it has recently been suggested that large-scale flooding of the Mississippi was dominant over the intervals 6500–4500 B.P. and 3000–2000 B.P. (Knox 1988). These intervals correspond to the peak moisture cycles and sedimentation values for the Delaware Valley.

Regional Correlations

In stream environments, attempts to structure the relationships between prehistoric site location, preservation, and landscape history are facilitated by an appreciation of fluvial geomorphic systems. Schumm's (1977) model of drainage basin dynamics stresses the transfer of sediment from upstream sources to downstream destinations to explain increasing accumulations of sediment in the direction of the drainage mouth. This model establishes three zones of sediment mobilization, each aligned with segments related to distance from the drainage head. The upstream portions of the drainageway (Zone 1) produce sediment for transport. Central segments transfer sediment along the basin (Zone 2) and drainage mouth portions are the sites of maximum deposition (Zone 3). Because stratified archaeological sites that are aligned with sectors of the drainage-way preserve variable thicknesses of sediment as well as vertically segregated archaeological components, the Schumm model may be appropriate for identifying patterned geoarchaeological correlations in the Middle Delaware stratigraphies and occupational records.

Table 13.5 summarizes the landscape histories of five

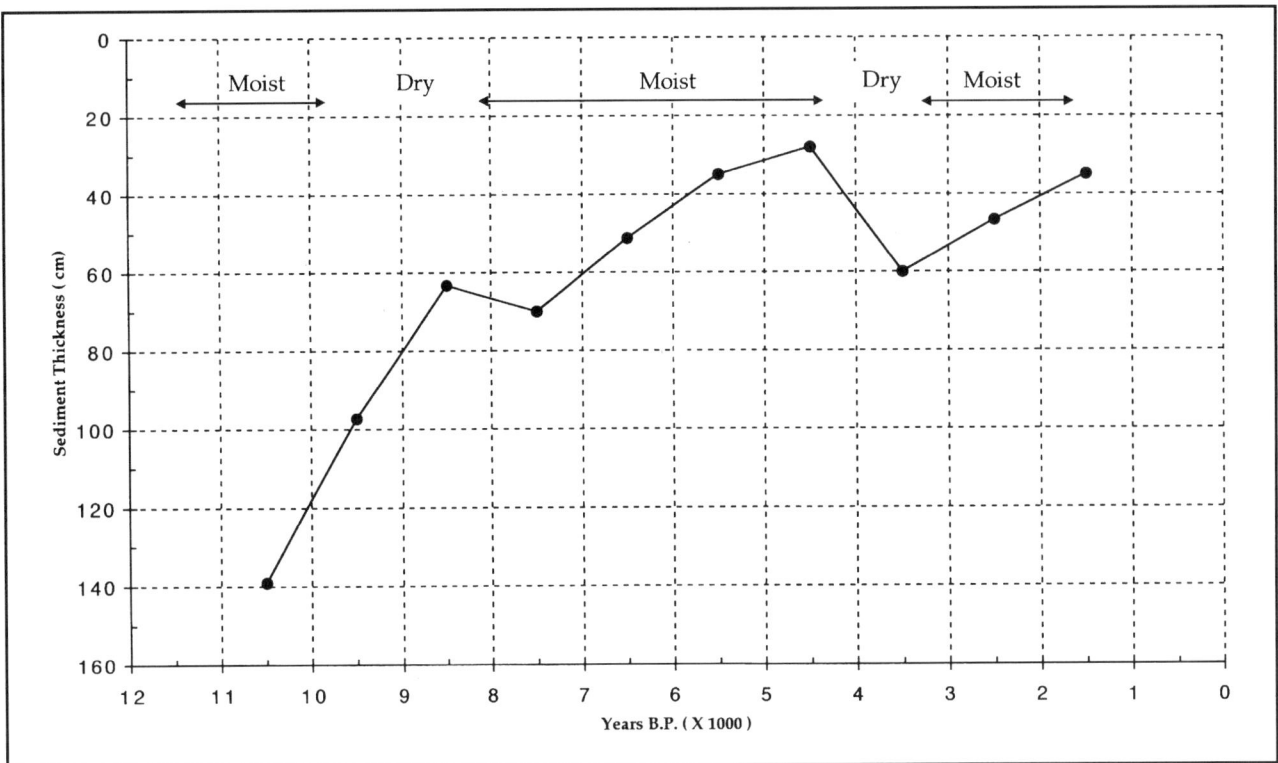

Figure 13.16. Holocene "moist-dry cycles" versus running mean sedimentation rates: Delaware Valley alluvial sites.

sites spanning the Middle Delaware Valley. They are grouped in order of increasing distance from the Delaware Water Gap, at the margin of the Upper and Middle Valley transition (see Figure 13.1). For each site (Column 1), the dominant landform is identified (Column 2), along with drainage segment (Column 3). In this case *transition zones* refers to segments of the Middle Valley stream trench that exhibit gradient changes intermediate between sediment transfer (Zone 2) and adjacent segments of either sediment production or deposition (Zones 1 and 3 respectively). Prehistoric components are noted in column 4; in cases of poor artifact context the component is deleted. Landscape history is presented in Column 5 and additional geoarchaeological observations are noted in Column 6.

Table 13.5 reveals that variability in the landform record (Columns 1 and 2) is paralleled by changes in the occupational and landscape histories (Columns 3, 4, and 5). Accordingly, in the downstream direction, the passage from the zone of sediment production to deposition is registered by a progressively subdued archaeological topography. The (prehistorically occupied) T-2 at Shawnee Minisink is equivalent to the 6–7 m high T-1 at Upper Shawnee Island; the latter surfaces grade down to a 4–6 m surface at Sandts Eddy, ultimately emerging as the 3–4 m terrace at Lower Blacks Eddy. The more pronounced upstream relief is a function first, of the differentiated glacial terrain in the vicinity of the Water Gap, and, second, of the incision of that terrain by the Delaware (Witte 1997). That incisional phase ultimately accounts for preservation of the only unequivocally identified Paleoindian component for the valley on the single T-2 outcrop (at Shawnee Minisink).

Relatively limited deposition in the glacial edge terrain also explains the rare accessibility of Early Holocene sediments—and by extension Early through Middle Archaic sites—on T-1 landforms (at Upper Shawnee Island and Sandts Eddy). Here deposits were encountered at depths of 3–6 m. In contrast, at the downstream end, near Lower Blacks Eddy, equivalent accumulations of sediment are fully 3000–4000 years younger. Sedimentation rates are 2.5 times higher (see Column 6 in Table 13.4). The protracted deposition as the stream graded to base level at the margins of the Coastal Plain is the reason for this variability. Near exponential surges in sedimentation along the downstream axis are systematic. Between Shawnee Minisink and Sandts Eddy aggradation rates increase on the order of 1.1–1.5, whereas below Lower Blacks Eddy they accelerate by a factor of 2–3.

Table 13.5. Correlation of Prehistoric Landscape Histories of Middle Delaware Valley Sites

Site	Land-form	Drainage Segment[a]	Prehistoric Components	Landscape History	Comments
Shawnee Minisink (36Mr43)	T-2	Transition Zone 1/2	Paleoindian, Early Archaic, Late Archaic, Woodland	Basal outwash succeeded by migrating stream that stabilizes to support Paleoindian horizon on silt capped T-2 surface; four cycles of fining upward, lateral accretion; overbanking regime after 5000 B.P. (Late Archaic); limited pedogenesis noted	Only site preserved on T-2 landform; 3 m cultural sequence from base (Paleoindian) to top (Woodland) of section; sedimentation rate drop from 6.7–1.5 cm/100 years
Upper Shawnee Island (36Mr45)	T-1	Transition Zone 1/2	Early Archaic, Middle Archaic (?), Late Archaic, Middle Woodland, Late Woodland	Stratified T-1 occupation; three cycles of sedimentation capped by soils and succeeded by erosion; "stacked" B horizon (lamellae?) to 6500-4500 B.P.; subsequent overbanking and more widely spaced cambric B horizons	6–7 m high T-1 surface; sedimentation rates 1.2–1.8 times that of Shawnee Minisink and increase for Late Holocene
Sandts Eddy (36Nm12)	T-1	Zone 2	Early Archaic, Middle Archaic, Late Archaic, Transitional Archaic, Early/Middle Woodland, Late Woodland	Channel migration & discrete "packages" of fining upward alluvium sealed by palleosols (Early-Middle Archaic); overbanking & lamellar profiles (Late/Transitional Archaic); stable channel, infrequent bankfull discharge (Woodland)	4–6 m high T-1 surface, morphologically identical with site landform at Upper Shawnee Island; preserves same cycles of deposition (laterally migrating to overbank stream regimen) sealed by developed soil horizons; sedimentation rates are 1.5 times greater than Upper Shawnee Island
Padula (36Nm15)	T-1 (Tributary)	Zone 2	Early-Middle Archaic (isolated), Late Archaic, Transitional Archaic, Early/Middle Woodland Late Woodland	Complex colluvial fan-tributary floodplain sequence; minimal terrace alluviation since 5000 B.P. (Late Archaic); ponded, aquatic environments during Middle Holocene; incision of contemporary channel ca. 5000 B.P.	Lateral segregation of Early Archaic through Woodland components; no vertical sedimentation atop T-1 for > 5000 years
Lower Blacks Eddy (36Bu23)	T-1	Transition Zone 2/3	Late Archaic, Transitional Archaic, Early/Middle Woodland, Late Woodland	Basal Mid-Holocene (?) lags buried by overbank silts, clays; 2 m of overbank Mid-Late Holocene alluvium capped by Late Archaic anthrosol (4500-3000 B.P.); 2-3 preserved sola (A-Bw-C) for Transitional-Late Woodland	3–4 m high T-1 terrace is extension of same landform preserved at upstream sites; pre-Mid Holocene sediments not identified; finer alluvium, deep Late Archaic soils & late Holocene sedimentation rates > 2.5 times higher than upstream suggest depositional locus

[a]After Schumm 1977.

The sedimentation patterns themselves vary as do sediment thicknesses by individual valley segments or sites. Fluvial mechanics regulate a "fining downstream" effect as stream discharge increases and sediment size decreases (Lane 1955). Although each of the site sequences demonstrated a time transgressive transition from lateral to vertical accretion in the Middle Holocene, the texture of the parent materials at Lower Blacks Eddy was typically finer than those at other sites by one or two size-grades. Here fine sands and silts constitute the parent materials for the Middle to Late Holocene alluvium whereas uniformly fine sands were more typical of the temporally equivalent upstream sediment "packages." Progressive downstream fining is accompanied

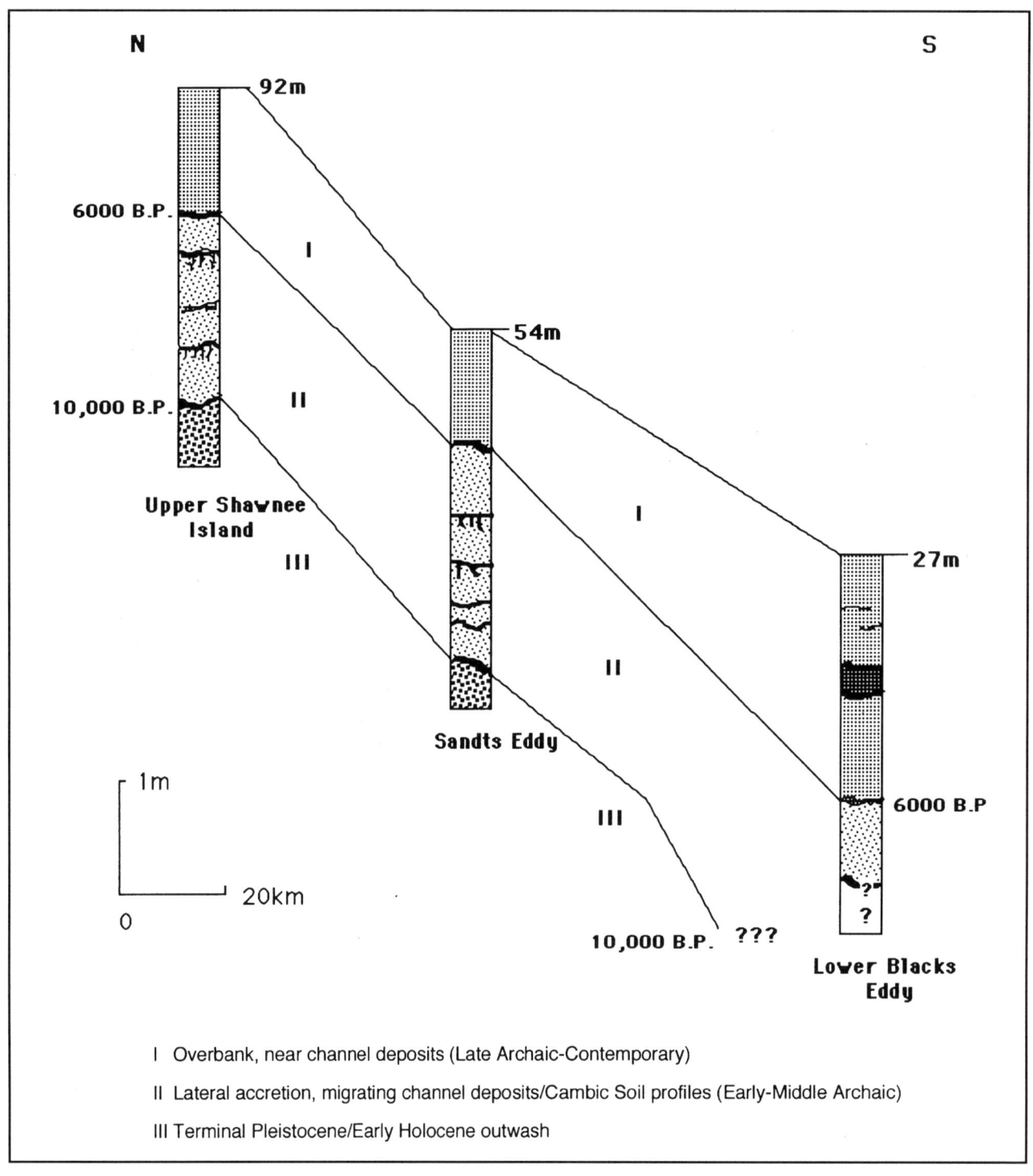

Figure 13.17. Schematic stratigraphy of the Middle Delaware Valley.

by impeded drainage and ponding; at Lower Blacks Eddy this is expressed by a preponderance of slackwater sedimentation and differentiated floodbasin features (Schuldenrein et al. 1991).

Reinforcing downstream trends to increased accretion, diminished sediment sizes, and impeded drainage are reflected in the archaeological stratigraphy of the sites, as illustrated in Figure 13.17. Because of limited sedimentation, the upstream sites preserved the most compressed, but perhaps best articulated, sequences.

Shawnee Minisink (not shown) housed all prehistoric components in a 3 m thick sequence. Farther downstream, Upper Shawnee Island and Sandts Eddy contained deeper, thicker, and younger deposits. They were also associated with a different landform (i.e., the T-1). Their successions featured Early Archaic through Woodland occupations in 4- to 6-m-thick accumulations. Finally, Lower Blacks Eddy contained no deposits older than Later Archaic in sediments extending more than 5 m in depth. As shown, depths of deposit are projected to be of such thickness in Zone 3 that the height of the water table would typically prevent access to pre-Late Archaic deposits even if preserved.

Significantly, however, the articulation of later prehistoric components is optimal at down-river locations because of the exposure of T-1 surfaces to bankfull discharge and attendant preservation of upper profiles. Such locations could conceivably showcase stratified Late Archaic through Woodland successions.

Despite variability in sedimentation mode and rates between valley segments and sites, several time-transgressive trends are characteristic of all Middle Delaware Valley profiles (Figure 13.17). First, three to four cycles of alluviation, soil formation, and erosion are typical. Second, a fluvial threshold is noted between 6000 and 5000 B.P. when sedimentation patterns passed from lateral accretion/fining upward to overbank regimes. These were paralleled by major transitions in soil-weathering sequences in which well-developed, but relatively thin, Bt-horizons (argillic) were superseded by deeper, but more subdued Bw (cambic) profiles that are often capped by preserved humic mats (Ab horizons). The overhaul in alluvial dynamics and weathering regimes may signal the appearance of generally modern environments, as this period coincides with the proliferation of sites across the landscape during Late to Terminal Archaic times. The evidence suggests that the contemporary Delaware has flowed in its present channel for about 6,000 years and that soil environments of today are enduring analogs to the cambic soil-forming environments that sustained the Late Archaic and Woodland populations of the Middle Delaware.

Overarching basin-wide trends in the geoarchaeological sequences of the Middle Delaware Valley suggest that the logistics of settlement are reflected in the morphology of the drainage basin and the variable sedimentation environments of the floodplain. This model refines our understanding of systematic change in floodplain sedimentation and site preservation across time and space. Additionally several basin-wide observations may be generated.

- Two-terrace (T-1, T-2) archaeological sequences are limited to the glaciated settings of the Upper and Middle Delaware (i.e., Shawnee Minisink).

- The most dominant and pervasive archaeological landform is the T-1 which begins in glacial terrain and extends the length of the Middle Valley. T-1 locations at the glacial margin contain deeply buried sites that have the highest potential for recovery of early occupations.

- Stratigraphic sequences, where preserved, will tend to thicken in a downstream direction. Shawnee Minisink, despite a 3-m-depth to the Paleoindian component, is a relatively compressed sequence compared to Sandts Eddy and Lower Blacks where the Pleistocene gravels may not have even been reached.

- By 5500 yr B.P. and certainly by the onset of Transitional-Late Archaic times, essentially contemporary floodplains were stabilized along the trunk stream. These promoted the emergence of differentiated riparian habitats in secondary valleys. Subsequent hydrographic and climatic changes along the Delaware were relatively minor by comparison.

CONCLUSIONS

The geoarchaeological research at Sandts Eddy explored the connections between patterned landscape change and human occupation for the duration of Holocene time at a key location on the northeastern glacial margin. Additionally the question of site preservation systematics was explored. Finally, the site was used as a baseline for modeling drainage-wide geological and archaeological successions the length of the Delaware Valley. These successions preserve the evidence and chronology for landscape dynamics (alluviation records), intervals of landform stability (soil sequences), and aboriginal settlement (archaeological records) in glacial, glacial-edge, and unglaciated segments of this very diverse valley.

In general, it was found that a braided channel setting gave way to a meandering stream in the Early Holocene. Fluvial transitions were registered sedimentologically in upward fining sequences that formed in parallel with the Early Archaic and lower Middle Archaic habitations. The floodplain and occupational geography of these settings was a series of point bars and islands that were intermittently settled by these earliest inhabitants of the valley. The artifacts and features containing their remains were preserved by the cohesive structures of argillic soils (Bt) and stronger cambic (Bw) horizons that were differentially sealed in by subsequent floods. After 6,000 B.P. the channel regime changed to one of overbanking and low-level accretion, as the Delaware was firmly entrenched in its present channel. Only the highest frequency floods (i.e., 100–500-yr intervals) added sediment to the T-1 terrace which already stood in excess of 3 m above the water line. This setting effectively mimics the contemporary terrain, such that Late Archaic and subsqunet Woodland occupations occurred across a landform that was somewhat lower but whose contours were broadly the same as those of the present. Evidence for soil formation (Bw horizons) is preserved in the Late Archaic levels, but the Woodland horizons are variously disturbed because of large-scale grading and landscaping activities of the Euroamerican period.

Because Sandts Eddy straddles the glacial margin, it marks a threshold for reconstructing human and landscape dynamics the length of the Delaware Valley. At stratified sites both upstream and downstream, major segments of the composite cultural and alluvial successions are preserved. Sandts Eddy orders these relationships across time and space because it contains one of the most complete composite sequences. It was shown that, upstream of the site, relict older terrace outcrops (T-2) have potential for containing Pleistocene archaeological remains (ie. at Shawnee Minisink). At the glacial margin itself (Sandts Eddy), the T-2 is largely gone, but the base of the T-1 may possibly contain evidence for Early and Middle Archaic components that often underlie thinly buried Woodland and Late Archaic horizons. Farther downstream, Late Holocene sediments achieve greater thicknesses whereas Early Holocene deposits are either eroded or inaccessible. Accordingly, deeper burial of Late Archaic and Woodland occupations is likely.

Prospects for developing comprehensive models of Holocene human paleoecology are greatly enhanced by the wealth of the Sandts Eddy database. This study demonstrates that interdisciplinary efforts merging approaches of the archaeological and natural sciences are key to unraveling complex human and landscape interactions.

Acknowledgments

The author wishes to thank Christopher Bergman (BHE Environmental) and John Doershuk (University of Iowa; Office of State Archaeologist) for the invitation to undertake the geoarchaeological analysis of the Sandts Eddy site. Their insights into site complexity and stratigraphic variability guided much of the analysis. Kurt Carr (Pennsylvania Bureau for Historic Preservation) recognized the critical significance of the site and its unique potential for Delaware Valley prehistory. Jim Bloemker (Williams Gas Pipeline-Transco) supported the effort on behalf of the project sponsor, Transcontinental Gas Pipeline. Don Thieme (GRA) offered key input, and Susan Malin-Boyce and Mark Smith (GRA) prepared the graphics with their usual care and expertise.

REFERENCES CITED

Anderson, D., and Schuldenrein, J. (1985). Prehistoric human ecology along the Upper Savannah River: Excavations at the Rucker's Bottom, Abbeville, and Bullard site groups. *Russell Papers 1985, Vol. I and Vol. II*. National Park Service, Archeological Services Branch, Atlanta.

Berg, T.M., Barnes, J.H., and Sevon, W.D. (1989). Physiographic Provinces of Pennsylvania (2nd ed.). *Pennsylvania Geologic Survey Map* No. 13, Fourth Series. Harrisburg.

Bergman, C., and Doershuk, J. (1994). *Archaeological Data Recovery for Transcontinental Gas Pipeline Corporation's 6.79 mile Leidy Natural Gas Expansion, Sandts Eddy Site (36-Nm-12), Northampton County, Pennsylvania*. Report submitted to Transcontinental Gas Pipeline Corporation. Houston, Tex.

Bergman, C., Doershuk, J., Moeller, R., LaPorta, P., and Schuldenrein, J. (1998). An Introduction to the Early and Middle Archaic occupations at Sandts Eddy. In *The Archaic Period in Pennsylvania*, edited by P.A. Raber, P.E. Miller, and S.M. Neusius, pp. 45–76. Pennsylvania Historic and Museum Commission, Harrisburg.

Bilzi, A.P., and Ciolkoscz, E.J. (1977). Time as a factor in the genesis of four soils developed in recent alluvium in Pennsylvania. *Soil Science Society of America Journal* 41:122–127.

Birkeland, P.W. (1999). *Soils and Geomorphology* (3rd ed.). Oxford University Press, New York.

Brakenridge, G.R. (1988). River flood regime and floodplain stratigraphy. In *Flood Geomorphology*, edited by V.R. Baker, R.C. Kochel, and P.C. Patton, pp. 157–168. John Wiley & Sons, New York.

Bull, W.B. (1988). Floods: Degradation and aggradation. In *Flood Geomorphology*, edited by V.R. Baker, R.C. Kochel, and P.C. Patton, pp. 139–156. John Wiley & Sons, New York.

Bullock, P., Fedoroff, N., Jongerius, A., Stoops, G. J., and Tursina, T. (1985). *Handbook for Soil Thin Section Description*. Waine Research Publishers, Wolverhampton, U.K.

Butzer, K.W. (1982). *Archaeology as Human Ecology: Method and Theory for a Contextual Approach*. Cambridge University Press, Cambridge.

Church, M., and Slaymaker, O. (1989). Disequilibrium of Holocene sediment yield in glaciated British Columbia. *Nature* 337:452–454.

Cook, S.F., and Heizer, R.F. (1965). *Studies on the Chemical Anaylsis of Archaeological Sites*. Publications in Anthropology No. 2. University of California Press, Berkeley.

Courty, M.A., Goldberg, P., and Macphail, R. (1989). *Soils and Micromorphology in Archaeology*. Cambridge University Press, Cambridge.

Crowl, G.H., and Sevon, W.D. (1980). Glacial border deposits of late Wisconsinan age in Northeastern Pennsylvania. *Pennsylvania Geologic Survey General Geology Report* G-71. Harrisburg.

Davis, R.E., Drake, A.A., Jr., and Epstein, J.B. (1967). Geologic map of the Bangor Quadrangle Map, Pennsylvania-New Jersey. *U.S. Geologic Survey Quadrangle Map* GQ-665. Washington, D.C.

Dijkerman, J.C., Kline, M.G., and Olson, G.W. (1967). Properties and genesis of textural subsoil lamellae. *Soil Science* 104:7–16.

Eidt, R.C. (1977). Detection and examination of anthrosols by phosphate analysis. *Science* **197**:1327–1333.

Epstein, J.B. (1969). Surficial Geology of the Stroudsburg Quadrangle, Pennsylvania-New Jersey. *Pennsylvania Geologic Survey General Geology Report* G-57, Fourth Series. Harrisburg.

Fehr, E., Freyermuth, D., Lopresti, J., Lopresti, V., and Kline, D. (1971). The Sandts Eddy site (36Nm12). *Pennsylvania Archaeologist* **41**(1-2):39–52.

Fenneman, N.M. (1938). *Physiography of the Eastern United States*. McGraw-Hill, New York.

Ferring, C.R. (1986). Rates of fluvial sedimentation: Implications for archaeological variability. *Geoarchaeology* **1**:259–274.

Ferring, C.R. (1992). Alluvial pedology and geoarchaeological research. In *Soils in Archaeology*, edited by V.T. Holliday, pp. 1–39. Smithsonian Institution Press, Washington, D.C.

FitzPatrick, E.A. (1993). *Soil Microscopy and Micromorphology*. John Wiley & Sons, Chichester, England.

Folk, R.L. (1974). *Petrology of Sedimentary Rocks*. Hemphill, Austin, Tex.

Foss, J.E., and Segovia, A.V. (1984). Rates of soil formation. In *Groundwater as a Geomorphic Agent*, edited by R.G. LaFleur, pp. 1–17. Allen & Unwin, Boston.

Friedman, G.M., and Sanders, J.E. (1978). *Principles of Sedimentology Rocks*. John Wiley & Sons, New York.

Gee, G.W., and Bauder, J.W. (1986). Particle size analysis. In *Methods of Soil Analysis (Part I)*, edited by A. Klute, pp. 383–411. American Society of Agronomy and Soil Science Society of America, Madison, Wisc.

Head, K.H. (1980). *Manual of Soil Laboratory Testing*. Pentech Press, London.

ISG (1997). *International Stratigraphic Guide: A Guide to Stratigraphic Classification, Terminology, and Procedure* (2nd ed.). Geological Society of America, Boulder, Colo.

Kinsey, W.F. (1972). *Archeology in the Upper Delaware Valley*. Pennsylvania Historical and Museum Commission Anthropological Series 2. Harrisburg.

Kinsey, W.F. (1975). Faucet and Bryam sites: Chronology and settlement in the Delaware Valley. *Pennsylvania Archaeologist* **45**:1–103.

Knox, J.C. (1983). Responses of river systems to Holocene climates. In *Late Quaternary Environments of the United States, Part 2: The Holocene*, edited by H.E. Wright, pp. 271–277. University of Minnesota, Minneapolis.

Knox, J.C. (1988). Climatic influence on Upper Mississippi Valley floods. In *Flood Geomorphology*, edited by V.R. Baker, R.C. Kochel, and P.C. Patton, pp. 279–300. John Wiley & Sons, New York.

Knox, J.C. (1999). Long-term episodic changes in magnitudes and frequencies of floods in the upper Mississippi River valley. In *Fluvial Processes and Environmental Change*, edited by A.G. Brown and T.A. Quine, pp. 255–282. British Geomorphological Research Group Symposia Series. John Wiley & Sons, Chichester, England.

Kolb, M.F., Lasca, N.P., and Goldstein, L. (1990). A soil-geomorphic analysis of the midden deposits of the Aztalan site, Wisconsin. In *Archaeological Geology of North America*, edited by N.P. Lasca and J. Donahue, pp. 199–218. Centennial Special Vol. 4. The Geological Society of America, Boulder, Colo.

Lane, E.W. (1955). Design of stable channels. *American Society of Engineers Transcripts* **20**:1234–1279.

Larsen, C., and Schuldenrein, J. (1990). Depositional history of an archaeologically dated flood plain, Haw River, North Carolina. In *Archaeological Geology of North America*, edited by N.P. Lasca and J. Donahue, pp. 525–540. Centennial Special Vol. 4. The Geological Society of America, Boulder, Colo.

Leverett, F. (1934). Glacial deposits outside the Wisconsin Terminal Moraine in Pennsylvania. *Pennsylvania Geologic Survey General Geology Report* G-7, Fourth Series. Harrisburg.

McNett, C.W. (editor). (1985). *Shawnee Minisink: A Stratified Paleoindian-Archaic Site in the Upper Delaware Valley of Pennsylvania*. Academic Press, Orlando, Fla.

Miller, B.L., Fraser, D.M., and Miller, R.L. (1939). Northampton County, Pennsylvania. *Pennsylvania Geologic Survey County Report* C-48, Fourth Series. Harrisburg.

North American Commission on Stratigraphic Nomenclature (NASCN). (1983). North American stratigraphic code. *American Association of Petroleum Geologists Bulletin* **67**:841–875.

Parker, G.G., Hely, A.G., Keighton, W.B., Olmstead, F.H. (1964). *Water Resources of the Delaware River Basin*. Geological Survey Professional Paper No. 381. U.S. Government Printing Office, Washington, D.C.

Rawling, J.E. (2000). A review of lamellae. *Geomorphology* **35**:1–9.

Ritter, D.F., Kinsey, W.F., and Kauffman, M.E. (1973). Overbank sedimentation in the Delaware River Valley during the last 6,000 years. *Science* **179**:374–375.

Schuldenrein, J. (1988). Implications of subsoil lamellae for reconstructing prehistoric occupation surfaces. Paper presented at the 53rd annual meeting of the Society for American Archaeology, Phoenix.

Schuldenrein, J. (1994). Alluvial site geoarchaeology of the middle Delaware Valley: A fluvial systems paradigm. *Journal of Middle Atlantic Archaeology* **10**:1–21.

Schuldenrein, J. (1995). Geochemistry, phosphate fractionation, and the detection of activity areas at prehistoric North American sites. In *Pedological Perspectives in Archaeological Research*, edited by M.E. Collins, B.J. Carter, B.G. Gladfelter, and R.J. Southard, pp. 107–131. Soil Science Society of America, Madison, Wisc.

Schuldenrein, J., Kingsley, R.G., Robertson, J.A., Cummings, L.S., and Hayes, D.R. (1991). Archaeology of the Lower Black's Eddy site, Buck's County, Pennsylvania: A preliminary report. *Pennsylvania Archaeologist* **61**(1):19–75.

Schumm, S.A. (1977). *The Fluvial System*. Wiley-Interscience, New York.

Stewart, M., Custer, J., and Kline, D. (1991). A deeply stratified archaeological and sedimentary sequence in the Delaware River Valley of the Middle Atlantic Region, United States. *Geoarchaeology* **6**:169–182.

Torrent, J., Nettleton, W.D., and Borst, G. (1980). Clay illuviation and lamella formation in a Psammentic Haploxeralf in southern California. *Soil Science Society of America Journal* **44**: 363–369.

U.S. Department of Agriculture. (1975). *Soil Taxonomy: A Basic System of Soil Classification for Making and Interpreting Soil Surveys*. Agriculture Handbook No. 436. U.S. Department of Agriculture, Soil Conservation Service, Washington, D.C.

Waters, M.R. (1992). *Principles of Geoarchaeology: A North American Perspective*. University of Arizona, Tucson.

Weed, C.S., Parish, C.K., Brett Cruse, J., Jones, P.S., Jones, J.L., and Cruse, M.E. (1990). *Cultural Resources Investigation of the Transcontinental Gas Pipe Line Corporation 6.79 Mile Leidy Natural Gas Pipeline Expansion, MP 29.51 - MP 36.30, Warren County, New Jersey and Northampton County, Pennsylavania*, Vol. 1. Report of Archaeological Investigations No. 24. Emanco, Inc., Houston. Submitted to Transcontinental Gas Pipeline Corporation, Houston.

Wendland, W.M., and Bryson, R.A. (1974). Dating climatic episodes of the Holocene. *Quaternary Research* **4**:9–24.

Witte, R. (1997). Late Quaternary deglaciation and fluvial evolution of Minisink Valley: Delaware Water Gap to Port Jervis, New York. In *Pliocene-Quaternary Geology of Northern New Jersey: Guidebook for 60th Annual Reunion of the Northeastern Friends of the Pleistocene*, edited by S. Stanford and R. Witte, pp. 5.1–5.23. New Jersey Geological Survey, Trenton.

CONTRIBUTORS

Steve Ahr
Fort Drum Public Works
Environmental Division
85 First Street West
Fort Drum, New York 13602

Heather Almquist
Institute for Quaternary Studies and
 Departments of Geological Sciences
 and Plant Biology
University of Maine
Orono, ME 04469

Randy Amici
Fort Drum Public Works
Environmental Division
85 First Street West
Fort Drum, New York 13602

Donald H. Cadwell
Research & Collections Division
New York State Museum
Albany, NY 12230

Carol Cady
St. Lawrence University
Library System
Canton, NY 13617

David L. Cremeens
GAI Consultants, Inc.
570 Beatty Road
Monroeville, PA 15146

Kathryn Curran
Department of Anthropology
University of Massachusetts
Amherst, MA 01003

P. Jay Fleisher
Earth Science Department
State University of New York
Oneonta, NY 13820

Daniel T. Forrest
Department of Anthropology
University of Connecticut
Beach Hall, Room 354
Mansfield Road, U-2176
Storrs, CT 06269-2176

Douglas Frink
Archaeology Consulting Team, Inc.
57 River Road
Suite 1020
Essex, VT 05452

John P. Hart
Research & Collections Division
New York State Museum
Albany, NY 12230

Allen Hathaway
Archaeology Consulting Team, Inc.
57 River Road
Suite 1020
Essex, VT 05452

Brian D. Jones
Mashantucket Pequot Museum
 and Research Center
110 Pequot Trail
P.O. Box 3180
Mashantucket, CT 06339-3180

Alice R. Kelley
5790 Bryand Global Sciences Center
University of Maine
Orono, ME 04469-5790

Ernest H. Muller
Department of Earth Sciences
Syracuse University
Syracuse, NY 13244

John C. Ridge
Department of Geology
Tufts University
Medford, MA 02155

continued

James Rapant
Fort Drum Public Works
Environmental Division
85 First Street West
Fort Drum, New York 13602

Laurie W. Rush
Fort Drum Public Works
Environmental Division
85 First Street West
Fort Drum, New York 13602

David Sanger
Department of Anthropology and
  Institute for Quaternary Studies
University of Maine
Orono, ME 04469-5790

Joseph Schuldenrein
Geoarchaeology Research Associates
5912 Spencer Avenue
Riverdale, NY 10471

Donald M. Thieme
Department of Geology
University of Georgia
Athens, GA 30601

Robert M. Thorson
Department of Geology and Geophysics
  and Department of Anthropology
University of Connecticut (U-2045)
Storrs, CT 06269-2045

Christian A. Tyron
Department of Anthropology
University of Connecticut
Storrs, CT 06269-2176